RELIABILITY-CENTERED MAINTENANCE: MANAGEMENT AND ENGINEERING METHODS

RELIABILITY-CENTERED MAINTENANCE: MANAGEMENT AND ENGINEERING METHODS

RONALD T. ANDERSON
Reliability Technology Associates

and

LEWIS NERI
US Army Aviation Systems Command

ELSEVIER APPLIED SCIENCE
LONDON and NEW YORK

ELSEVIER SCIENCE PUBLISHERS LTD
Crown House, Linton Road, Barking, Essex IG11 8JU, England

Sole Distributor in the USA and Canada
ELSEVIER SCIENCE PUBLISHING CO., INC.
655 Avenue of the Americas, New York,
NY 10010, USA

WITH 20 TABLES AND 76 ILLUSTRATIONS

Softcover reprint of the hardcover 1st edition 1990

British Library Cataloguing in Publication Data

Anderson, Ronald T.
Reliability-centred maintenance.
1. Reliability engineering & maintainability engineering.
Management aspects
I. Title II. Neri, Lewis
658.562

ISBN-13: 978-94-010-6826-0 e-ISBN-13: 978-94-009-0757-7
DOI: 10.1007/978-94-009-0757-7

Library of Congress Cataloging-in-Publication Data

Anderson, Ronald T.
Reliability-centered maintenance : management and engineering methods / Ronald T. Anderson, Lewis Neri.
p. cm.
Includes bibliographical references.

1. Reliability (Engineering) 2. Maintainability (Engineering)
I. Neri, Lewis. II. Title.
TA169.A63 1990
620′.0045—dc20 89-77190

Disclaimer: The views expressed in this book are those of the authors and do not reflect the official policy or position of the Department of the Army, Department of Defense, or the US Government.

Foreword

In this book the authors provide a fresh look at basic reliability and maintainability engineering techniques and management tools for application to the system maintenance planning and implementation process. The essential life-cycle reliability centered maintenance (RCM) activities are focused on maintenance planning and the prevention of failure. The premise is that more efficient, and therefore effective, life-cycle maintenance programs can be established using a well disciplined decision logic analysis process that addresses individual part failure modes, their consequences, and the actual preventive maintenance tasks. This premise and the techniques and tools described emphasize preventive, not corrective, maintenance.

The authors also describe the techniques and tools fundamental to maintenance engineering. They provide an understanding of the interrelationships of the elements of a complete RCM program (which are applicable to any complex system or component and are not limited only to the aircraft industry). They describe special methodologies for improving the maintenance process. These include an on-condition maintenance (OCM) methodology to identify defects and potential deterioration which can determine what is needed as a maintenance action in order to prevent failure during use.

Additionally, the authors describe an aircraft flight safety prediction model that can be used to facilitate application of the RCM analysis process. The model takes into account defects and failure mechanisms that may be introduced during production, storage, operation, and maintenance. Standardized functional fault tree diagrams are presented which serve as templates to facilitate application or tailoring to specific systems or major components.

This book can be a valuable reference document for continuous improvement of the system maintenance planning and implementation process.

JOHN F. ZUGSCHWERT
Executive Director,
American Helicopter Society

Preface

This book describes a broad based management and system engineering approach to the maintenance planning process for complex systems and components. It emphasizes a preventive approach to maintenance, focusing on maintenance engineering as a methodology completely integrated with reliability and maintainability (R&M) engineering. It explicitly describes all essential life-cycle activities as well as the basic elements of a total reliability-centered maintenance (RCM) program.

The book is based on the premise that a more efficient and cost-effective life-time maintenance and logistic support program can be established using a well disciplined, knowledge-based RCM decision logic analysis process. This process is based on the identification of safety critical failure modes and deterioration mechanisms through engineering analyses and evaluation of experience data to determine failure consequence severity levels and the most effective apportionment of maintenance tasks for each level. The book recognizes that a well planned, reliability and safety driven maintenance program leads to lower cost, lower risk and more effective maintenance tasks.

Although this book is based on work done by the authors for US Army aircraft systems and components, it should have wide use in the engineering community; the methods and techniques described are applicable to systems used in transportation, military, nuclear, gas and other high technology industries and, thus, are not limited only to the aircraft industry. The book is addressed to the maintenance engineer, logistic engineer, design engineer, reliability and maintainability engineer, safety engineer, project engineer and manager, as well as graduate students and others who are involved in maintenance, logistics, system engineering and management and are concerned with the establishment of cost-effective maintenance programs that are specifically geared to the preservation of the inherent level of reliability and safety designed and built into hardware systems.

This book provides practical up-to-date information, guidelines and engineering data pertaining to maintenance engineering and the closely related R&M engineering disciplines. It is a complete reference document covering all essential aspects of maintenance planning, specification and implementation. The book describes how to specify maintenance tasks and requirements for: (1) detecting and correcting incipient failures either before they occur or before they develop into major defects, (2) reducing the probability of failure, (3) detecting hidden failures that have occurred and (4) increasing the cost-effectiveness of a system's maintenance program. It describes how to determine and implement the most effective mix of:

(1) *Scheduled inspection or tests* that are designed to measure aging or deterioration of a component or structure — based on the deterioration found, the hardware item either undergoes maintenance or remains in service.
(2) *Scheduled removal tasks* at predetermined fixed intervals of age or usage.
(3) *Unscheduled tasks* consisting of routine monitoring during normal operation whereby components are allowed to fail or where impending failure can be detected prior to occurrence.

This book is organized into six chapters, each covering an essential aspect in the planning and establishment of cost-effective, RCM derived maintenance support programs. *Chapter* 1 provides general information related to the purpose and scope of the book; it discusses the establishment of complete and cost-effective maintenance programs from an overall life-cycle standpoint. It also presents a brief discussion on the maintenance cost and reliability characteristics of US Army helicopters. *Chapter* 2 covers the basic theory and engineering foundation for RCM including a description of airline maintenance and US Army aircraft maintenance programs. It describes the RCM decision logic process and how it is applied to determine hard-time replacement, on-condition and condition monitoring maintenance requirements and the basic maintenance tasks, i.e., servicing, monitoring, rework (repair, overhaul and rebuild) and replacement. It also discusses how the results of an RCM analysis are used to determine logistic support plans and requirements. *Chapter* 3 covers R&M theory including basic hardware life characteristics, reliability degradation and growth, and reliability, maintainability and availability analysis concepts. It describes how to plan a life-cycle R&M program. It discusses the determination and evaluation of R&M parameters and how they are used as input to RCM and logistics planning. *Chapter* 4 covers RCM engineering practice including R&M modeling and prediction, failure mode, effects and criticality analysis (FMECA) and fault tree analysis (FTA), and shows how they can be applied to identify safety critical parts and to develop criticality data essential for RCM logic analysis. *Chapter* 5 discusses the application of RCM within the depot maintenance process. It describes the depot maintenance work requirements including the procedures and standards for processing a component or complete aircraft system through the depot. It discusses the failure modes encountered at the depot and methods for their detection and repair. It also describes an on-condition maintenance (OCM) program designed specifically for US Army aircraft. This special maintenance engineering technique involves assessing the actual condition of in-service aircraft in order to improve the availability of the overall fleet by identifying specific aircraft, in rank order, that need repair or reconditioning to prevent degradation of reliability and safety. *Chapter* 6 presents a special aircraft flight safety prediction model, based on the fault tree analysis procedure, that can be used to develop criticality data for application with the RCM analysis process. It describes how the model takes into account defects and deterioration mechanisms that can be introduced during production, storage and maintenance. It presents generic functional fault tree diagrams (FTD's) which serve as templates for the subsequent evaluation of specific aircraft or major components and the computation of criticality data. Application of the model with the RCM process is illustrated by applying it and the applicable FTD to a major dynamic component. Sample criticality and decision logic data are developed for direct input to logistic support analysis records and the preparation of a cost-effective RCM based maintenance program.

Acknowledgments

The authors must thank a number of people for their help and advice in the preparation of this book, in particular, Dr. Daniel Henry III for his valuable assistance in the preparation of the material in Chapters 5 and 6 and in the development of the Army aircraft safety analysis model. Also, Mr. Robert Ladner of the US Army Aviation Systems Command provided much help in the preparation of material on the application of RCM within the depot maintenance process at the Corpus Christi Army Depot.

The authors also acknowledge Dr. Frank Iddings, Director of the Nondestructive Testing Information Analysis Center (NTIAC) at Southwest Research Institute (SwRI) for his help in the preparation of the aircraft failure modes and nondestructive inspection practices described in Chapter 5. Many of the tables and figures included in this book were prepared as part of work done with NTIAC for the US Army and presented previously in various military handbooks and publications.

Mr. Dominic Vaughan of Elsevier Applied Science Publishers provided editorial comments and suggestions which added greatly to the quality of the final manuscript.

Acknowledgement is also given to Ms. Karen Lipkin for developing the aircraft engine criticality data and for editing the entire manuscript and preparing graphical and tabular material.

Finally, the authors would like to acknowledge Ms. Janet Anderson and Ms. Betty Kinney for their rapid and accurate typing of all draft material as well as for their efficient typing of the final manuscript.

Contents

List of Figures

List of Tables

CHAPTER 1

Introduction

The large, complex, high technology systems and components, such as those used in aircraft, require much more than high performance and versatility. They must be reliable and maintainable in order to render their operation both safe and cost-effective and they must be supported by an efficient, responsive maintenance program. The attainment of reliable and maintainable systems requires the application of sound engineering effort, starting early in the research and development process to design-in high reliability and ease of maintenance features and continuing after deployment, during field operation, to implement well planned maintenance support tasks. Only through a total life-cycle program that incorporates proven reliability and maintainability (R&M) engineering practice and well planned, properly executed maintenance procedures can safe, reliable and affordable hardware systems be achieved and maintained.

Maintainability and maintenance are distinct, yet related, disciplines covering the design, restoration and preservation of systems and components in their operational state. *Maintainability* is a specialized branch of the systems engineering discipline. It can be carried out together with reliability engineering analyses and actions as part of a total reliability-centered maintenance (RCM) program in support of a hardware system's design, development, production and operation and maintenance life-cycle process. It includes the engineering activities necessary to incorporate ease of maintenance features into the design and to establish appropriate repair and logistic requirements to maintain the designed-in R&M characteristics during service at an acceptable cost. Design emphasis is placed on the use of proven long-life component parts, the use of easily accessible and interchangeable modules and units, the incorporation of ease of inspectability and maintenance features and the use of sophisticated built-in diagnostic systems.

Maintenance deals with the specific procedures, tasks, instructions, personnel qualifications, equipment and resources needed to satisfy the system maintainability requirement within an actual use environment. It is the action necessary to retain a system in or to restore it to a serviceable condition. It includes tasks for servicing, repairing, removing and replacing, modifying, overhauling, inspecting and verifying hardware condition. Preventive maintenance is performed to retain a system in a satisfactory operational condition by inspection, and subsequent repair or replacement, and by scheduled overhaul, lubrication, calibration, etc. Corrective maintenance is performed to restore an item to a satisfactory condition after failure or after its performance has degraded below that which was specified.

Planning for maintenance also involves establishing requirements for logistic support. Plans must be formulated that integrate the various elements of logistics, e.g., test and support equipment, spare/repair parts and training requirements, with the system's R&M design features. The establishment of the maintenance and logistics requirements is accomplished through analysis of the system's designed-in maintainability and reliability features and attributes.

Although maintenance is performed to retain a system in or restore it to a serviceable condition, poor maintenance practice will actually degrade the condition of the system. Foreign objects left in assemblies, bolts not tightened sufficiently or overtightened, dirt not removed, parts replaced improperly, lubricant improperly installed and other problems brought about by frequent and poorly executed preventive and corrective maintenance will result in significantly lower operational reliability and safety, higher maintenance cost and unnecessary downtime.

Also, the actual field maintenance environment may be other than what was anticipated during development. For instance, a subassembly removed for repair in a desert area may be placed in direct sunlight while awaiting transfer. Component temperatures may exceed those experienced during normal operation for an extended period, thus reducing their life expectancy. Mechanical stresses imposed on components during removal, repair and replacement may exceed that designed for a given environment.

Therefore, critical to effective maintenance, and the assurance that R&M and safety will not be degraded during service, is to integrate into the system's R&M and maintenance engineering program appropriate activities for determining those requirements, tasks and controls necessary to prevent (or significantly reduce) maintenance-induced defects.

The integration of maintenance and R&M engineering practices into an overall RCM program was first done, by the airlines, in the early 1970s. The airlines structured RCM as a broadly applicable new philosophy of maintenance endorsed by the Air Transport Association (ATA), the Aerospace Manufacturers' Associates (AMA), and the US Federal Aviation Administration (FAA). The concept has been referred to (by the airlines) as MSG-2 and more recently, in a revised form, as MSG-3. MSG refers to Maintenance Steering Group, the airline industry body which formulated the RCM concept.

RCM concepts are now extensively applied by the US Department of Defense (DoD) to derive optimum maintenance plans. These RCM derived maintenance plans greatly extend the useful life of a hardware system, prevent a decrease of reliability and/or deterioration of safety, and reduce support cost as well as total life-cycle cost (LCC). Also, a well planned RCM program leads to the establishment of better diagnostics and failure indicators which facilitate those maintenance tasks involving the detection of impending failures and the determination and verification of the condition of the hardware.

RCM is based on the premise that more efficient life-time maintenance and logistic support programs can be developed using a well disciplined decision logic analysis process which focuses on the consequences of failure and the actual preventive maintenance tasks. RCM techniques are applied during the system design and development process — and re-applied after deployment during operation as part of a sustaining engineering activity.

The RCM logic process considers maintenance tasks relative to three areas:

(1) Hard-time replacement (HTR) where degradation, because of age or usage, prior to functional failure can be prevented by a replace/overhaul task at a predetermined, fixed interval (generally in terms of operating time or, for aircraft systems, in flying hours)
(2) On-condition maintenance (OCM) where degradation prior to functional failure can be detected by periodic inspections and evaluations
(3) Condition monitoring (CM) where degradation prior to functional failure can be detected in sufficient time by instrumentation (e.g., temperature, pressure, vibration indicators)

Replacing an item at a fixed interval must be supported by a statistically sound analysis of failures occurring during test or field operation. The failure distribution is determined in addition to the mean time or usage to failure. Hard-time replacement is most effective when there is a close correlation between reliability and age, i.e., the variance about mean time to failure is narrow. Items in this class are generally parts which are subject to metal fatigue or other kinds of mechanical wear which are well correlated with time, or time in use.

Recognition of the limitations of hard-time maintenance approaches has led to emphasis on OCM concepts based on inspection. However, OCM is effective only when potential failure can be ascertained reliably and inexpensively. In defining the OCM tasks, it must be kept in mind that detection prior to failure can be accomplished only if the inspection occurs during the period between the onset of noticeable and unacceptable deterioration and the occurrence of actual failure. The inspection interval selected should be the largest that provides an acceptable likelihood of successful detection.

The most sophisticated maintenance process is based on concepts of 'condition monitoring', in which failure potential is under constant surveillance using built-in test equipment. This is a technology limited approach which previously appeared only in very high-value (safety, cost or mission) items. With the emergence of new classes of effective and inexpensive sensors and microprocessors, condition-monitoring is much more prevalent in modern complex, high technology systems and components.

The complete RCM logic analysis process involves four major steps:

Step 1 Perform failure mode analysis to identify the critical items in a particular end-item, e.g., engine, transmission, rotor system, airframe, etc.

Step 2 Apply the RCM decision logic to each critical item in order to select the optimum combination of HTR, OCM and CM maintenance task requirements or to determine if redesign is needed

Step 3 Implement the RCM decisions by defining specific maintenance tasks, requirements and appropriate intervals for their implementation and by developing necessary data needed for logistics analysis

Step 4 Apply a sustaining engineering effort using actual hardware reliability–age experience data to optimize the process

The key to the process is application of the RCM decision logic (Step 2). As indicated, the logic process is applied to each critical item and judgments are made as to the necessity of various maintenance tasks. The tasks deemed to be necessary, together with the intervals determined to be appropriate, form the total scheduled maintenance program. The RCM decision logic is described in Chapter 2.

Use of the RCM decision logic facilitates the development of high quality maintenance plans in less time and at lower costs. It establishes a system maintenance history and knowledge-base that allows correlation of the preventive maintenance task to the actual experience of specific parts and their failure modes and criticalities. It helps ensure that all safety critical parts and their failure modes are considered in the development of maintenance requirements. Furthermore, the process allows for routine, on-line information exchange among engineering and maintenance staff and management. It provides the capability to manage and audit the RCM process over the entire life-cycle of the hardware item.

The flow of information into and within an RCM program is a continuing process where analyses performed during development are based on reliability design data and analyses performed during field service are based on operational experience data.

During development, the primary determinant of reliability is the design itself. Many systems are designed with redundant or backup equipment in order to function after a failure has occurred. Modern system design, particularly aircraft system design, also incorporates a great deal of built-in test equipment which greatly reduces the chance that a hidden failure will precipitate a sudden, catastrophic event.

The realization of the reliability and safety inherent in the system (or subsystem) design is dependent on the selection and use of properly rated and screened components and the use of proper reliability management and control techniques during manufacture and assembly. Data on component reliability are obtained from the vendors as part of their item descriptions and from such centralized government funded sources as the US Government–Industry Data Exchange Program (GIDEP) and the US DoD Reliability Analysis Center (RAC), where reliability experience data are collected. Data from the equipment manufacturer reliability control plan, and especially the documented results of various screening and burn-in processes, also constitute a rather basic input to the early phases of the RCM process in terms of establishing initial reliability projections.

The design configuration (and all of its associated reliability data) and the system mission description (and its associated stress environment) constitute the basis on which a failure mode, effects and criticality analysis (FMECA) is performed. The resultant FMECA provides a strong foundation for the RCM decision logic and is, therefore, one of most important data sources during development.

In addition to using these well defined data sources and procedures, it is often necessary to fill certain gaps by relying on the seasoned judgment of those who are exercising the RCM decision logic and to rely heavily on their collective experience with similar items.

During the in-service period, as system operational experience is gained, a broad inventory and logistic planning data base comes into existence. This period not only constitutes the end-item environment, but is also the time frame in which some of the most vital RCM data and information are generated and collected. It provides information on the occurrence of specific failures, required removals, repair costs, spares requirements, and much of the economic data needed for the refinement of the initial RCM-derived maintenance program.

The RCM infrastructure in place for an operating system is, in fact, an information management system with the following principal components:

- a system for reporting failures and their consequences
- a system for continuous assessment of reliability–age behavior of significant items
- a system for rigorously controlling the introduction of new maintenance tasks
- a system for periodic reassessment of on-going tasks and for purging those which no longer provide an RCM supportable value
- a system for expeditiously dealing with unanticipated failures at all levels of criticality

The information collected within this system is critical to a number of functions. It constitutes, first of all, the primary data base for essential corrective feedback to the RCM decision process. It is also extremely important, especially in the early operational period, that the RCM logic decisions be reassessed in the light of any new available data in order to eliminate, on a sound basis, any unnecessary or excessive maintenance tasks. It is equally important that any newly encountered (unanticipated) failures be fed back into the system and subjected to the RCM decision process. The experience base is also used for a continuing logical re-assessment of task intervals from a cost-reduction point of view. The experience data base is also used as an input to provisioning and logistic plans and as an R&M corporate memory for the design of new systems and components.

One of the underlying precepts of the overall RCM process is to provide the most cost-effective maintenance program that will yield certain pre-established levels of safety and reliability. The relevant economic decisions are made, to a large extent, in terms of known and projected life-cycle maintenance and logistic support costs. Logistic support sensitivity, therefore, influences the RCM process from the very beginning.

Many of the more useful RCM data are derived from information systems which were put in place for other operational and logistic purposes. The extraction of relevant data is done through a carefully established process governed by priorities which favor safety-dictated necessity over economic desirability. In the case of aircraft, the log book is one of the most important of these existing information systems. Aside from dealing with routine operational and performance parameters, these logs also document any unsatisfactory conditions observed or reported during flight. These observations are organized to alert maintenance crews who, in turn, annotate the repairs made as a result of these reports. After initial use and review, the log books are kept on file for an extended period.

Another important routine source for aircraft is the maintenance information system, which is a record of all maintenance tasks performed. It provides useful information on maintenance frequency and cost for specific items. Identification and routing tags are attached to units removed for repair as a result of scheduled maintenance, or for other reasons. These tags often contain a great deal of information on the nature of the failure, or potential failure, as well as on the conditions under which the problem developed.

When the initial maintenance program is implemented and operational experience begins to accrue, the documentation and reporting system, integral to both the maintenance and logistic support functions, serves as a major source of RCM information for program revision and product improvement. Even with the maturity of any given system development this is clearly a continuing process which intimately links RCM and logistic support.

Many of the techniques and examples presented in this book have been developed and applied by the authors to military systems, and, in particular, to US Army helicopters, where efficient and economical maintenance is essential in achieving operational readiness objectives. In order to provide a background and historical perspective for the techniques and examples, a brief discussion of helicopter development and their reliability characteristics is given in the following paragraphs.

Helicopters can be defined as those aircraft which derive both lift and propulsive force from a powered rotary wing and have the capability to hover and to fly rearward and sideward, as well as forward. The theoretical basis for rotary-wing flight was first established in 1926. Analysis was at first confined to the autogyro, but by 1927 a theory of helicopter performance during vertical ascent was developed, which was then extended in 1928 to cover horizontal flight with the rotor axis vertical. By 1935 the analysis was extended to flight with the rotor axis inclined forward to give a component of rotor thrust. Early experimental work centered around the autogyro; however, by 1938, the era of the helicopter began to emerge when adequate controllability was first demonstrated by a helicopter in the hover mode. At this point it was clear that there were three main categories of rotary-wing aircraft.

1. The classic or 'pure' helicopter, which had no separate means of propulsion, i.e., all of the power was supplied to the rotor or rotors.
2. The autogyro, whose rotor was kept in rotation during flight by aerodynamic forces only, the engine power being supplied to a propeller that provided a forward thrust component for translational flight. The rotor, thus, was only a lifting device.
3. The compound or hybrid helicopter, in which part of the power was supplied to the rotor for producing lift and part to a propellor for providing propulsion. The addition of a fixed wing was used to reduce the lift component provided by the rotor in translational flight. This enabled higher forward speeds to be achieved without encountering severe fluctuations in rotor lift (periodic fluctuations had, in the past, been responsible for high rotor drag and inherent vibrational problems).

From 1940, up until the early 1960's, the overall performance capabilities of helicopters were relatively limited. However, beginning in the late 1950's, technological improvements, including reduction of parasitic drag, improved rotor systems, auxiliary propulsion, and lighter weight structures and engines, resulted in considerable growth in almost all aspects of helicopter operational capability.

The increase in the spectrum of obtainable performance has since then had a major impact on military planning. New operational applications such as attack and heavy lift missions have become feasible and it is now possible to optimize configurations for particular classes of missions, rather than to use only one or two available helicopter types for a complete range of applications, as used to be the practice.

US Army helicopters today are classified according to the general mission they are designed and developed to accomplish. Included are the following helicopter types:

- Attack — A fast, highly maneuverable heavily armed helicopter for combat fire support and escort missions. It can typically be a compound vehicle, i.e., with auxiliary forward propulsion and/or a stub wing used to unload the main rotor in high-speed flight.
- Cargo — A medium or heavy lift class of helicopter that is intended primarily for heavy load-carrying missions. The loads may be carried internally or externally. These helicopters generally have a wide range of center of gravity travel.
- Observation — A small, light helicopter that can be used for a variety of missions including surveillance, target acquisition, command and control, etc. Light armament may be installed.
- Training — A small helicopter usually with seating only for instructor and student-pilot, or a helicopter of one of the other mission classes specifically assigned to the training mission.
- Utility — A class of helicopter that is assigned a wide variety of missions such as medical evacuation, transporting personnel, and/or light cargo loads. Speed and maneuverability are required in order to minimize vulnerability when operating over hostile territory.

Some of the advanced technologies being incorporated into the development of helicopters include:

- Composite materials — The use of composites might permit low cost tailoring of shape vs span, greatly increased tolerance to damage, whether from gunfire or impact, and reduce the complexity and, hence, the cost of such traditionally high cost components as tail rotor systems and main rotor blades.
- Metallurgical developments — The development and successful adaptation of high hardness materials to such components as transmission gearing could permit helicopter main transmission assemblies to handle approximately 20% more power at approximately 10% less weight. Such assemblies and other dynamic components also are being used, which will need little or no

lubrication and which, in emergency situations, will be able to operate for periods without any lubrication.

- Maintenance warning systems — Such systems already appear in present day helicopters and are scheduled for increased use in the next generation of rotary-wing aircraft. These systems are self-checking systems that will warn the operator when they have reached the end of their useful life. This will aid the trend to major subsystems that can be monitored for condition or removed and overhauled on an 'on-condition' basis, rather than on a specific timetable.
- Increased overhaul periods — Significantly increase the time between overhaul periods for dynamic components such as rotors, transmissions, controls and drive shafts, with the trend to eliminating specific periods altogether and going to an 'on-condition' basis for overhaul.
- Noise and vibration reduction — The use of tailored composite materials may even permit drastic reductions in the rotor noise, and possibly the total elimination of the familiar rotor slap. Developments in dynamic isolation might permit reductions of vibration by up to 60% over present day helicopters. This in turn could lead to substantial reductions in total maintenance man hours.
- High-lift airfoils — High-lift rotor airfoils have been derived primarily from the super-critical wing technology and subsequently tailored for helicopter use. These show promise of increasing the coefficient of lift from 10 to 50% over present helicopters.

Regardless of the type, helicopter systems and components can be grouped into standard management segments or functional groups against which data can be compiled and R&M can be evaluated. These functional groups along with a rough indication of their percent contribution to direct maintenance cost are shown in Figure 1–1.

Clearly the dynamic components and the power plant are the primary contributors to maintenance costs. It should be pointed out that maintenance data usually include maintenance and operator damage, equipment scavenging and failures due to environmental causes. During the initial deployment phase, as many as 50% of the maintenance removals of some components have actually been proven to be good units. The accessibility of components has considerable impact on their removal

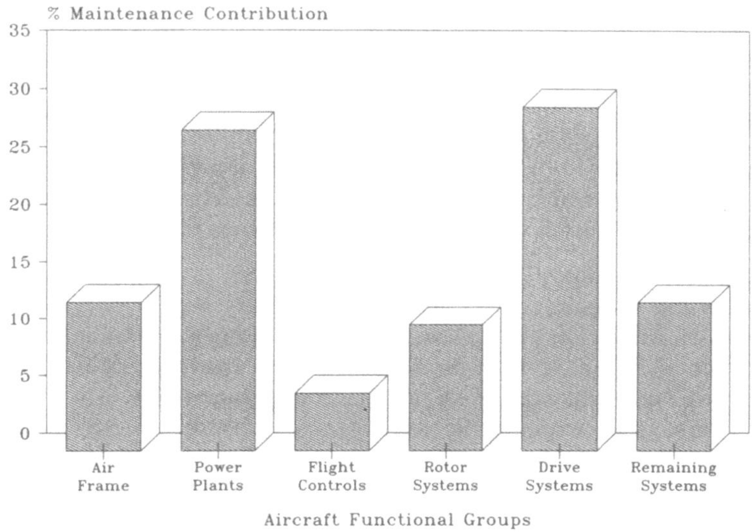

Figure 1–1 Maintenance Contribution: Aircraft Functional Groups

or repair rate. This has been observed on fuel subsystems, for example, where direct maintenance on the fuel control unit is difficult when the engine is installed in the aircraft.

The parts and their failure modes associated with the dynamic components which give rise to their high maintenance support costs are:

- Bearings — Bearing failures contribute significantly to engine and transmission unreliability.
- Gear Teeth (Spalling) — While seldom catastrophic, gear spalling is recognized as potentially being a nucleus for a more serious tooth fatigue failure, if not discovered and corrected.
- Gear Mountings — Gear mountings may be prone to fretting deterioration, particularly in bevel gearing where the attachment of the gear to the shaft is made through splines of bolts. Fretting is a time dependent phenomenon and exists to some degree of severity at nearly every unlubricated interface.
- Housings — Cracks have occurred in magnesium cases. Occasionally they are the result of random flaws in material and processing, but more often they occur in unflawed castings as the result of vibratory stresses introduced externally.

- Seals — Seals exhibit a wearout failure mode that results in leakage, and are additionally sensitive to handling and external environment.
- Spacers, Bearing Liners and Retention Hardware — Spacers, liners and other components required to locate bearings have proven to have high failure rate wear problems. Bearing locknuts and other retention hardware have occasionally backed off, sometimes with catastrophic results. A high proportion of locknut failures may be related to the maintenance interval.

The primary objective of the RCM process is to determine that combination of maintenance tasks for a complex system or component item which will significantly reduce the major contributors to unreliability and maintenance cost, such as those described above for helicopters, in light of the consequences of failure, particularly on safety. This process, which first uses engineering data during development to establish preliminary maintenance support requirements and then actual experience data after deployment to refine and update the requirements, is fully described in Chapter 2. The special engineering techniques needed to implement an effective life-cycle RCM program are described in Chapter 4.

CHAPTER 2

The RCM Program

While oriented originally to the needs of the commercial air transport industry, RCM is a discipline which has application and acceptance within the US Department of Defense (DoD) and is now being applied, as well, to nuclear power systems and other complex high technology systems. Each service within the US DoD was tasked to adapt the airline maintenance concept to military equipment. Although the RCM concept is structured somewhat differently by each service, it has since been applied to a number of US military aircraft programs, either in part or in total.

Application of RCM involves evaluating maintenance based on the design characteristics and operational functions of the system and its failure modes and consequences. It requires the application of a detailed knowledge-based logical analysis to select those condition monitoring (CM), on-condition maintenance (OCM) and hard-time replacement (HTR) maintenance tasks that are most effective in preventing the system's significant part failure modes.

Using available system safety and reliability (historical failure mode and effects) data, RCM identifies those parts which are critical in terms of mission accomplishment and operating safety. It determines the feasibility and desirability of maintenance, it highlights maintenance problem areas for redesign consideration, and it provides supporting justification for maintenance. The driving force in RCM analysis is to reduce the scheduled maintenance burden and to eliminate excessive support costs while preserving the inherent level of reliability and safety designed and built into the hardware system.

RCM is based upon the premise that maintenance cannot improve upon the safety or reliability inherent in the design of a hardware system. Good maintenance can only preserve these characteristics. The RCM philosophy dictates that maintenance shall be performed on critical parts only when it will prevent a decrease in reliability and/or deterioration of safety to unacceptable levels or when it will reduce life-cycle cost. It further dictates that maintenance shall not be performed on noncritical parts unless it will reduce life-cycle cost.

The process addresses individual part failure modes. Thus, for a large, complex component, different maintenance tasks could be specified, because of the various possible failure modes that may exist. As an example, a given component might undergo condition monitoring during normal operations to detect a particular failure mode, while still having an on-condition inspection or a hard-time replacement requirement for a failure mode that is not detectable during routine operator monitoring.

The logic process is applied first during design and development to establish the system maintenance requirements. Because it may be necessary during early development to make decisions with only limited or predicted information, the logic process allows the selection of the most conservative maintenance approach 'by default'. The logic process is again applied during operation, as available data move from a predicted state to actual operational experience values, to re-examine the RCM decisions and to modify or to confirm default-based decisions.

Once all parts and their significant failure modes have been subjected to the RCM logic process, the resultant data are evaluated to arrive at the maintenance tasks requirements. The individual part requirements are ultimately merged into a complete maintenance plan that includes the intervals and sequences for performing the individual scheduled maintenance tasks.

2.1 AIRLINE INDUSTRY RCM PROGRAMS

Much of the pioneering work on RCM was done by the commercial airline operators and manufacturers through the US Air Transport Association (ATA). One of the earliest formal treatments of the subject can be found in the ATA's 1968 Handbook MSG-1, 'Maintenance Evaluation and Program Development', which was developed jointly by several airlines for use with the Boeing 747. The decision logic was later updated and the procedures were generalized for application to other emerging wide-body aircraft (DC-10, L-1011) expected to enter the inventory during the 1970's. This version of the handbook was designated MSG-2. It has been the practical guide to RCM practice for more than a decade.

In 1980, the ATA published the 'Airline/Manufacturers Maintenance Planning Document', MSG-3.[1] Maintenance programs for the Boeing 756 and 767 were developed under MSG-3 and for the A-300 and Concorde under the 'European MSG-3'. Although MSG-3 represents a major revision of MSG-2, the fundamental concept has not been abandoned. However, implementation and understanding have been greatly facilitated by a task-orientation in the decision logic in lieu of the original maintenance process approach. Like its predecessors, MSG-3 is organized as a planning document to coordinate the participation of specialists from the aircraft operators and manufacturers, and from the appropriate national regulatory authorities. This approach is designed to achieve a broadly acceptable program in an efficient manner. The essence of the process remains the disciplined use of a decision logic to yield a maintenance program which preserves the designed-in levels of safety and reliability at the lowest possible cost.

The equipment already in service has not been exempted from the RCM approach. A number of airlines have introduced RCM on an in-service basis for their fleets of 727's, 737's and DC-8's, and in fact these aircrafts served as test beds for the early development of RCM.

RCM is now accepted as a permanent part of the airlines' aircraft development, operating and maintenance practice. It is also expected to continue to evolve as experience indicates ways in which it can be improved. The greatest testimonials to RCM suitability are its introduction into the US DoD aviation services and, now, an increasing indication of its adaptation to non-aviation products, devices and equipments through the commercial and industrial segments of our society.

RCM was developed within the commercial airline industry when it became apparent that traditional maintenance practice, with its emphasis on rigidly scheduled inspection, servicing and removal, was not impacting reliability as expected. This became increasingly obvious as more complex aircraft systems were introduced and the relationship between reliability and preventive maintenance tended to disappear. In contrast to the traditional approach, current RCM practice within the airline industry focuses on the consequences of failure in a prioritized hierarchical structure (top-down approach) and uses a decision logic process to develop an optimum minimum cost maintenance program. The RCM development process begins with design and extends throughout the system's service life. During this latter period, information from a well organized experience base ('age exploration') is used to effect product improvements and to serve as a guide to the design of new systems.

While RCM development is generally thought of as commencing with the implementation of the decision process, this step is really dependent on the prior establishment of certain objectives and the organization of a considerable amount of preliminary data.

The process really begins with the enumeration of the aircraft operational performance requirements including both economic and operational factors. The available technology determines whether or not the users' expectations can be met and, if so, in what general form. These characteristics begin to emerge as the design proceeds from conceptual to specific and detailed levels. The process eventually defines specific components, devices, subsystems and equipment for inclusion in the overall aircraft design. Each of these elements has a reliability profile consisting of some combination of empirical experience, including test results, and analytically derived, predictive, failure rate information. In addition, each design iteration is subjected to a failure mode, effects, and criticality analysis (FMECA) in order to understand how each element is likely to contribute to system unreliability within the proposed configuration. When the design process proceeds in this manner the essential information for the RCM decision process is automatically provided. Note, however, that the design is not frozen when the RCM decision process begins, but is allowed to evolve. The maintenance program is optimized within this process and the interaction continues throughout system life as more information (age exploration) becomes available for product improvement and redesign.

The RCM decision process for systems and power plant items (non-structural elements) begins with an identification of maintenance significant items (MSI's), the failure modes and effects associated with each, and their known or predicted failure rates. This information is already at hand if FMECA and other reliability engineering tasks have previously been implemented within the design phase.

The RCM decision process itself employs a progressive logic diagram for each identified failure. The output is a set of tasks and intervals which constitute the total scheduled maintenance program. Task intervals are initially set on the basis of relevant prior experience and available data, or, when there is little firm basis, reliance is placed on the intuitive judgment of an experienced working group.

A radically different RCM decision process from the one used for power plants and systems is used for structures. This uniqueness derives from the different way in which structures fail and the much heavier reliance on visual inspection to determine condition. The overall objective, however, remains the same: the development of a maintenance program keyed to the consequence of failure and designed so as to produce the required level of reliability and safety at the lowest possible life-cycle cost. The primary inputs to the process of developing the program are the known, or predicted, susceptibility of various components to damage and the degree of difficulty involved in the detection of this damage.

The decision process for structures begins with the identification of structurally significant items (SSI's): those whose failure would result in a reduction of aircraft residual strength or in direct loss of structural function. The initial maintenance program for non-SSI's is developed rather directly from past experience with similar items and from the manufacturer's recommendations.

The SSI's, on the other hand, are subjected to a much more detailed analysis, similar in complexity to that conducted for aircraft systems and power plants. The first step in this procedure categorizes items as either damage tolerant or safe life. An item is considered to be damage tolerant if the remaining structure retains its essential integrity until the damage can be detected by ordinary scheduled maintenance. Safe-life items are not damage tolerant and reliability is achieved through removal of such items from service before failure is likely to occur.

Damage-tolerant SSI's are subjected sequentially to a full decision process which assesses susceptibility to deterioration from fatigue, environmental effects and accidental damage. Any item which fails to survive this process is either reclassified as safe life or is subjected to redesign. The level, method, threshold and frequency of inspection are then established and a program to explore the age characteristics of the SSI's is applied.

Safe-life SSI's are subjected to a more limited process which assesses susceptibility to deterioration from environmental effects and accidental damage. The inspection parameters are determined for these SSI's, but an age-exploration program is not applied.

The overall program is then assembled from an overlay and collation of the functions developed for the two classes and the preparation of a consolidated, documented program plan. Even after a properly structured maintenance program for a given system is put in place by the using airline it remains a dynamic element within the on-going RCM process. From a RCM point of view, it is mandatory that a system be maintained for monitoring and compiling the reliability characteristics of the equipment under actual operating conditions. This infrastructure is to be in place from the time an aircraft system first enters the operating inventory of a user until it is retired from service.

2.2 THE US ARMY AIRCRAFT RCM PROGRAM

The US Army's RCM Program is conducted in accordance with AMC-P 750–2, 'Guide to Reliability-Centered Maintenance'.[2] This document explains in detail how to use the RCM decision logic in conjunction with the failure mode, effects and criticality analysis (FMECA) to develop a scheduled maintenance plan including the tasks and intervals for preventive maintenance checks and services, and provides information for overhaul, age exploration, economic analysis and redesign. The document is used with US MIL-STDs-1338–1A and -2A[3,4] for establishing and implementing a Logistic Support Analysis (LSA) program.

The RCM program identifies the maintenance needed to ensure the preservation and/or restoration of the inherent reliability, safety and mission accomplishment of aircraft systems and components at a minimum expenditure of resources and to prevent indiscriminate maintenance which is not cost-effective. Its specific objectives, as delineated in AMC-P 750–2, are to:

- Establish design priorities which facilitate preventive maintenance
- Plan preventive maintenance tasks that will restore safety and reliability to their inherent levels when equipment/system deterioration has occurred
- Obtain the information necessary for design improvement of those items whose inherent reliability proves inadequate
- Accomplish these goals at a minimum total cost, including maintenance costs and the costs of residual failures

The essential activities in the RCM process include: (1) developing required part criticality input data, (2) applying the part criticality data to the RCM decision logic process to select effective and applicable maintenance tasks, (3) recording the decisions, (4) determining task intervals, (5) implementing the RCM maintenance task decisions and intervals and (6) applying a sustaining engineering activity effort, after deployment, to assess, verify and adjust the RCM decisions and the subsequent maintenance program and LSA data and outputs.

The key to the complete process is the use of the RCM decision logic. The logic used for US Army aircraft is based on the airlines' MSG-3 decision logic and is described in Section 2.3. The decisions resulting from the use of the logic form the maintenance program plan and also dictate any changes that are to be made to the technical data package for the hardware item. Backup information covering the changes is maintained on file in hard copy to provide the necessary audit trail.

A failure mode analysis is performed to identify the critical parts and to obtain the criticality data used as input to the RCM decision logic. Procedures for performing failure mode analysis, including FMECA, are given in Chapter 4. Each critical part failure mode is subjected to the questions in the logic diagram and the answers are recorded on a worksheet provided for this purpose or are entered directly into a computer system. The logic process forces the selection of the most effective and applicable hard-time, on-condition and condition monitoring maintenance tasks. Intervals for each on-condition and hard-time preventive maintenance task are individually determined. The maintenance tasks are then combined into common intervals wherever possible to reduce overall costs and scheduling complexity.

The RCM logic decisions are tracked into the field to establish the success of the program. All of the RCM decisions are constantly refined and updated through a sustaining engineering activity during field operation. This is accomplished through continuous monitoring of failure experience and subsequent failure analysis and product improvements. Chapter 6 describes a safety prediction model developed specifically for Army aircraft that can be used with field experience data to reapply the RCM decision logic as part of the sustaining engineering activity.

A complete RCM program has been established for US Army aircraft. The program provides organized methods for planning, managing and conducting the required life-cycle RCM analysis. The program includes six major on-going functions:

(1) Review Team Monitoring

A review team is established for each development program to monitor the adequacy of the RCM effort and, in particular, to ensure an effective interface with the integrated logistic support (ILS) activity.

(2) RCM Technique Development and Improvement

The specific on-condition, condition monitoring and hard-time maintenance tasks are reviewed continually and improved, as necessary, and integrated into the system maintenance programs. The RCM decision logic methodology is also continually reviewed and updated, if necessary, to reflect the changing needs and characteristics of Army aircraft systems and components.

(3) RCM Requirement Preparation

RCM requirements, i.e., R&M logistic support concepts/parameters, are prepared for inclusion in contractual and program documents. Detailed schedules for RCM analysis and planning for the system development cycle as well as for the sustaining engineering activity performed during the operational phase are prepared and reflected in the program requirement documents. Also, contractual requirements for RCM are cross-referenced to ILS requirements.

(4) RCM Application for Developmental Systems

RCM tasks are applied throughout the system's life-cycle.

During Concept Exploration

The RCM tasks and how they interface with the ILS process are planned during this phase. The planning addresses the interrelated objectives of R&M and the logistic support and maintenance plan. The RCM program plans are prepared in accordance with the approach, milestones and funds required to implement RCM as specified in the appropriate requirement documentation. Historical data from existing systems are reviewed to relate past experience to the logistic support requirements of the new system. Logistic support alternatives are evaluated, including the anticipated scheduled maintenance burden of each alternative and any anticipated advancement in reliability and safety design techniques which would impact the projected maintenance burden. The maintenance concept is established, including identification of the capability required to retain those safety and reliability parameters incorporated during design.

During Demonstration and Validation

- An FMECA (functional) is performed, criticality thresholds are allocated to the system and subsystems consistent with R&M and safety objectives, and any needed design improvements are determined and corrections identified. Failure modes remaining in the nonacceptable range are designated as design deficiencies to be corrected during the development phase.
- RCM decisions are recorded in the system logistic support analysis records (LSAR) and updated (by application of RCM logic analysis) as the design matures.
- Maintenance task requirements, including contract requirements, based on the maintenance concept and on RCM logic considerations are developed and incorporated into the ILS requirements.
- RCM implementation is monitored by the review team.

During this phase sufficient data are documented to support the development of firm R&M/logistics support objectives for inclusion in the subsequent requirement documents.

During Full-Scale Development

- RCM documentation is updated (by application of RCM logic analysis) at the beginning of this phase and thereafter as required.
- FMECA's (hardware) are performed to an indenture level one higher than the lowest level at which corrective and preventive maintenance are prescribed. An analysis of the failure modes is performed at the lowest level where corrective and preventive maintenance are prescribed. Also, historical data from existing systems are reviewed to assure the adequacy of the data base prior to the subsequent RCM logic analysis.
- A complete RCM logic analysis is performed to define specific condition monitoring, on-condition and hard-time replacement maintenance tasks or to identify the need for design improvements.
- Maintenance tasks are specified for each maintenance significant component failure mode in accordance with the maintenance task(s) identified by the RCM logic analysis.
- Corrective maintenance analysis is performed to determine maintenance tasks that are required for each repairable item. This analysis takes into account the design characteristics of the equipment as reflected in the R&M analyses.

- The adequacy of the overall maintenance plan, including the preventive maintenance checks and services and assigned maintenance levels, is evaluated during this phase in accordance with the appropriate test plans.
- RCM implementation is monitored by the review team.

During Production and Deployment

- Maintenance procedures are developed for maintenance significant items in accordance with the specified maintenance tasks derived from the RCM logic analysis.
- Maintenance requirements derived from RCM analyses are incorporated into the appropriate technical publications. This includes scheduling and preparing/procuring changes or revisions of the applicable technical publications.
- Appropriate intervals are determined for each maintenance task. Also, the time required for military personnel to perform scheduled maintenance is reviewed to determine if it is reasonable and is the minimum essential for the retention of safety and reliability that was designed into the system (i.e., a burdensome maintenance system cannot compensate for an inadequate design).
- The maintenance concept/plan is fully demonstrated.
- A sustaining engineering program is instituted.
- Feedback (both field and engineering) data are assessed to determine the extent that the field is actually performing the scheduled maintenance services and to identify needed adjustments to the maintenance and reliability improvement programs.

(5) RCM Application for Fielded Systems

The RCM tasks for fielded aircraft systems are:

- Performing a failure mode analysis based on field experience data.
- Performing RCM logic analysis using the Army aircraft decision logic described in Section 2.3.
- Selecting maintenance tasks and intervals, developing logistics data from the RCM logic analysis and preparing improved maintenance plans that integrate the RCM tasks and ILS data into a complete, cost-effective maintenance program.

- Performing reviews of existing Depot Maintenance Work Requirements (DMWR's) and supplemental documentation and assuring that the requirements reflect the RCM philosophy and data requirements. The reviews include assessing the savings, i.e., man hour savings, man hours cost savings, parts cost savings, material cost savings and net cost savings resulting from the application of the RCM logic process.
- Planning and implementing the airframe condition evaluation (ACE) program.
- Performing age-exploration analysis on those items identified by the RCM logic analysis where a failure relationship between age and reliability must be established.

(6) RCM Data Bank

A complete data base is maintained to support the performance of RCM logic analyses as well as other RCM/ILS tasks.

An essential function of the RCM program is its application to aircraft that have been in service for many years and are being supported by scheduled maintenance programs that were not developed by the RCM logic process. For in-service aircraft, the RCM logic is based on actual experience data from the RCM data bank and, thus, there will be less need for default answers. The application of the decision logic should result in an increased use of on-condition maintenance tasks and a significant reduction in the overall scheduled maintenance workload.

The RCM program for in-service aircraft involves many tasks as delineated in item 5 above. A key task is the updating of the DMWR's to include preshop analysis (PSA) as an integral part of the maintenance process. PSA is performed on aircraft systems and components upon their induction into the depot. Through PSA the condition of an item is determined prior to performing maintenance and used as the basis to determine only those repair or overhaul procedures which are essential to return the item to a satisfactory serviceable condition.

PSA is a logical inspection process with the inspection focusing on the reason(s) the item was sent to the depot, the component operating times and the condition of the hardware. The inspector physically and functionally inspects for condition, based on guidelines, in the DMWR, and then specifies the extent of disassembly, repair and part replacement required to return the item to a serviceable status. As a result, unnecessary overhaul is eliminated and maintenance is performed only when the condition warrants it. The guidelines for assessing the condition of aircraft components are reviewed continually. Improved PSA guidelines and criteria are included in revisions to the applicable DMWR. The process to review and revise the DMWR's to reflect the latest on-condition PSA techniques is described in Chapter 5.

Another key task for in-service aircraft is planning and evaluating the airframe condition evaluation (ACE) program. The ACE program involves an annual evaluation of aircraft to determine their condition. It uses a profiling technique for evaluating aircraft condition and for identifying items most in need of depot maintenance. The evaluation is based on representative indicators of structural condition specifically selected for each aircraft type. Typical indicators include the condition of the main lift beam, the nose fuselage skin and the upper bulkhead. Also, the external areas of the aircraft and its components, both structural and dynamic, are examined for deterioration caused by corrosion. Record keeping and the manipulation and analysis of data are accomplished by automated means. Portable automated structure inspection equipment, including a device to measure the depth of corrosion, are used by the inspection team for quick, efficient and accurate on-site evaluations.

The ACE program is continually reviewed and updated in order to increase its efficiency. Emphasis is placed on improving data/trend analysis and improving feedback of deficiency information to more readily identify maintenance problems for timely resolution and on using the data to correct problems at the maintenance organizational levels. The methodology and mathematical/statistical formulae are reviewed to incorporate the latest experimental design and statistical process control (SPC) techniques. The diagnostic equipment and on-site inspection techniques are also reviewed and updated to further improve the efficiency and accuracy of the on-site evaluations. The ACE program is described in more detail in Chapter 5.

Effort is also made to establish the methods for predicting the remaining safe and useful life of aviation components which are used during PSA and ACE to assess the economic benefits and the safety concerns that impact replace or repair decisions and to establish specific criteria for assessing remaining life. This includes evaluating techniques used to measure the amount of degradation caused by various material failure mechanisms, i.e., fatigue, creep, creep/fatigue corrosion, stress corrosion, erosion, corrosion fatigue, and thermal embrittlement. It includes evaluating measurement techniques (both nondestructive and destructive) such as metallographic replication and testing of miniature specimens. It also includes evaluating analytical procedures used to model and estimate damage accumulation and remaining life which integrate materials' properties, component configurations and operating parameters.

To help prevent and control corrosion in Army aircraft, the RCM program also includes an on-going task to assess the extent of corrosion and its cost, investigate corrosion detection and prevention techniques, and formulate specific recommendations to prevent or reduce corrosion in both developmental and fielded Army aircraft. The corrosion control effort emphasizes early detection and correction as well as data/trend analysis and timely feedback of deficiency information. It involves identifying the most significant corrosion critical items by taking into account the type and stage of corrosion in an item in light of its structural function and criticality. It includes developing special repair procedures, specifying wear limits, and establishing other actions, including application of ion implantation and plasma coating techniques, to correct the problems.

In addition to establishing the on-condition maintenance tasks for in-service aircraft, the RCM program also emphasizes the development of on-board condition monitoring (CM) maintenance systems for the detection of system degradation prior to functional failure. The intent of these systems is to identify safety critical conditions (and the appropriate actions to be taken) or any degradation in performance to the pilot early enough to safely terminate the flight and to effect repairs. They would provide fleet wide component trend data, on-board fault annunciation and prioritization and automated records.

A complete CM maintenance system would involve the use of an on-board monitoring system with sensors, a flight data recorder, fault analyzer and cockpit display, and would be supported by a ground-based computing system. It would use artificial intelligence and integrated diagnostics to detect system faults and predict future maintenance requirements, and expert system software to process and analyze the performance data in order to perform trend analysis and fault isolation and diagnosis and to provide a prediction of the remaining useful life of the monitored system.

These maintenance systems are also continually evaluated as part of the RCM program to determine if improvements can be incorporated that would further increase their effectiveness and decrease the cost of maintaining an aircraft. Emphasis is placed on improving the fault isolation capabilities as well as the diagnostic data obtained on each aircraft to identify hidden problems for correction during scheduled maintenance. The early discovery of hidden deterioration allows repair before secondary damage occurs on other related parts, thus further lowering the cost of maintenance. It also reduces the number of unnecessary removals and replacements of serviceable components and thus helps in reducing spare part shortages.

The performance of RCM logic analyses as well as other RCM/ILS tasks requires the availability of an extensive and cumulative base of data and information. Consequently, data are compiled for this purpose and a complete, user-oriented on-going RCM data bank is maintained. The data bank is continually expanded to include the most recent field experience information as well as component repair data, and data from the depot, the field and other overhaul facilities.The data are used to update the part failure factors used as the basis for RCM/ILS analysis and to perform reliability–age-exploration analysis.

It should be noted that in the early stages of a system life-cycle, the reliability–age relationship may not be perfectly understood, causing the selection of conservative maintenance tasks and intervals. Data in the RCM data bank are used as the basis to perform age-exploration analysis and ultimately to adjust the intervals for scheduled maintenance as well as to validate the overall maintenance program.

Once data for inclusion in the data bank have been validated, codified and analyzed, they are entered into the bank and used to generate reports, compute composite failure factors, support special projects, support determination of maintenance requirements, determine component reliability–age characteristics and resulting scheduled maintenance changes, and provide direct data for spare part provisioning and/or to revise applicable LSA records as part of a sustaining engineering activity.

The data output products are constantly refined to improve their adequacy in supporting reliability analysis, variability analysis, logistics analysis, RCM program effectivity analysis, age exploration, critical characteristic identification and in meeting other user needs. Emphasis is placed on providing improved part failure rates, FMECA and reliability–age data in support of the RCM/ILS process.

In addition, the RCM program procedures are revised and refined to improve the process of developing uniform and complete RCM-based maintenance programs using data from the RCM data bank, facilitate application of the logic analysis and the use of reliability–age experience data within the overall RCM/ILS process, and establish a more effective RCM and ILS interface. The intent is to improve and speed up the process of performing failure mode analysis and applying the RCM decision logic process as well as to reduce cost, enhance the uniformity of treatment and provide a more organized, accessible data and procedural audit trail.

Training is also provided to develop the necessary engineering skills and knowledge required to plan and implement the RCM methodology on aircraft systems and components. The training program covers FMECA, RCM decision logic analysis, maintenance task and interval specification, age-exploration and maintenance planning in accordance with the RCM driven ILS process. Also, maintenance personnel are trained in the latest repair procedures and practices and the use of special testing and diagnostic equipment.

The following section describes the RCM decision logic process and how it is applied to select appropriate hard-time replacement, on-condition and condition monitoring maintenance tasks. Section 2.4 discusses how the results of an RCM analysis are used to determine logistic support plans and requirements.

2.3 THE DECISION LOGIC PROCESS FOR US ARMY AIRCRAFT

A key step in the RCM process is the application of the decision logic to facilitate the rapid development of uniform, complete and cost-effective RCM based maintenance programs. This section describes the RCM decision logic used by the US Army to develop maintenance support programs for aircraft.

Specific questions are asked in an established sequence for those parts considered critical. The RCM logic is first applied during development to the critical parts and their failure modes identified by failure mode analysis (failure mode analysis techniques are described in Chapter 4, Section 4.3). Actual experience data are used later, after deployment, as part of a sustained engineering effort to eliminate default decisions and to optimize the program. The driving force is to reduce the scheduled maintenance burden and support cost while maintaining the necessary operational readiness state.

Implementation of the logic process requires a working familiarity with the RCM process, the design characteristics and functions of the system or component under evaluation and the structure, organization and capabilities of the overall maintenance support system that is in place. Certain data elements must also be available to the engineer for his use in applying the decision logic. These data elements include a complete description of the current maintenance program (or the maintenance program used on a similiar hardware system) and the aforementioned failure mode data. These data help the engineer in making the proper decisions and in selecting the maintenance tasks. The data are also used later to help in establishing the maintenance task intervals and determining the necessary logistic support requirements.

The RCM logic forces a progressive determination to be made on how impending failures can be detected and corrected, before they develop into major defects. The intent is to preserve, to the maximum degree possible, the inherent levels of reliability and safety designed into the system or component and to increase the cost-effectiveness of the overall maintenance support program by reducing the probability of failure and by detecting any hidden failures that may have occurred.

The RCM decisions are implemented by defining the specific requirements for each selected task and determining appropriate intervals for their implementation. The tasks cover lubrication and servicing, operational checks, inspection and functional checks, rework (repair, overhaul and rebuild) and replacement, as necessary, to reflect the decisions and actions resulting from the logic analysis. For example, if an OCM task is determined by the logic process to be applicable and effective in detecting a potential part failure mode, the actual task may involve examining the condition of a hardware item using a specific checklist, inspection procedure, standard or Army regulation. It may include a functional check to determine if one or more functions of the item performs within specified limits.

If a crew monitoring task is applicable and effective, the task may involve monitoring by the crew of instrumentation and recognition of potential failures through the use of normal physical senses (e.g., odor, noise, vibration, temperature, visual observation, changes in physical input force requirements, etc.). For a monitoring task to be effective, a reduced resistance to failure must be detectable and the rate of reduction in failure resistance must be predictable. Indicators that annunciate failures at the time of occurrence are not applicable. Examples include the detection of leaky seals through noting excessive oil consumption or smoky engine operation, the detection of clogged fuel nozzles by difficult engine starting, and the detection of minor cracks in engine components by a decrease in available engine power.

Similarly, if a replacement task is applicable and effective, the task may involve substituting a serviceable part, subassembly or module for an unserviceable counterpart. It requires removal from service of an item at a specified life limit. This task is normally applied to so-called single celled parts such as cylinders, engine disks, safe-life structural members, etc. The item must show functional degradation characteristics at an identifiable age and a large proportion of units must survive to that age. Component items with hard-time replacement limits are replaced during the overhaul process.

As with other reliability and logistic analyses and tasks, the logic process is reapplied as available data move from a predicted state to actual operational experience values with a higher degree of certainty. In developing the initial maintenance program within the RCM process, it is frequently necessary to make action decisions without adequate information. The RCM decision logic is structured to yield sure-safe practices in these situations through the selection, 'by default', of the most conservative course.

This practice of employing default logic is the safest course, but it is also the most expensive. Consequently, one of the most urgent steps to be accomplished, once a system is operational and an experience base begins to accumulate, is to reassess all RCM default decisions. The objective is to eliminate excessive maintenance costs while retaining established and required levels of reliability and safety.

The US Army's RCM decision logic is presented in Figure 2–1 following and in Figures 2–3 through 2–6 presented later in this section. It is based on the airlines' MSG-3 decision logic, but specifically tailored for Army aircraft systems and components. Each identified part failure mode is processed through the logic so that a judgment will be made as to the necessity of a task. The resultant tasks and intervals form the total scheduled maintenance program.

There are two major steps in the logic process. The *first* step (Figure 2–1) is to evaluate each part failure mode for determination of the appropriate consequence category which prioritizes the failures to be prevented. The *second* step (Figures 2–3 through 2–6) is to evaluate and select for each consequence category the most effective and applicable maintenance task combination that can prevent the failures or reduce their likelihood of occurrence.

The logic diagram is designed to lead, through the use of standard assessment questions, to the most effective preventive maintenance task combinations. Simple 'YES' or 'NO' answers are recorded on a worksheet, such as that illustrated in Figure 2–2, or entered directly into a computer system. This decision logic is applied to each critical part

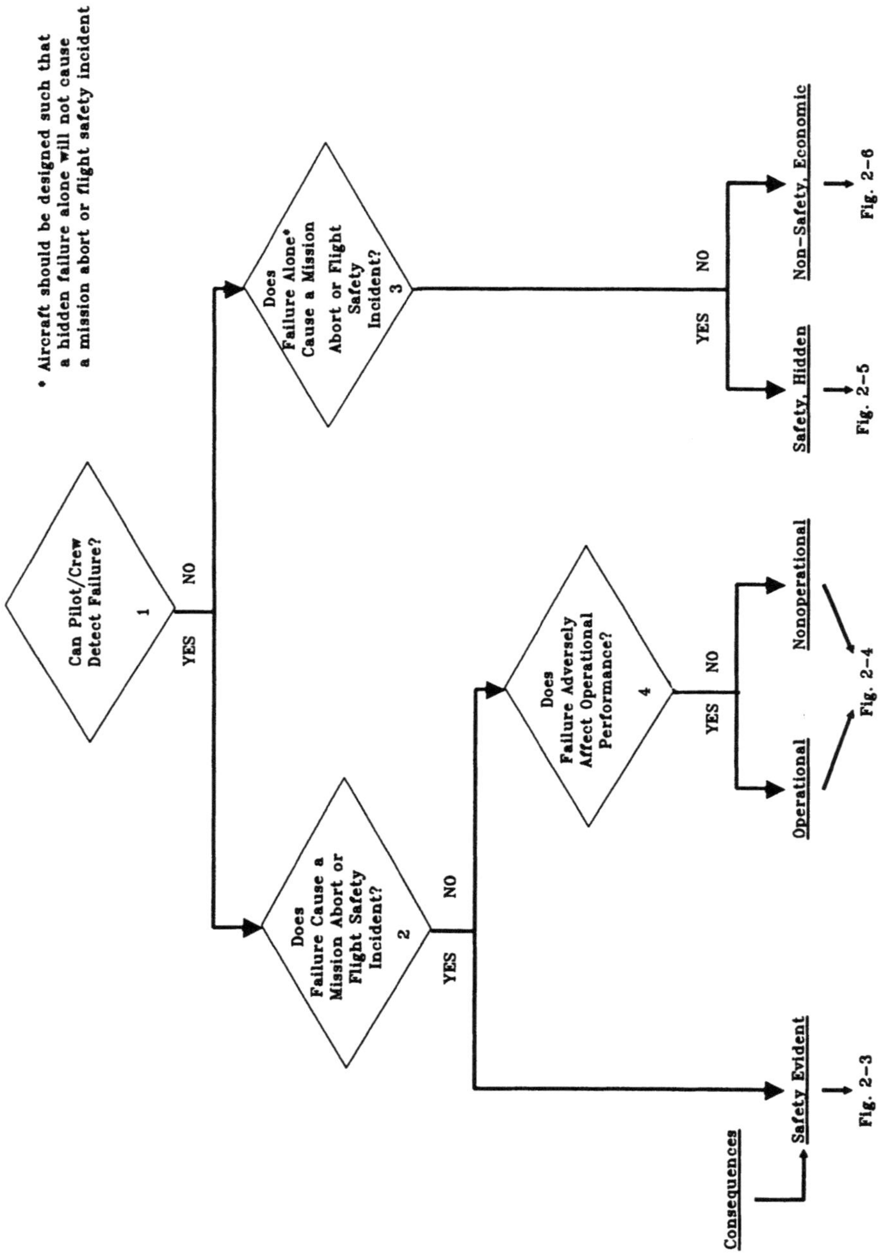

Figure 2-1 Army Aircraft RCM Decision Logic: Consequence Categories

MAINTENANCE PROCESS ANALYSIS WORKSHEET				
MAJOR ITEM	PREPARED BY		PREPARING ORGANIZATION	
NOMENCLATURE	PART NO.	DATE	REVISION NO.	
FAILURE MODES	A	B	C	D

NO.	LOGIC QUESTION	FM	Y	N	INFORMATION SUMMARY
1	Can Pilot/Crew Detect Failure?	A			
		B			
		C			
		D			
2	Does Failure Cause a Mission Abort or Flight Safety Incident?	A			
		B			
		C			
		D			
3	Does Failure Alone Cause a Mission Abort or Flight Safety Incident?	A			
		B			
		C			
		D			
4	Does Failure Adversely Affect Operational Performance?	A			
		B			
		C			
		D			

FAILURE MODES CONSEQUENCE CATEGORY	A	B	C	D

NO.	LOGIC QUESTION	FM	Y	N	INFORMATION SUMMARY
* 5, 11, 16, 22	A Servicing Task?	A			
		B			
		C			
		D			
6, 12	A Crew Monitoring Task?	A			
		B			
		C			
		D			
17, 23	Verify Operation?	A			
		B			
		C			
		D			
7, 13, 18, 24	An On-Condition Task?	A			
		B			
		C			
		D			
8, 14, 19, 25	A Rework Task?	A			
		B			
		C			
		D			
9, 15, 20, 26	Replacement?	A			
		B			
		C			
		D			
10, 21	A Combination of Tasks?	A			
		B			
		C			
		D			

* Identify Applicable PM Task Question Numbers

Figure 2–2 Maintenance Process Analysis Worksheet

failure mode and judgments are made as to their consequences and the necessity of various maintenance tasks. The tasks deemed to be necessary, together with the intervals determined to be appropriate, form the total scheduled preventive maintenance program.

The determination of the proper consequence category (step 1) requires answering the following questions:

Question 1: *Can pilot/crew detect failure?*

This question asks if the operating crew will be aware of the failures during performance of their normal operating duties, i.e., monitoring of instrumentation and through the use of normal physical senses (e.g., odor, noise, vibration, temperature, visual observation, changes in physical input force requirements, etc.). If the answer is 'YES', the failure is evident and the process proceeds to question 2. If the answer is 'NO', the failure is hidden and the process proceeds to question 3.

Question 2: *Does failure cause a mission abort or flight safety incident?*

This question asks if the failure by itself, not in combination with other functional failures (i.e., no redundancy exists and it is a primary item), will cause a mission abort or flight safety incident. A 'YES' answer indicates that the maintenance tasks are to be developed in accordance with the safety evident consequence category and that task development proceeds in accordance with questions 5 through 10 (Figure 2–3). A 'NO' answer indicates an economic effect and that question 4 must be asked.

Question 3: *Does the failure alone cause a mission abort or flight safety incident?*

This question asks if the failure alone, or in combination with an additional functional failure, will cause a mission abort or flight safety incident. Note that a design objective for an aircraft is that a hidden failure alone should not cause a mission abort or flight safety incident. A 'YES' answer to this question indicates that the maintenance tasks are to be developed in accordance with the safety hidden consequence category and that task development proceeds in accordance with questions 16 through 21 (Figure 2–5). A 'NO' answer indicates a non safety economic effect and that maintenance task development proceeds in accordance with questions 22 through 26 (Figure 2–6).

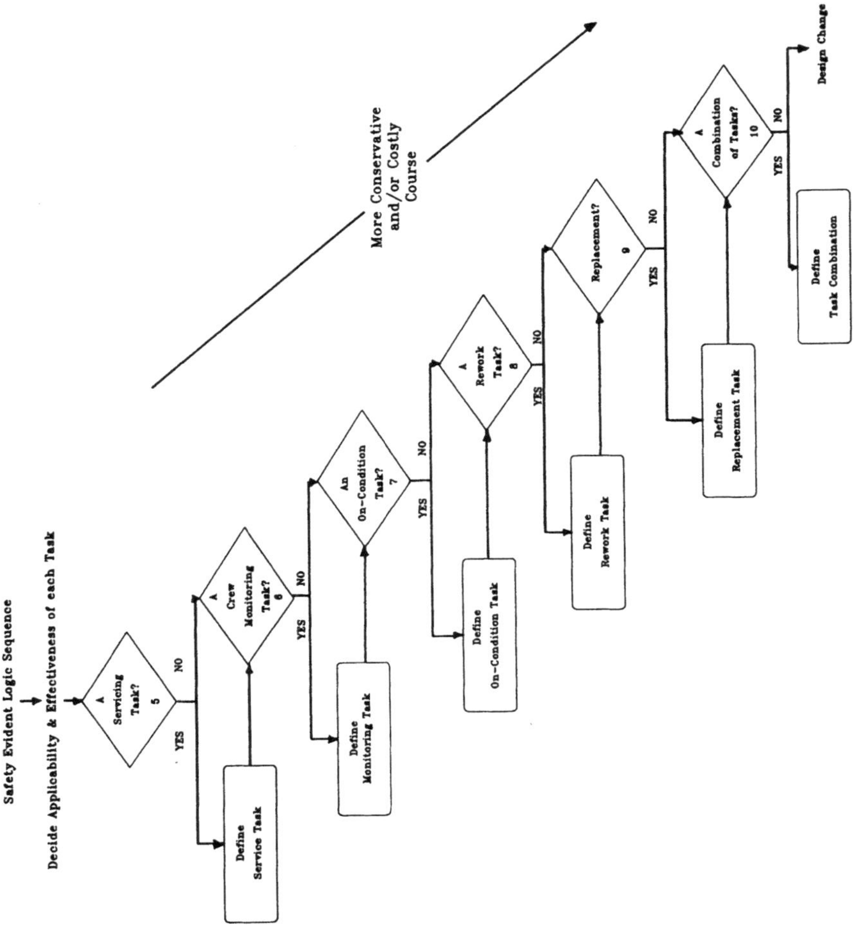

Figure 2–3 Army Aircraft RCM Decision Logic: Safety Evident Logic Sequence

Question 4: Does failure adversely affect operational performance?

This question asks if the failure could compromise the mission flexibility; e.g., aircraft altitude restriction, non-icing restriction, weight restriction, etc., and requires correction prior to use. If the answer to this question is 'YES' or 'NO', task selection will be handled in accordance with the operational/non-operational consequence category and task development will proceed in accordance with questions 11 through 15 (Figure 2–4).

Each identified failure processed through the above logic sequence will then be directed into one of the five consequence categories described below:

(1) Safety, Evident

This category is for failures whose occurrence results in a loss of function and has a direct adverse effect on safety. The logic questions and sequence for this consequence category are presented in Figure 2–3. All questions (5–10) in this category must be asked. Preventive maintenance is required to assure safe operation. If no effective or applicable task results from this evaluation, then a design change must be implemented to eliminate the part failure mode.

(2) Economic, Operational

This category is for failures whose occurrence affects operational performance only. The logic questions and sequence for this category are presented in Figure 2–4. Preventive maintenance is generally performed if the cost is less than the combined cost of the operational loss and the cost of repair. Either a 'YES' or 'NO' answer to question 11 requires proceeding to the next question. A 'YES' answer to any of the remaining questions will complete the analysis and the resultant task will satisfy the requirements. If all answers are 'NO', i.e., no task is considered effective and applicable, and if the economic penalties are severe, a design change should be implemented.

(3) Economic, Non-Operational

This category is for failures whose occurrence affects non-operational performance alone. The logic questions and sequence for this category are also presented in Figure 2–4. However, in this case preventive maintenance is generally performed only if the cost is less than the cost of repair.

(4) Safety Hidden

This category is for hidden failures whose occurrence alone or in combination with another failure results in a loss of function and has a

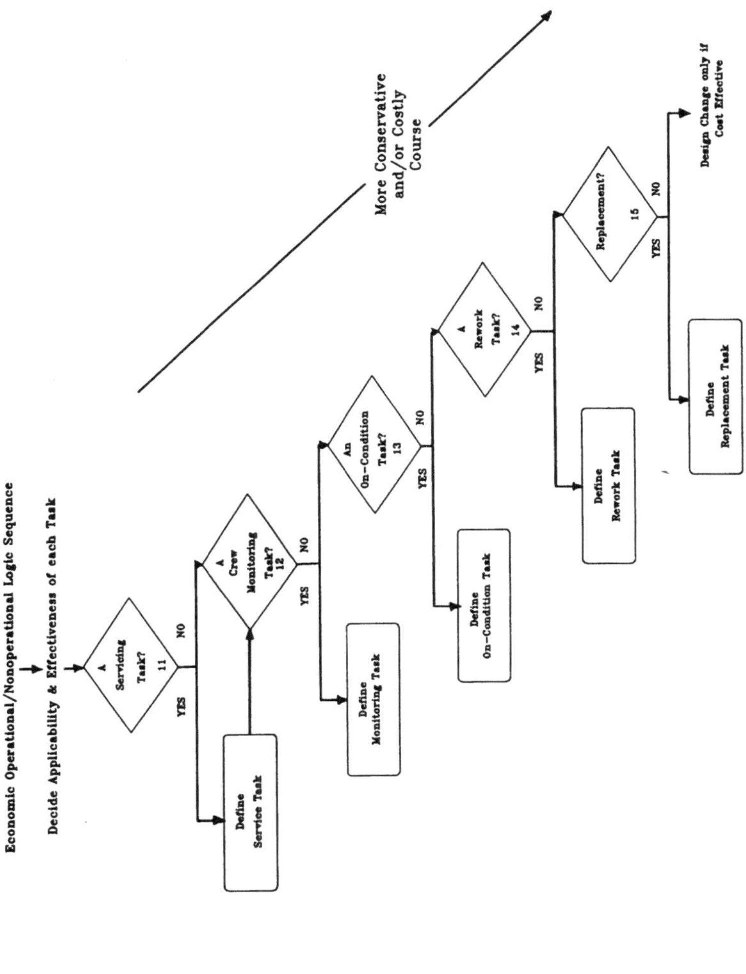

Figure 2–4 Army Aircraft RCM Decision Logic: Economic Operational/Non-Operational Logic Sequence

direct adverse effect on safety. The logic questions and sequence for this category are presented in Figure 2–5. Preventive maintenance is required to assure the operating reliability of the backup function(s) in order to avoid the safety effects of multiple failures. All questions (16–21) must be asked. If no tasks are found effective or applicable, then a design change must be implemented.

(5) Non-Safety Economic

This category is for hidden failures whose occurrence alone or in combination with other failures does not affect safety. The logic questions and sequence for this category are presented in Figure 2–6. Either a 'YES' or 'NO' answer to question 22 requires proceeding to the next question. A 'YES' answer to any of the remaining questions will complete the analysis and the resultant task will satisfy the requirement. If all answers are 'NO', i.e., no task is considered effective and applicable, and if the economic penalties are severe, a design change should be implemented.

It should be emphasized that regardless of the answer to the questions regarding servicing, the next task selection question must be asked in all cases. When following the safety effects (hidden or evident) paths, all subsequent questions must be asked. In the other paths, subsequent to the first question, a 'YES' answer will allow exiting the logic. At the user's option, advancement to subsequent questions after a 'YES' answer is allowable, but only until the cost of the task is equal to the cost of the failure prevented.

It also should be emphasized that default logic is reflected in paths outside the safety effects areas by the arrangement of the task selection logic. In the absence of adequate information, logic dictates that a 'NO' answer be given and the subsequent question be asked. As 'NO' answers are generated, the only choice available is the next question, which in most cases provides a more conservative, stringent and/or costly route.

Selecting the most effective and applicable preventive maintenance task or combination of tasks is handled in a similar manner for each of the five consequence categories. Following are the questions asked for each kind of preventive maintenance task, identified to the decision logic question sequence, along with some guidance in making the correct decisions.

Questions 5, 11, 16 & 22: *Is a servicing task applicable and effective?*

This task covers any act of servicing to maintain the inherent design capabilities. It includes activities performed at regular intervals to keep an

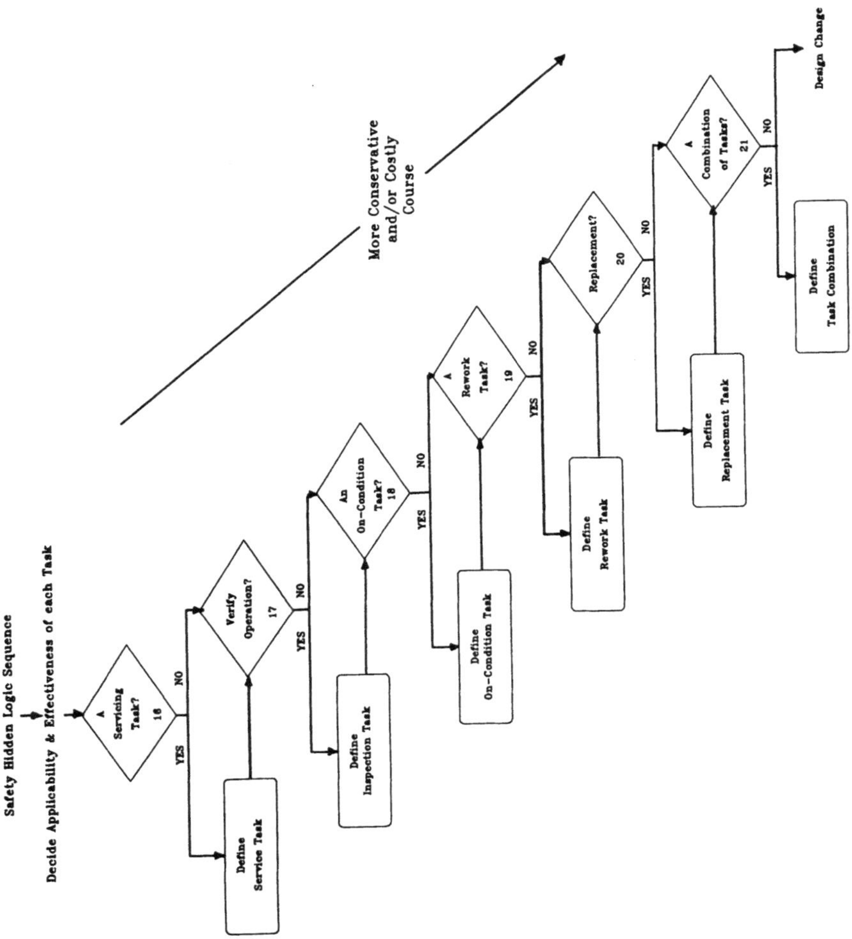

Figure 2–5 Army Aircraft RCM Decision Logic: Safety Hidden Logic Sequence

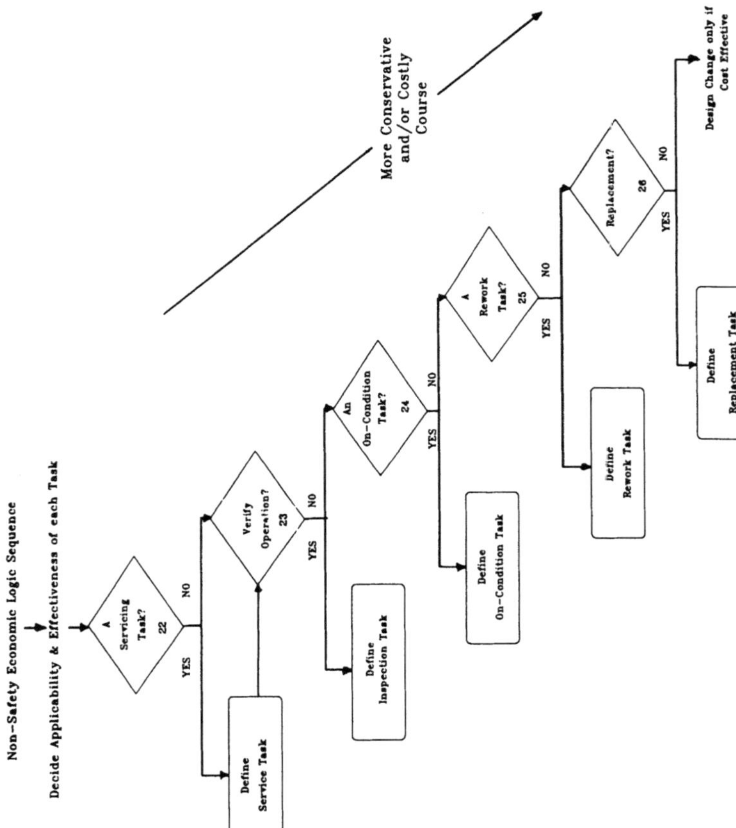

Figure 2-6 Army Aircraft RCM Decision Logic: Non-Safety Economic Logic Sequence

item in proper operating condition, i.e., to clean (decontaminate), preserve, drain, paint, or to replenish fuel, lubricants, hydraulic fluids or compressed air supplies. To be applicable, the replenishment of the consumable must reduce the rate of functional deterioration. For safety categories this task must reduce the risk of failure and for economic categories the task must be cost-effective.

Questions 6 & 12: *Is a crew monitoring task applicable and effective*?

This task consists of any monitoring of system operation by the crew members during their normal duties. This includes monitoring of instrumentation and recognition of potential failures by the operating crew through the use of normal physical senses (e.g., odor, noise, vibration, temperature, visual observation, changes in physical input force requirements, etc.). To be applicable, reduced resistance to failure must be detectable and rate of reduction in failure resistance must be predictable. Indicators that annunciate failures at the time of occurrence are not applicable. For the safety category this task must be part of the normal duties of the operating crew and must reduce the risk of failure to assure safe operation. For the economic categories this task must be part of the normal duties of the operating crew.

Questions 17 & 23: *Is a check to verify operation applicable and effective*?

This check is to verify operation qualitatively that an item is fulfilling its intended purpose and to detect any impending failure. To be applicable, verification of operation must be possible. For the safety category this task must ensure the operating reliability of the hidden function to reduce the risk of a multiple failure. For the economic categories this task must ensure the operating reliability of the hidden function in order to avoid economic effects of multiple failures and must be cost-effective.

Questions 7, 13, 18 & 24: *Is an on-condition task applicable and effective*?

This task is to detect degradation of the function by an on-aircraft or off-aircraft task, e.g., a functional checkout, preflight inspection or through ACE or PSA. It may be performed against a specific checklist or standard and may include a functional check to determine if one or more functions of an item perform within specified limits. To be applicable, reduced resistance to failure must be detectable and rate of reduction in failure resistance must be predictable. For safety categories this task must reduce the risk of failure to assure safe operation. For economic categories this task must be cost-effective, i.e., the cost of the task must be less than the cost of the failure.

Questions 8, 14, 19 & 25: *Is a rework task applicable and effective*?

This task is to repair, overhaul or rebuild an item in order to reduce its failure rate to an acceptable level. *Repair* is the application of maintenance services or other maintenance actions to restore serviceability to an item by correcting specific damage, fault, malfunctions or failure in a part, subassembly, module (component or assembly), end item or system. *Overhaul* is the maintenance effort necessary to restore an item to a completely serviceable/operational condition as prescribed by the appropriate technical publication. Overhaul is normally the highest degree of maintenance performed by the Army. Overhaul does not normally return an item to like-new condition. *Rebuild* consists of those actions necessary for the restoration of unserviceable equipment to a like-new condition in accordance with original manufacturing standards. Rebuild is the highest degree of material maintenance applied to Army equipment. The rebuild operation includes the act of returning to zero those age measurements (i.e., hours) considered in classifying Army aircraft systems and components.

This task involves that work (on/off the aircraft) necessary to return the item to an acceptable level of performance, quality and reliability. Since this task may vary from the repair of single parts up to a complete overhaul or rebuild, the extent of each rework task has to be defined and the task must be accomplished in accordance with the proper DMWR. To be applicable the item must show functional degradation characteristics at an identifiable age and a large proportion of units must survive to that age. For safety categories this task must reduce the risk of failure to assure safe operation. For economic categories this task must be cost-effective, i.e., the cost of the task must be less than the cost of the failures prevented.

Questions 9, 15, 20 & 26: *Is a replacement task applicable and effective*?

This task involves substituting a serviceable part, subassembly or module (component or assembly) for an unserviceable counterpart. It requires removing the item from service at a specified life limit. It is generally applicable to engine parts, structural members and other parts which show functional degradation at an identifiable age. For safety categories replacement at a safe-life limit must reduce the risk of failure to assure safe operation. For economic categories replacement at an economic-life limit must be cost-effective, i.e., the cost of the task must be less than the cost of the failures prevented.

Questions 10 & 21: *Is there a task or combination of tasks which is applicable and effective?*

This question is applicable only to the safety consequence categories; its answer is established by the responses to the questions in those categories. It must be emphasized that for these categories a task or task combination must be defined to prevent the failure mode, or a design change must be implemented. This question requires that the tasks are reviewed to assure that the most effective and applicable maintenance program will be put in place.

Once the maintenance tasks have been selected, the next step is to set the appropriate intervals and to properly phase the maintenance to minimize downtime and optimize labor expenditures. Actual data showing the hardware reliability–age relationships are used to determine the most effective intervals for accomplishment of the maintenance tasks. Also, data from other hardware systems may be used, particularly if the data show that the accomplished maintenance tasks are effective and economically worthwhile. In those cases where no data exist, or there is very little similarity to other systems, data from reliability engineering analysis performed during development can be used as the basis to set the task intervals.

The reliability–age data are reviewed in light of maintenance cost and downtime considerations to establish the most cost-effective intervals and phase-maintenance schedule. This involves listing the maintenance tasks identified by the RCM logic analysis on a time line and then grouping the tasks into appropriate modular intervals. The establishment of appropriate maintenance task intervals and phase-maintenance schedules is normally done as part of the system's integrated logistic support program.

The US Army's RCM decision logic described in this section uses a standard, well supported microcomputer and an existing, mature, reliable and comprehensive data base management system. The logic provides:

(1) A well organized, readily accessible life-cycle document and procedural audit trail.
(2) The compilation of output data in real time.
(3) A means for routine, on-line information exchange among the engineering staff and management.

(4) A compatible interface with logistic support analysis records (LSAR) for eventual overall integration within a computerized ILS system.

As experience is gained through application of the logic, high quality maintenance plans are developed in less time and at lower cost. A maintenance history is provided for each aircraft system or component that can be correlated with specific parts and their failure modes and criticalities. The process increases the probability that all safety critical parts and their failure modes are considered in the development of maintenance requirements and that the level and content of the requirements are optimally specified.

2.4 THE RCM & ILS INTERFACE

The RCM logic provides essential data for determining maintenance support requirements as part of a system's integrated logistic support (ILS) program. The ILS program includes the management and technical activities to define maintenance support requirements based on the R&M characteristics of the system and the RCM logic data and to acquire the required support at minimum cost during the operational phase. This section describes the interfaces between R&M, RCM and ILS program activities.

When the ILS program is properly executed, support requirements are defined in terms of the system design and relative to each other. The program includes actions to identify, define, analyze, quantify and process ILS requirements in all phases of a system acquisition program. An effective ILS program provides:

- A unified, structured and interactive means of establishing the maintenance program and identifying logistics support requirements.
- Continual information exchange between the system designers, R&M engineers and maintenance/logistics analysts, such that support considerations are made to influence design.
- A data base for the performance of logistic support analysis (LSA) and trade-off studies.
- A method of identifying deviations from anticipated behavior/ operational goals so that corrective action may be taken.

An ILS program is generally conducted in accordance with US MIL-

STD-1388–1A, 'Logistic Support Analysis — Program Requirements'. This standard provides a focus for prioritizing and directing efforts to address R&M, RCM and logistic factors. It describes the LSA process and provides guidelines and rationale for the selection and tailoring of ILS program tasks. The standard defines five major task areas. The following paragraphs provide a brief abstract of these areas and some of the specific tasks within that are based on R&M and RCM data:

(1) 100 — Program Planning & Control
Provides for formal program planning and review actions. The R&M history of existing or similiar systems or gross R&M estimates and other data and information are used to aid in tailoring the LSA activities for subsequent tasks.
(2) 200 — Mission & Support Systems Definition
Establishes support objectives and related design goals, thresholds, and constraints through comparison with existing or similar systems and analyses of support and cost drivers. A use study is performed which documents the mission and operating requirements, allowable maintenance periods, projected environmental requirements and other system parameters that are essential to performing various R&M analyses. The R&M history on any support, test or mission equipments being considered for use with the new system is evaluated. A comparative analysis may be performed between an existing or similiar baseline system and the new system's R&M and support parameters. R&M predictions are used to help evaluate alternative design solutions, which have maintenance support improvement potential, to problems with the current system. The system design characteristics serve as a basis for the R&M predictions and are used to help set logistic support goals and to perform system sensitivity analysis.
(3) 300 — Preparation & Evaluation of Alternative
Optimizes the support requirements for the new system in order to develop hardware which achieves the best balance between cost, schedule, performance and supportability. R&M prediction, failure mode effects and criticality analysis, and RCM data are developed and entered into the LSA record system. Evaluations are performed including support trade-offs, system trade-offs, training trade-offs, repair level analyses, diagnostic trade-offs, comparative evaluations, energy trade-offs, survivability trade-offs and transportability trade-offs.

(4) 400 — Determination of Logistic Support Resource Requirements

Identifies the logistic support resource requirements of the new equipment in its operational environment(s) in order to develop plans for post production support. LSA data records are completed and validated and all required data are entered into the LSA record system.

(5) 500 — Supportability Assessment

Assures that specified requirements are achieved and deficiencies corrected. Actual test and initial field data are used for final assessment of the support requirements. This is a necessary step for complete assessment of all logistic elements, since some of the elements, such as the final technical manuals and the depot maintenance work requirements, are not available until after development testing or fielding.

Data produced by the R&M, RCM and ILS activities form the basis for design trade-offs while alternative concepts are being explored. The ILS requirements are documented in a series of worksheets known as Logistic Support Analysis Records (LSAR) in accordance with US MIL-STD-1388–2A 'DoD Requirements for a Logistic Support Analysis Record' and used subsequently to establish the support resource requirements. The LSAR is the permanent file of all documents and data created during the ILS process.

RCM logic data are major inputs to the ILS process and appear on the LSAR R&M data record along with failure modes, effects and criticality analysis data from R&M studies. The R&M and RCM data on this record form the basis for preparing the other MIL-STD-1388 data records used to establish support resource requirements. The R&M record facilitates the application of RCM to the LSA process by providing for:

(1) The recording of answers to applicable logic questions for failure modes, identified in a failure mode analysis.
(2) The recording of the disposition of each failure mode processed through the logic. This is the first indication, within the ILS program, that a potential RCM scheduled maintenance task may be required in the maintenance program.
(3) The documenting of the unique task code for each identified RCM task.

Implementation of the RCM decisions is accomplished through the completion of task analysis summary and maintenance and operator task

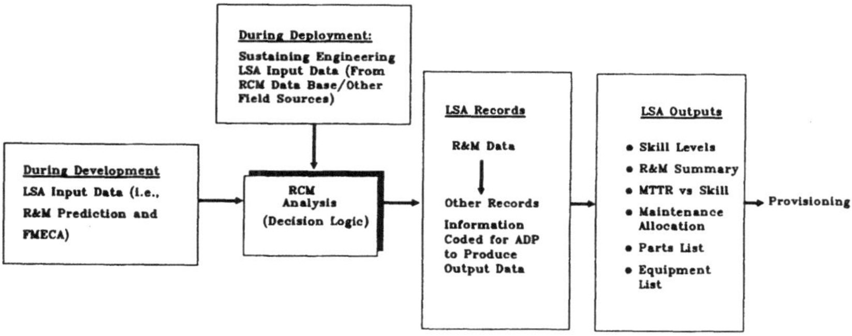

Figure 2–7 The RCM/ILS Process

analysis records for all the identified scheduled maintenance tasks. Identification of a task is established and tracked by coded entry. The records show task and personnel requirement summaries and contain a list of the sequential task steps, descriptive information for publications use, information for training requirement determination, and a statement of requirements for support equipment and repair parts. The result of the complete RCM/ILS process is the compilation of a provisioning master record (PMR) from which procurement of support items is derived. Resulting data are also used as direct input into, or as source information for, the development of other ILS data products including technical manuals and personnel and training requirements.

The LSAR's also allow for routine incorporation into the maintenance support program of new, more efficient and less costly maintenance concepts as they are defined. For fielded systems this can be readily implemented through reapplication of the RCM logic using field experience data and compilation of revised LSAR's.

Figure 2–7 depicts the process by which development data and, later after deployment, actual experience data are used to establish ILS requirements through application of RCM analysis.

The RCM-driven ILS process is initiated early in the development of a new hardware system to impact design and operational concepts, to identify the gross logistic resource requirements of alternative concepts, and to relate design, operational, manpower and support characteristics to readiness objectives and goals. It includes reviewing comparable existing hardware systems and determining major manpower and cost drivers. System trade-offs are made between the support, operational and design concept and between alternative support concepts such as organic vs contractor support, built-in vs external test capability and varying numbers of maintenance levels. These trade-offs are to take into account existing support resources, maintenance policies and results of trade-off studies. They are based on gross engineering data derived from preliminary performance information, R&M requirements and support planning and on data derived from similar fielded systems, support criteria and existing support resources for such similar systems as they apply to the various maintenance organizations.

The early logistic planning and analysis activities, undertaken while alternative system design concepts are being explored, are directed towards the identification of ILS alternatives, establishment of R&M and cost goals, identification of equipment deficiencies in deployed systems, identification of potential logistic problems and identification of qualitative and quantitative personnel requirements.

Once system level trade-offs are made, the analysis shifts to lower level optimization. All elements of ILS, including task frequencies, task times, personnel and skill requirements, supply support requirements, and support and test equipment requirements, are defined through an integrated assessment of operator and maintenance functions and tasks. Optimization is achieved through allocating specific tasks to maintenance organizational levels, performing repair vs discard analysis of components and parts, performing RCM logic data analysis and formulating design recommendations to reduce maintenance times or to eliminate special support requirements. Data resulting from these analyses are used as direct input into, or as source information for, the preparation of specific logistic support elements such as the actual provisioning list, technical manuals, and personnel and training requirement information.

The logistic support analyses are based on detailed R&M engineering data and information supplied by the system contractor. This information is generally provided in the form of data items and delivered as part of the contract. The analyses take into account existing support resources, maintenance policies and results of trade-off studies based on engineering data derived from performance information, R&M predictions, application of RCM logic analysis and support planning. When the initial support levels are developed, certain R&M characteristics essential to the ILS process are unknown. As an example, although reliability degradation can be estimated, there are no data on the actual degradation as various hardware items age. Similarly, the information required to evaluate cost-effectiveness and reliability–age relationships becomes available only after the item has been in service for some time. After deployment, when the maintenance plan is implemented and as operating experience is gained, activities focus on:

(1) determining the actual reliability–age characteristics and re-applying the RCM decision logic to respond to failures not anticipated during development,
(2) assessing the desirability of additional maintenance tasks,
(3) adjusting the maintenance task intervals, and
(4) eliminating the cost of unnecessary and over intensive maintenance resulting from the use of default answers in the initial RCM logic analysis.

A key activity after deployment is to gradually replace the default decisions from the RCM logic analysis, that were originally made because of the absence of definitive information, with well founded decisions based on actual operating experience.

The RCM/ILS process requires an extensive and cumulative base of data and information. The data must be continually refined and updated to include the most recent information. Also, Bayesian statistics can be used to apply prior development information (e.g., predictive data) to the analysis of more recent information, such as test or actual field data. The results achieved, therefore, would utilize the widest spectrum of information available. In this fashion, not only could R&M prediction estimates and the latest test information on an item be used exclusively and independently of one another for LSA, but also more detailed R&M data can be provided, reflecting both information sources. This is particularly useful for assessing operational reliability of newly deployed systems or components when only limited operational data exist.

As indicated, the ILS activities are based on data derived from R&M and RCM logic analyses. During early development, the failure factor estimates needed as input to ILS activities are derived from existing or similiar equipment. Later in development (and after deployment), the failure factors are derived from detailed R&M analysis.

The R&M data include mean time between failures (MTBF); mean time to repair (MTTR); and failure mode, effects and criticality analysis (FMECA). The MTBF, MTTR and FMECA data elements are major inputs to the RCM/ILS process and appear on the R&M record in accordance with MIL-STD-1388–2A. These data elements are developed as part of the R&M program activities which, in general, are conducted in accordance with the requirements of US MIL-STD-785, 'Reliability Program for Systems and Equipment' and US MIL-STD-470, 'Maintainability Program for Systems and Equipment'.[5,6] MIL-STD-785 includes relevant tasks for reliability modeling, reliability allocations and predictions, and failure modes, effects and criticality analysis. MIL-STD-470 includes tasks for maintainability modeling, maintainability allocations and predictions, and failure modes, effects and criticality analysis.

A system reliability prediction performed during development establishes the MTBF and part failure mode/rate numerics. The prediction and its associated mathematical models are generally derived from MIL-HDBK-217 and from other sources.[7] The prediction establishes the basic part/component replacement rates of the design and is used as input to logistics support analysis (LSA) and trade-off studies of alternative design concepts.

Similarly, MIL-HDBK-472 is used for maintainability prediction and particularly for deriving factors for repair time, maintenance frequency per operating hour, preventive maintenance time and other maintainability factors.[8] The techniques given in MIL-HDBK-472, in general, involve the determination of MTTR using failure rates obtained from the reliability prediction and maintenance time factors derived from a review of the system design characteristics. Conceptually, the repair of hardware items after the occurrence of a failure necessitates the initiation of a corrective maintenance task which ultimately results in the interchange of a replaceable part or assembly. In order to achieve a complete 'repair', various activities both before and after the actual interchange are necessary. This includes activities for localization, isolation, disassembly, interchange, reassembly, alignment and checkout.

The composite time for all repair activities is called the repair time. The part/assembly failure rates and repair times are combined to arrive at a weighted corrective maintenance action rate. The prediction process also involves preparing a functional-level diagram for the system and determining the repair time for each replaceable item. The functional-level diagram reflects the overall maintenance concept and the complete replacement breakdown for all items that comprise the system.

MIL-HDBK-472 information allows the analyst to determine the number of people required to maintain a given number of systems within a specified operating/calendar time period. In conjunction with maintainability predictions, additional maintenance data supplied by the contractor allow decisions to be made regarding difficulty of maintenance (which translates into skill levels of personnel), tools and equipment required, consumable items used while performing maintenance and facilities required.

Critical to RCM and ultimately logistic support analysis is FMECA. The FMECA identifies potential failure modes, thus establishing the initial basis for formulating maintenance task requirements. It systematically identifies the likely modes of failure, the possible effects of each failure, and the criticality of each effect on equipment function, safety or some other outcome of significance.

The R&M predictions and the FMECA for a development system are generally performed by the prime contractor as part of the reliability, or product, assurance program; however, the analyses must be coordinated with the RCM/ILS program and the results must be made available as essential input to RCM logic analysis. This coordination should address timing of the analyses, their level of detail and the specific documentation requirements.

In cases where historical data are inadequate or where support experience with a new system is needed prior to deployment, supportability tests and evaluations may be conducted. The supportability test and evaluation program serves three objectives: (1) to expose supportability problems so that they can be corrected prior to deployment; (2) to provide measured data for logistic and support related design parameters for input into system level estimates of readiness, operating and support costs, manpower requirements, and logistic support resource requirements; and (3) to demonstrate contractual compliance with quantitative logistic and support related design requirements. Test and evaluation planning, scheduling, and cost investment conform to this order of priorities to maximize the return on investment in supportability test and evaluation.

An effective test and evaluation program requires close coordination of efforts between all systems engineering disciplines to prevent duplication of tests and to maximize test program effectiveness. Maximum use should be made of data available from reliability tests, maintainability demonstrations, publications, validation/verification efforts, environmental tests, endurance/durability tests, and other tests to satisfy supportability assessment requirements. A well integrated test program involves establishing test conditions that maximize the utility of the test results. This is an important factor considering that the availability of hardware and time to conduct tests and evaluations are generally at a premium for most system acquisitions.

One of the major factors that determines the utility of test results in satisfying the objectives of the supportability test and evaluation program is the test environment. Historically, there has been a large discrepancy between test results and field-observed supportability parameters. This gap is to a large degree caused by the conduct of tests in ideal environments, the use of contractor technicians to perform maintenance during tests, the selective deletion of some test results (i.e., non-chargeable failures), and not using the planned support resources (technical manuals, tools, test equipment, etc.). Realistic test environments for supportability assessment must be established, reflecting the system's intended operational environment and the intended support resources (all elements of ILS) that will be available to operate and maintain the system after deployment. While a total simulation of the field environment may not be practical or cost-effective, the test environments should be as close as possible to the anticipated field environment and any differences between the test and field environments should be known and their impact understood. These differences must then be accounted for in using the test results to update system level projections for readiness, operating and support costs, manpower requirements and logistic support resource requirements.

Once the systems have been delivered and assigned to their organizational units, the RCM decisions and the ILS outputs are reassessed through a sustaining engineering process that takes into account actual operating experience. Analysis of logistic and support related data on the system in its operational environment is necessary in verifying that the system has met its supportability objective. This assessment can be made using field feedback data that are available from the readiness, supply and maintenance reporting systems. In some cases, data from standard reporting systems must be supplemented in order to meet the supportability verification objective within acceptable risks. Requirements for supplemental data must be weighed against the cost and resources to obtain such data plus the impact upon using units to gather the data.

Analysis of supportability data on a new system in its operational environment provides for verification of the achievement of logistic and support related design goals, identification of support problems not foreseen or encountered during test and evaluation, and information that can lead to supportability enhancements on the new system and future system acquisition programs. To realize these benefits, this analysis must be effectively planned, managed and conducted. In those cases where existing standard field reporting systems will not provide the necessary data or accuracy to conduct this analysis, then supplemental data collection programs must be planned, approved, budgeted for and implemented. Care should be exercised in planning this activity to assure that field results are collected during 'normal' field operations. Collecting data immediately after deployment may be biased if any of the following situations are in effect:

a. New equipment fielding teams are still with the system.
b. Operator and maintenance personnel received training from other than the intended normal training sources.
c. Initial supply support was obtained from other than standard supply systems.
d. Interim support resources are being used pending deployment of other items (e.g., support and test equipment).

Analysis of data obtained from field reporting systems can provide significant information for system enhancements through support system modifications, product improvement programs or modification of operating tactics. Furthermore, comparative analysis field results, test and evaluation results, and engineering estimates can provide information for use on future acquisition programs to better project manpower, cost and readiness parameters.

After deployment, as operating experience is gained, activities focus on determining the actual reliability–age characteristics and applying the RCM decision logic to respond to failures not anticipated during development, to assess the desirability of additional maintenance tasks, and to eliminate the cost of unnecessary and over-intensive maintenance resulting from the use of default answers in the initial RCM logic analysis. Actual experience data are used to update the failure factors which are used as the basis for these activities.

The cornerstone of an effective ILS process is the RCM logic analysis, which, as described earlier, is conducted to identify maintenance problem areas for design consideration and to establish the most effective maintenance support program. The logic is applied to the individual failure modes of each safety critical part identified by a failure mode analysis. Once the RCM logic process has been completed, the tasks in the scheduled maintenance program are then specified. This includes preparing specific requirements for servicing, condition monitoring, on-condition adjust/align/calibrate, rework (repair)/overhaul and replacement (at life-limits) in accordance with the decision logic results and output data.

To facilitate application of the RCM program (and extend useful life), design emphasis is placed on the use of proven long-life component parts, the use of easily accessible and interchangeable modules and units, the incorporation of ease of inspectability features and the addition of more sophisticated on-board diagnostic systems. This leads to more cost-effective maintenance and to the establishment of better indicators or precursors of failure, which greatly improve the on-condition maintenance tasks, improve the condition monitoring tasks performed by the operating crew and extend the time limits for hard-time replacement tasks at the depot.

The RCM/ILS program integrates many of the relevant reliability, maintainability and safety program tasks and other special studies in order to achieve the common objective of orienting the development and operational phases toward a practical, supportable and affordable hardware system. It provides output data for trade-off analysis with the design engineering function and for preparation of a complete cost-effective maintenance support plan.

The maintenance support plan describes how the system will be maintained and the relationship to the overall maintenance concept. It documents the requirements and tasks to be accomplished for restoring or preserving the operational capability of the complete system. It provides definition as to what constitutes a repair action and the scope of maintenance activities planned for execution at the various levels of repair. It specifically identifies and defines logistic support requirements including:

a. Maintenance tasks including requirements to support the system at each level of repair
b. Spare provisioning
c. Tool and test equipment, including calibration equipment and calibration requirements
d. Manpower-training and skill levels
e. Maintenance manuals and data
f. Training manuals
g. Support equipment/facilities
h. Shipping and transportation
i. Quality control
j. Configuration management

The plan is prepared during the design phase, based on the initial RCM decision logic data, and is updated, as necessary, during development and production phases and reassessed after deployment during operation in light of the revised RCM analysis that reflects actual field experience data.

The plan also establishes requirements and/or interfaces for R&M engineering tasks, LCC analysis and other related hardware development activity. In addition to the scheduled maintenance task requirements identified during application of the RCM logic, any scheduled tasks that were assumed in establishing the inherent R&M characteristics of the equipment or component must either be included in the maintenance plan or identified as being omitted from the maintenance plan. R&M estimates and projected failure rates and failure modes and effects may need adjusting if an assumed scheduled maintenance action is omitted from the maintenance plan after application of the RCM logic.

As previously mentioned, maintenance planning starts during the early concept phase and the initial requirements are formulated during the demonstration and validation phase based on the R&M and RCM analyses. The plans are updated as necessary during the course of the development program and reassessed after the system is fielded as part of a sustaining engineering activity to reflect revised R&M and RCM logic analysis data derived from actual field experience information.

CHAPTER 3

R&M Theory and Fundamental Concepts

This chapter provides a framework for the RCM engineering techniques (given in Chapter 4) and provides a brief summary of basic R&M concepts upon which the techniques are based. Hardware reliability–age characteristics are described, followed by a description of reliability degradation and growth; basic reliability, maintainability and availability analysis concepts; and life-cycle RCM activities.

3.1 RELIABILITY–AGE CHARACTERISTICS

The term reliability is defined as the probability that a hardware item will satisfy its performance requirements for a specified time interval under operational conditions. The reliability definition stresses four elements, namely, probability, performance requirements, time and use conditions. Probability is the likelihood that an event will or will not occur. It is a quantitative term expressed as a value between zero and one. Performance requirements indicate that criteria must exist which clearly specify, describe or define what is considered to be satisfactory operation. Time represents a measure of a period during which we can expect satisfactory performance. Operational conditions represent the environmental conditions under which we expect the item to function.

Determining reliability involves understanding concepts pertaining to failure rate as a function of age. A failure rate is a measurement of the

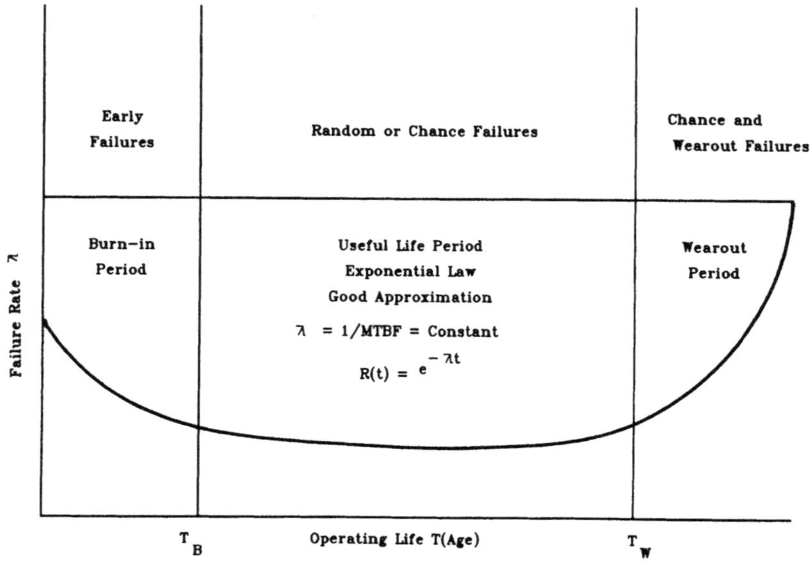

Figure 3–1 Life Characteristic Curve

number of malfunctions occurring per unit of time. Separate consideration is given to three discrete periods when viewing the failure characteristics of a complex hardware item or system over its life span (and when considering a large sample of its population). These periods are shown in Figure 3–1.

The time periods shown in the figure are characterized as follows:

(1) Initially, the item population exhibits a high but decreasing failure rate that stabilizes at an approximate value (at time T_B) when the weak units have died out. This initial failure rate is unusually pronounced in new hardware systems. Many manufacturers provide a 'burn-in' period for their product prior to delivery which helps to eliminate a high portion of the initial failures.

(2) The item population, after having been burned-in, reaches its lowest failure rate level, which is normally characterized by a relatively constant failure rate, accompanied by negligible or very gradual changes due to wear. This second period (between T_B and T_W) is called the useful life period, characterized mainly by the

occurrence of chance or random failures. The exponential failure distribution is widely used as a mathematical model to approximate this time period.

(3) The third and final period occurs when the item population reaches the point where the failure rate starts to increase noticeably (T_W). This point is identified as the end of useful life or the start of wearout. Beyond this point on the time axis, the failure rate increases rapidly. When the hardware failure rate due to wearout (i.e., accumulated damage, fatigue and/or degenerative factors) becomes unacceptably high, replacement or repair of the item should be made. Hard-time maintenance replacement schedules (of critical short-life components) are based on the recognition of this time period.

Optimizing reliability involves eliminating early failures by application of controlled environmental stress screening (ESS) and burn-in during manufacturing and extending the start of wearout by long-life component design and/or timely preventive replacement of short-life component parts during field operation. The general technique considers only the useful life period in reliability design efforts in which reliability is predicted by means of the single parameter exponential distribution:

$$R(t) = e^{-\lambda t}$$

where $R(t)$ is the probability that the item will operate without failure for the time period t (usually expressed in hours) under stated operating conditions, and λ is the item failure rate (usually expressed in failures per hour) and is a constant for any given set of stress, temperature and quality level conditions. It is determined for parts and components from large scale data collection and/or test programs.

When appropriate values of λ and t are inserted into the above expression, the probability of success (i.e., reliability) is obtained for that time period.

The reciprocal of the failure rate is defined as the mean time between failures (MTBF):

$$\text{MTBF} = 1/\lambda$$

The MTBF is primarily a figure of merit by which one hardware item can be compared with another. It is a measure of the failure rate (λ) during the useful life period.

The concepts associated with the three time periods shown in Figure 3–1, when implemented through appropriate techniques described later, can be used to establish (and ultimately control) the reliability of the system under consideration.

3.2 RELIABILITY DEGRADATION AND GROWTH

It must be emphasized that a reliability (or MTBF) estimate, in general, reflects the reliability potential of a system during its useful life period (i.e., the period after early production, where quality defects are dominant, and prior to the time when wearout becomes dominant). It does not necessarily reflect the reliability of the system when first released to the field, after initial manufacturing, and operated and maintained in its service environment. It reflects the inherent reliability of the system, as it is defined by its engineering documentation, and by taking into account the designed-in stress–strength derating factors and gross environmental application, manufacturing and quality factors.

Experience has shown that the reliability of a system or component as it first leaves production is much less than its inherent reliability. In order to assess the magnitude of the reliability degradation due to manufacturing, the impact of the manufacturing process, i.e., the efficiency of quality controls and inspections, and the effectiveness of any applied environmental stress screen (ESS) must be evaluated. This includes estimating the number of both intrinsic and induced defects. Intrinsic defects arise from the basic limitations of the constituent parts used in the item and are a function of process maturity and inspection and test methods used by the suppliers. Induced defects are those which enter the item as a result of manufacturing process stresses, handling damage, and workmanship and inspector errors.

The role of inspection is to weed out these defects and to determine compliance with engineering specifications. A well planned inspection performed in accordance with documented instructions, clear acceptability criteria, proper instrumentation and trained personnel will have a high efficiency and, consequently, weed out a large number of defects. Note, however, that no inspection is perfect; a 100% error-free inspection is impossible to attain. More importantly, these actual defects can be overshadowed by the presence of latent defects, the results of weakened parts, which fail only under the proper conditions of stress—usually during field operation.

The purpose of an ESS is to convert possible latent defects into actual defects which can then be removed by inspection. It involves applying controlled time or cyclic-stress procedures, derived from failure mode studies to identify both latent design and workmanship failure mechanisms. The ability of an ESS to convert latent defects into actual defects can be evaluated based on the type of stress, its stress level and the length of the screen or the number of cycles.

Hardware wearout, with aging as the dominant failure mechanism, can degrade reliability and significantly shorten or reduce the useful life period of a system, particularly a helicopter system. Also, situations occur in which a helicopter, for example, may be called upon to operate beyond its design capabilities because of an unusual mission requirement, or to avoid a ground threat. These situations could damage (or weaken) its structure or its dynamic components. Operational abuses due to rough handling, heavy loads or neglected maintenance, can contribute materially to reliability degradation which eventually results in failure. Degradation also occurs as a result of poor maintenance and, particularly, as a result of poorly trained or unskilled technicians. Furthermore, excessive handling brought about by frequent or poorly executed maintenance can also degrade system reliability. However, with the emphasis on condition monitoring systems this degradation can be reduced.

Thus, it is evident that the actual in-service reliability for any given system is a function of both its designed-in inherent level and the degradation which occurs during manufacturing and after deployment during operation and maintenance. A reliability growth process can be applied during development and manufacturing to force a system to grow from the expected degraded level, if the process was not applied, back to a value which approaches its designed-in reliability level.

The basic concepts associated with the reliability growth process and its application to new development systems involve consideration of test, failure, correction and retest activities. Specifically, reliability growth is an iterative test–fail–correct process. There are three essential elements involved in achieving reliability growth, namely:

(1) Detection and analysis of hardware failures
(2) Feedback and redesign of problem areas
(3) Implementation of corrective action and retest

The rate at which reliability grows during development and manufacturing is dependent on how rapidly these three elements can be

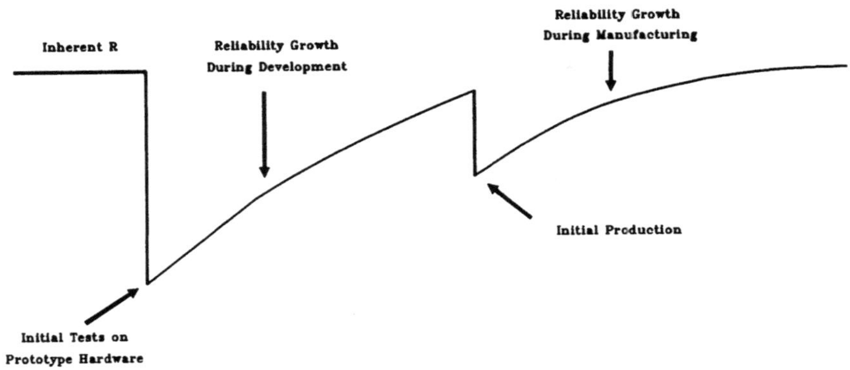

Figure 3–2 Reliability Growth Process

accomplished and, most importantly, how well the corrective action effort solves the problem identified. Figure 3–2 depicts the reliability growth process. During early development and test, reliability (or MTBF) is well below that predicted based on the system design characteristics. As development and test efforts progress and further problem areas become resolved, reliability approaches the inherent (design based) level.

As production begins, a decrease in reliability occurs due primarily to workmanship errors. As production continues, and skill increases, reliability again approaches the inherent value.

The application of reliability growth concepts, particularly reliability growth testing and ESS during system development and manufacturing, will facilitate the meeting of system requirements and resource allocation goals. It provides a quantitative means for determining the time and costs required to grow to a given level of reliability under varying degrees of corrective action rigor. Reliability growth testing and ESS are described in Chapter 4.

3.3 RELIABILITY CONCEPTS AND THE EXPONENTIAL DISTRIBUTION

The exponential formula for reliability introduced in Section 3.1 can be derived from the basic definition of probability.

When a fixed number, N_0, of components are repeatedly tested, there will be, after a time t, N_s components which survive the test and N_f components which fail. The reliability or probability of survival is at any time t during the test:

$$R(t) = \frac{N_s}{N_0}$$

Since $N_s = N_0 - N_f$, reliability can be written:

$$R(t) = \frac{N_0 - N_f}{N_0} = 1 - \frac{N_f}{N_0}$$

and

$$\frac{dR}{dt} = \frac{-1}{N_0}\frac{dN_f}{dt} = f(t)_i$$

where $f(t)_i$ is the failure density function, i.e., the probability that a failure will occur in the next time increment dt.

Let: $z(t)_i$ be the hazard rate, or the probability that a failure will occur in the next instant of time assuming previous survival; then:

$$z(t)_i = \frac{f(t)_i}{R(t)_i}$$

The quantity $z(t)_i$ can be defined as the hazard rate of element i at time t. In general, it can be assumed that the hazard rate of complex systems remains constant over practical intervals of time and that $z(t)_i = \lambda_i$, the expected number of random failures per unit of operating time of the ith element, i.e., the failure rate. Thus, when a constant failure rate is assumed:

$$z(t)_i = \lambda_i = \frac{f(t)_i}{R(t)_i} = \frac{\dfrac{-dR(t)_i}{dt}}{R(t)_i}$$

Solving this differential equation for $R(t)_i$ gives the exponential distribution function commonly used in reliability prediction:

$$R(t) = \exp(-\lambda_i t)$$

Also, the mean time to failure can be determined by:

$$\text{MTBF} = \int_0^\infty R(t)\mathrm{d}t$$

when a constant failure rate is assumed:

$$\text{MTBF} = \int_0^\infty \exp(-\lambda_i t)\,\mathrm{d}t = \frac{1}{\lambda_i}$$

The above expressions for $R(t)$ and MTBF are the basic mathematical relationships used in reliability prediction. It must be noted, however, that these expressions were derived based on the assumption that the failure rate of the item under consideration is a constant. When the failure rate is not constant, the more general hazard rate must be considered, in which case reliability is obtained using the more general expression:

$$R(t)_i = \exp - \int_0^\infty z(t)_i\,\mathrm{d}t$$

The emphasis on the exponential distribution in reliability analysis makes a discussion of the use of this function as a failure-probability model worthwhile. The mechanism underlying the exponential reliability function is that the hazard rate (or the conditional probability of failure in an interval, given survival at the beginning of the interval) is independent of the accumulated life.

The use of this type of 'failure law' for complex systems is usually justified because of the many forces that can act upon the system and produce failure. For example, different deterioration mechanisms, different part hazard-rate functions, and varying environmental conditions often result in effectively random system failures.

Another justification for assuming the exponential distribution in long-life complex systems is the so-called 'approach to a stable state', wherein the system hazard rate is effectively constant regardless of the failure pattern of individual parts. This state results from the mixing of part ages when failed elements in the system are replaced or repaired. Over a period of time, the system hazard rate oscillates, but this cyclic movement diminishes in time and approaches a stable state with a constant hazard rate.

The life characteristic curve, shown in Figure 3–1, can be further defined by the following components of failure:

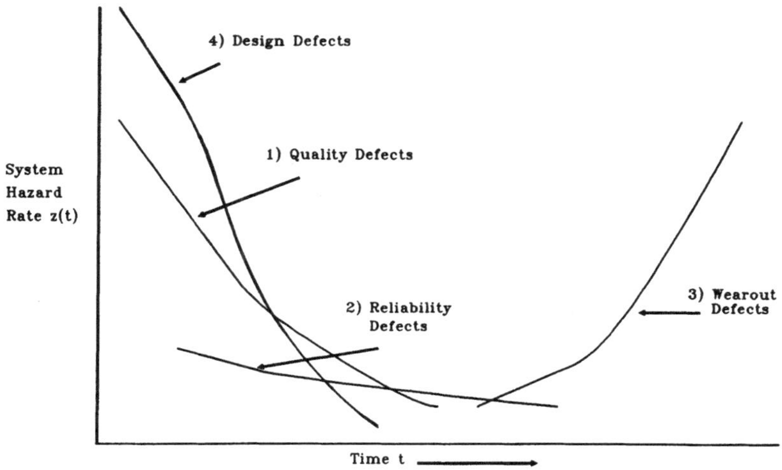

Figure 3–3 System Failure Components

(1) Quality defects—represent early failures and have a decreasing hazard rate.
(2) Reliability (or stress related) defects—represent failures during the early and useful life period; have a constant (or slightly decreasing) hazard rate.
(3) Wearout defects—represent failures during the normal and end-of-life period; have an increasing hazard rate.
(4) Engineering (or design) defects—normally represent early failures and have a decreasing hazard rate; however, an immature design can allow these defects to dominate all other defects.

These components of failure are shown pictorially in Figure 3–3.

As previously mentioned, the basic assumption normally made in reliability prediction is that, during the useful life period, the sum of the above components would result in a constant hazard (or failure) rate that can be described by the exponential failure distribution. This means that the hardware item must reflect a mature design where *design failures* have been eliminated or minimized, *quality defects* have been minimized, and *wearout* is not noticeable or is beyond the period of concern.

It should be noted that for many systems that are primarily comprised of mechanical components and parts, the sum of the above failure characteristics will not necessarily result in a constant failure rate. Failure analysis studies indicate that while electrical parts normally exhibit a long useful life period with a relatively higher constant or random failure rate, mechanical parts are characterized by a short useful life period with a relatively lower random failure rate. Both part categories exhibit similar early failure characteristics.

Thus, in general, wearout failure is the dominant characteristic with respect to mechanical parts and random failure is the dominant characteristic with respect to electrical parts. As indicated earlier, optimizing reliability requires systematic and deliberate actions to be taken during development and manufacturing. These actions include:

1. Designing mechanical components to extend their life and to minimize the effects of wear and/or identifying short-life components for timely replacement during field operation.
2. Minimizing design failures through FMECA, reliability testing and design review.
3. Minimizing workmanship defects through the application of a controlled ESS and burn-in process.

3.4 MAINTAINABILITY CONCEPTS

Maintainability is defined as the probability that a hardware item will be retained in or restored to a specified operating condition, within allowable time limits, using available test equipment, facilities, personnel, spare parts and prescribed procedures. Maintainability prediction, as with reliability prediction, is an analytical process of estimating the parameters that describe this probability. It accounts for the design characteristics and maintenance features of the system (i.e., test points, self check features, accessibility, modularization, adjustment, etc.) and provides a measure of the ease and speed with which maintenance operations can be performed and failures can be diagnosed and corrected.

Repair is quantitatively evaluated in terms of times required to perform elementary maintenance activities. These time elements are mathematically combined to form a statistically meaningful measure of system downtime through several conventional techniques described in the following discussion. Maintainability can be expressed in terms of mean time to repair (MTTR).

Conceptually, the repair of hardware items after the occurrence of a failure necessitates the initiation of a corrective maintenance task which ultimately results in the interchange of a replaceable part of assembly. In order to achieve a complete 'repair', various activities both before and after the actual interchange are necessary. These activities can be subdivided into the following time elements:

LOCALIZATION TIME — The time associated with eliminating as many as possible of the unfailed functions from further consideration by performing rapid tests (frequently involving only operating controls, displays and/or monitoring devices) before proceeding with the more difficult diagnostic techniques of fault isolation.

ISOLATION TIME — The time associated with tracing a failure down to a replaceable item through the use of test equipment.

INTERCHANGE TIME — The time associated with the physical removal of a failed item and its replacement with a new item.

DISASSEMBLY TIME — The time associated with gaining access to the replaceable item(s) identified during fault isolation.

REASSEMBLY TIME — The time associated with disassembly, except that the steps are performed in reverse order.

ALIGNMENT TIME — The time associated with the manipulation of operating and maintenance controls and mechanical parts so as to bring the equipment within its specified operating ranges.

CHECKOUT TIME — The time associated with the verification that the repair has restored the equipment's normal performance.

The composite time for all the above activities is called the repair time, R_p. In order to provide weight factors for the expected number of corrective maintenance actions, the failure rate of each replaceable component/part/assembly, λ_p, is used. The failure rate and repair time are combined to arrive at a corrective maintenance action rate. This process is repeated for each replaceable component/assembly in the system. From the maintenance actions rates (R_p) derived for each replaceable item, the MTTR can be determined using the following expression:

$$\text{MTTR} = \frac{\Sigma(R_p)(\lambda_p)}{\Sigma\lambda_p}$$

3.5 AVAILABILITY ANALYSIS

Availability is the probability of a hardware system or component item being in service when required. It provides a single combined measure of

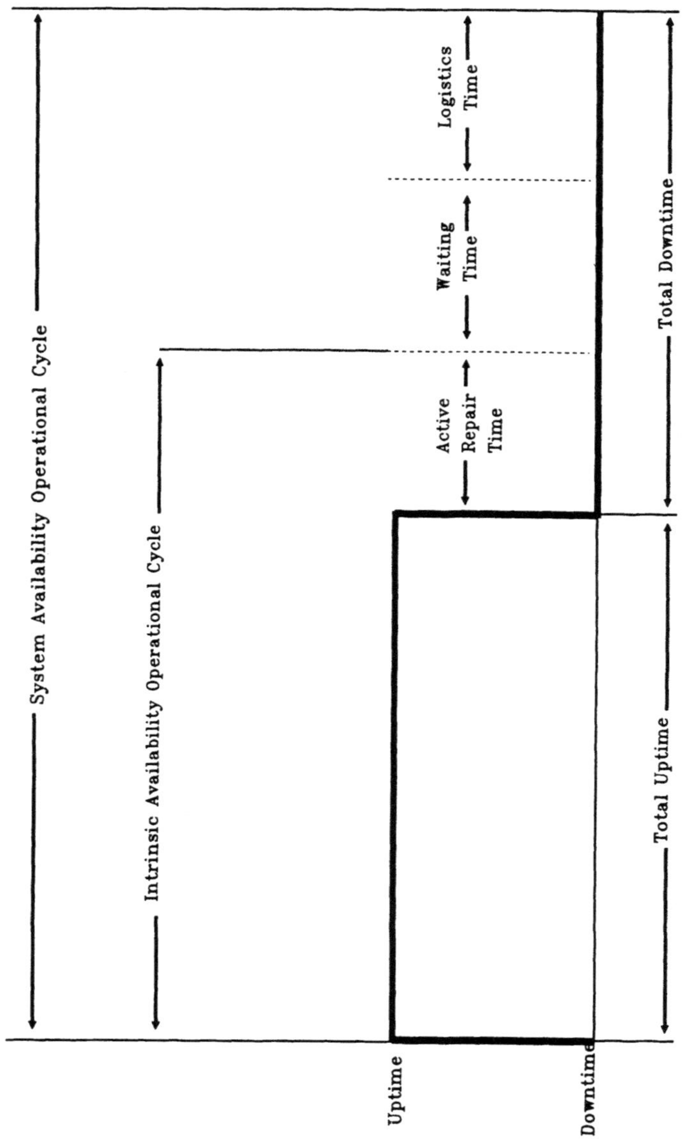

Figure 3–4 Operational Cycles for Intrinsic and System Availability

the reliable operation of the system and its ability to be efficiently maintained. It has a similar meaning for repairable equipments to that of reliability for non-repairable equipments. The difference is that reliability only accounts for the single event failure and availability accounts for both failure and repair events.

It is defined mathematically as:

$$A = \frac{\text{total system uptime}}{\text{total system uptime} + \text{total system downtime}}$$

Various other definitions of availability have been established based on the time elements included in total system downtime. Intrinsic availability is defined as consisting only of the actual active repair time and neglects any other logistic or personnel factors. On the other hand, operational or system availability is defined to include active repair time, waiting or administrative time (including shipping time) and logistic time. Figure 3–4 helps to clarify the distinction between intrinsic and operational availability, by showing their operational cycles and the various time elements.

From the figure it can be seen that system availability takes into consideration all delay factors and hence provides a realistic picture of the actual time the system will be available to perform its intended function. Because of difficulty in evaluating total downtime, care must be exercised in assessing system availability, due to the large number of factors that will affect its actual value.

The relationship of availability and reliability with time is shown in Figure 3–5.

Instantaneous availability, $A(t)$, is defined as the probability that a system will perform a specified function under given conditions at a prescribed time.

The instantaneous availability is bounded such that

$$R(t) \leqslant A(t) \leqslant 1$$

since $A(t) = R(t)$ for an item that does not undergo repair. An important difference between $A(t)$ and $R(t)$ is their behavior for large times. As t becomes large, $R(t)$ approaches zero, whereas availability functions reach some steady-state value.

A system consisting of one or more identical channels (or components), each having a constant hazard and repair rate (i.e., failure and repair are random), is a multiple-state system since each channel can be either in operation or under repair (or even in standby awaiting repair). The

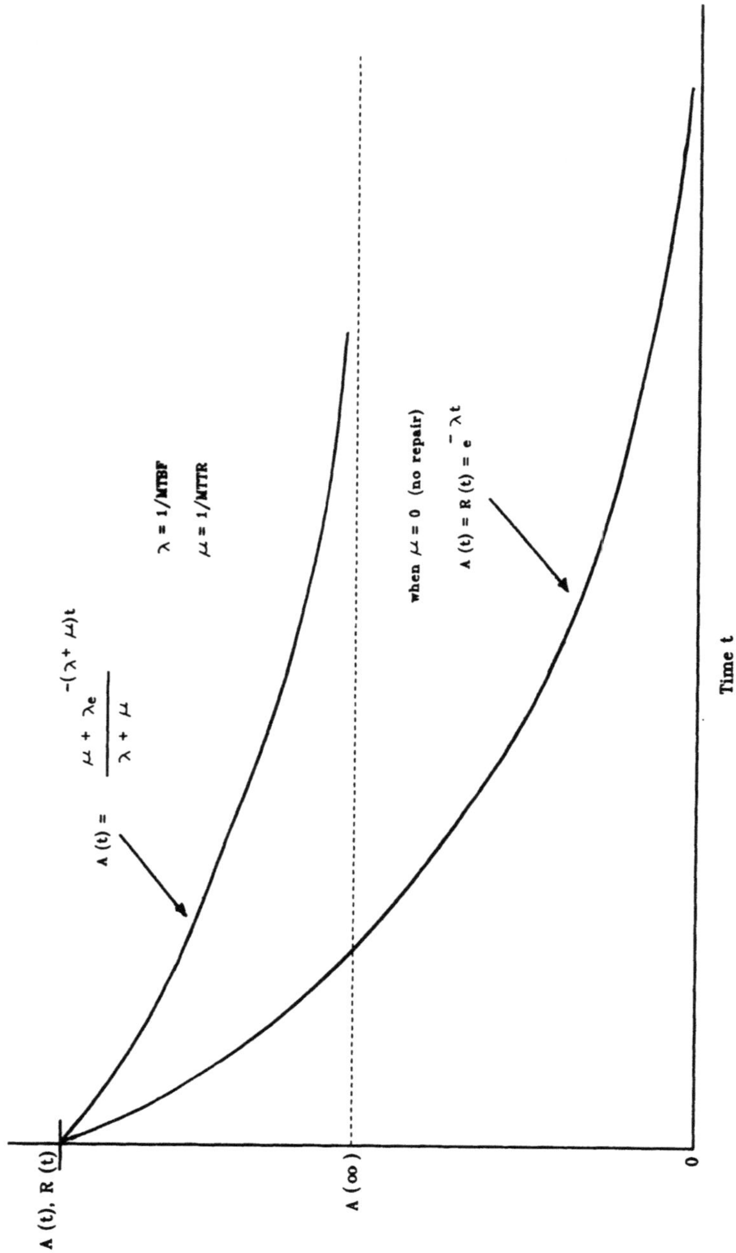

Figure 3–5 Availability and Reliability of a Single Element

steady-state availability of this typical equipment is, generally, evaluated using a Markov-chain model. A Markov-chain model is a discrete-time stochastic process in which it is assumed that the occurrence of any future equipment state is independent of any past state and depends only on the present state. The following steady-state availability equations are derived from the Markov process:

For a single channel:

$$A(\infty) = \frac{\mu}{\lambda + \mu}$$

For a two channel active redundant system:

$$A(\infty) = \frac{\mu^2 + 2\mu\lambda}{\mu^2 + 2\mu\lambda + 2\lambda^2}$$

where:

$$\lambda = \frac{1}{\text{MTBF}}$$

$$\mu = \frac{1}{\text{MTTR}}$$

Also, from the Markov process, steady-state availability can be approximated for n redundant channels, where k must be operable for system success, by

$$A_{k,n} \simeq 1 - \frac{n!}{(k-1)!} \cdot \gamma^{n-k+1}$$

where $\gamma = \lambda/\mu$ and MTBF can be approximated for n redundant channels, where k must be operable, by

$$\text{MTBF} \simeq \frac{(k-1)!}{n!} \cdot \frac{\mu^{n-k}}{\lambda^{n-k+1}}$$

The above formulae can be used to determine the appropriate R&M parameter values (i.e., MTBF, MTTR) to meet a functional availability requirement; however, in using these formulae the following assumptions and limitations must be kept in mind:

(1) The redundant channels are identical where both units are initially operating, one on-line the other off-line; each unit fails at the rate λ and is repaired at the time t.

(2) Repair is done on only one element at a time, and any element under repair remains so until the complete repair has been accomplished.
(3) Each element is completely independent. This means that redundant channels are designed such that they are completely isolated from each other, that failures do not propagate from one channel to the other and that common failure modes do not exist.
(4) Perfect switchover exists, i.e., the reliability of any components used to sense failure of the on-line channel and to switch to the off-line channel is assumed to be one.
(5) Both logistic time and repair time are considered random. This means that the repair is completed at a random time after it is started and also that the repair is initiated at a random time after the failure occurs. It should be noted that, although the assumptions of random failure and repair (constant λ and μ) are not always correct, both assumptions are necessary to avoid extensive mathematical complications. Furthermore, assuming constant repair rates does not usually introduce serious limitations when availability calculations are performed. The steady-state availability of a repairable component is dependent only on the MTTR and the MTBF and is usually reached after only a few repair times. Thus, assuming constant component repair rates, generally, affects system results only during the early transient, and even this effect is slight in most cases.
(6) The MTTR of each element (serial and redundant channels) and any logistic delay times are identical.
(7) The failure rate of a redundant channel is less than or equal to ten times its repair rate.
(8) The product of a channel MTTR (plus logistic delay time) and its failure rate is much less than one.

It should also be kept in mind that, although the equations indicate that the use of redundancy provides a significant increase in availability, it does impose a penalty by adding an additional serial element in the unscheduled maintenance chain. Furthermore, all redundancy applications are governed by such limiting factors as feasibility, cost, weight and complexity. Detailed analyses of the system and its requirements with careful consideration of the factors that will limit the effectiveness of redundancy must be performed.

Since both MTBF and MTTR are influenced by the design strategies, cost trade-off analyses are performed to converge on the optimum MTBF–MTTR mix. The complete trade-off process is an iterative process which begins with the application of minimal R&M design attributes or features and extends to the application of state-of-the-art technology.

Figure 3–6 depicts the trade-off between MTBF and MTTR which can be made to achieve a level of inherent availability. The figure shows that in order to maximize inherent availability it is desirable to make the ratio of MTBF to MTTR as high as possible. An item can be designed and built having high reliability with respect to maintainability, or ease of maintenance can be designed into the item, resulting in high maintainability with respect to reliability. Frequently, the most practical way to achieve high availability is to supplement the design for reliability with a design for efficient, rapid repair and a high degree of maintainability.

In general, designing for high reliability means: (1) selecting parts and components with proven reliability and life characteristics; (2) derating parts to reduce deterioration and to provide a margin of safety; (3) using carefully designed-in redundancy of a form most appropriate to the hardware under consideration; (4) carefully planning, performing and documenting reliability tests including reliability development and growth tests, reliability qualification tests, environmental stress screens and acceptance tests; (5) using extensive and effective controls, disciplines, and provisions employed in a well-designed reliability program; and (6) requiring that all failures be analyzed when they occur, with rapid feedback of test and failure analysis results to the designers for correction of inadequate design.

Designing for a high degree of maintainability, in general, means: (1) incorporating easily accessible and interchangeable units, assemblies and modules; (2) providing automated and continuous scanning of selected measures of performance, within acceptable limits of variation of each parameter; (3) providing automatic alarm systems that warn when tolerance limits have been exceeded and possibly automatic logging of selected performance parameters to permit early trend detection; (4) providing features for the automatic detection, location and diagnosis of failures to the maximum extent possible; and (5) using extensive and effective controls, disciplines and provisions employed in a well-designed maintainability program.

From the preceding discussion it can be seen that system availability provides a useful tool in determining how anticipated improvements in R&M will affect the actual time the system can be used, at any phase in its operational life. Application of availability analysis during design

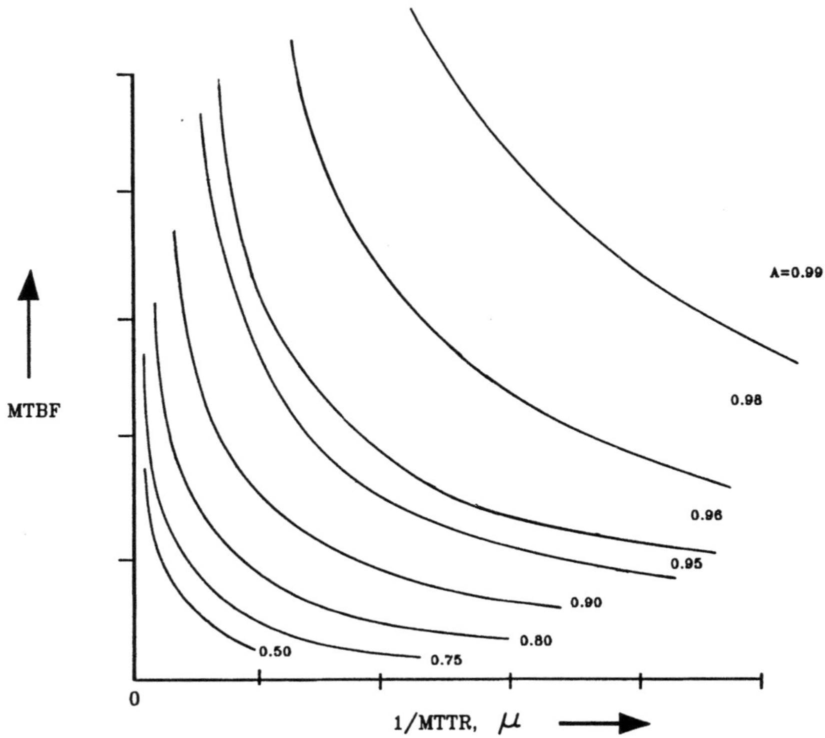

Figure 3–6 Availability as a Function of MTBF and 1/MTTR

provides a basis to perform trade-offs and sensitivity analyses and to force the design to be iterated to optimize R&M. Application during development and manufacturing facilitates improvement and growth. Application during field use provides a basis for making operational and maintenance management decisions as well as assessing achieved availability levels.

3.6 R&M MODELING

Mathematical evaluation models are used to apportion R&M requirements to various elements of hardware within the total system and to predict the design's inherent reliability and maintainability levels. Estimates based on evaluation models then become benchmarks for subsequent R&M assessment efforts.

In order to evaluate the reliability of a hardware system, a method is needed to reflect the reliability connectivity of its parts. This is accomplished by establishing a mathematical relationship between system reliability and the individual components and parts and their failure rates that make up the system. For most hardware systems, failure is a reflection of component failure, i.e., the system fails when any individual component fails. This is known as a serial reliability configuration. Failure of any one part in the series would result in failure of the system. Further, it may be assumed that failure of any part would occur independently of the operation of the others.

The series configuration may be represented by the following block diagram:

Input → $R_1(t)$ — $R_2(t)$ … $R_n(t)$ → Output

and system reliability is the product of the reliabilities of the individual blocks:

$$R_s(t) = (R_1(t) \cdot R_2(t) \ldots R_i(t) \ldots R_n(t)$$

where $R_s(t)$ is system reliability and $R_i(t)$ is the reliability of the ith block for the time t.

It should be noted that as a hardware development program progresses from conceptual to detailed design, the hardware definition also progresses to a much lower level of the assembly. To illustrate, Figure 3–7 provides a partial list of a helicopter system hierarchy identified to the development phases and the corresponding reliability techniques that are applicable to the level of definition. Reliability must be evaluated during all development program phases; consequently, the techniques required to predict and evaluate reliability must also be more detailed to reflect the greater level of hardware definition.

The concept of equipment hierarchy is also useful when combining lower level component failure rates to obtain an estimate of the system failure rate. The constant failure rate allows the computation of system reliability as a function of the reliability of a lower component to be accomplished in the following manner:

$$R(t) = \prod_{i-1}^{n} \exp(-\lambda_i t) = [\exp(-\lambda_1 t)] \cdot [\exp(-\lambda_2 t)] \ldots [\exp(-\lambda_n t)]$$

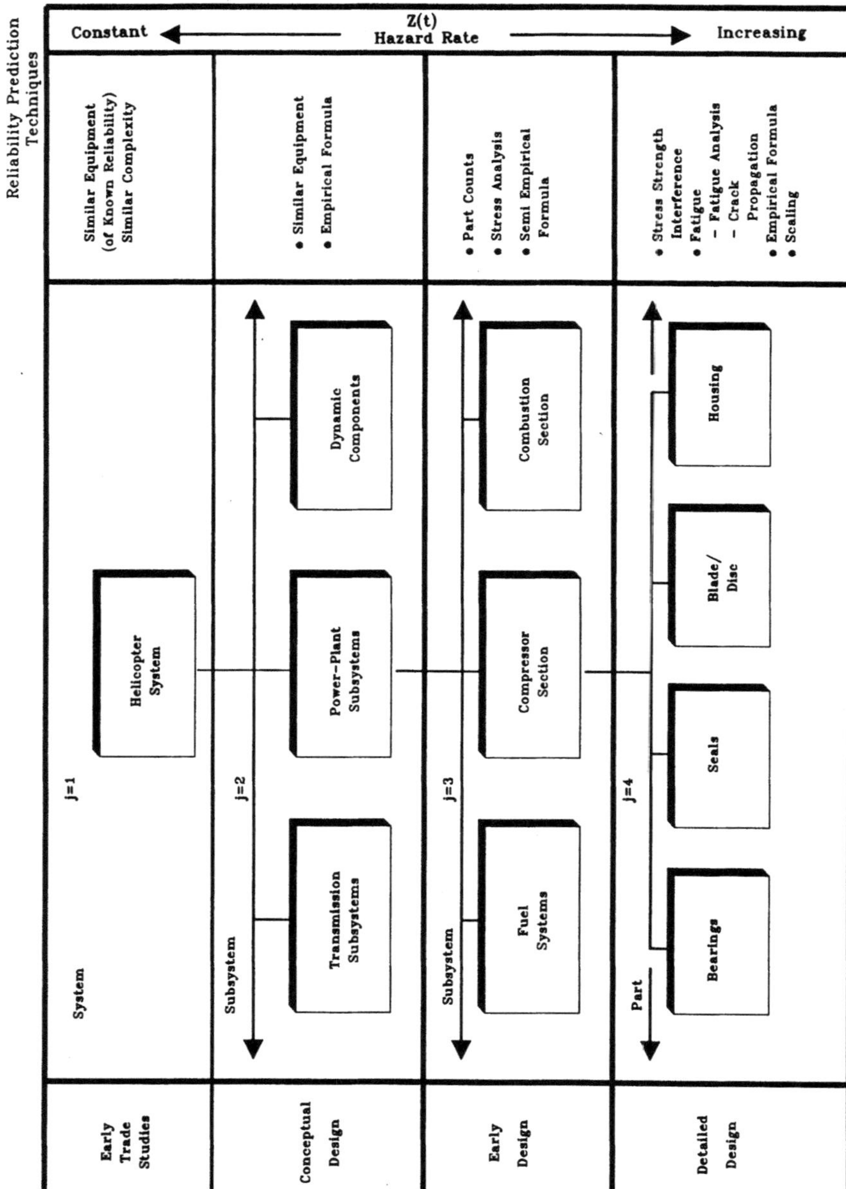

Figure 3–7 Helicopter Equipment Hierarchy (partial listing)

This can be simplified:

$$R(t) = \exp[-(\lambda_1 t + \lambda_2 t + \ldots + \lambda_n t)] = \exp[-\lambda_1 + \lambda_2 \ldots + \lambda_n)t$$

The general form of this expression can be written:

$$R(t) = \exp\left(-t\sum_{i=1}^{n}\lambda_i\right)$$

Another important relationship is obtained by considering the *j*th subsystem failure rate (λ_j) to be equal to the sum of the individual failure rate of *n* independent elements of the subsystems such that:

$$\lambda_j = \sum_{i=1}^{n} \lambda_i$$

Revising the MTBF formulae to refer to the system rather than an individual element gives the mean time between failures of the system:

$$\text{MTBF} = \frac{1}{\lambda_j} = \frac{1}{\sum_{i=1}^{n} \lambda_i}$$

Successive estimates of the *i*th subsystem failure rate can be made by combining lower level failure rates using

$$\lambda_j = \sum_{i=1}^{n} \lambda_{ij} (j = 1, \ldots, m)$$

where λ_{ij} is the failure rate of the *i*th component in the *j*th level subsystem and λ_j is the failure rate of the *j*th level subsystem.

As the system progresses to a greater level of detail, simple elements (parts) are designed, and it becomes increasingly difficult to justify the constant failure rate assumption for the reliability analysis. However, as previously discussed, non-constant part hazard rates of many parts and components will combine and a constant failure rate can be considered a valid approximation for the higher level assembly:

$$z_j(t) = \lambda_j = z_{1,j}(t) + z_{2,j}(t) + \ldots + z_{n,j}(t)$$

where $z_{1,j}(t)$, $(i = 1, 2 \ldots, n)$ are the non-constant hazard rates associated with part failure modes at time *t*, and λ_j is the higher assembly level where the constant failure rate is valid.

For the non-constant hazard rate assumption it is recommended that further detailed reliability analysis at the part (failure mode) level employs probability theory to compute higher level equipment reliability. The equation can be written:

$$R_j = \exp(-\lambda_j t) = \prod_i^n (Ps_i) \cdot (1 - P_j)$$

where

$$Ps_i = \exp[-\int_0^\infty z(t)\,\mathrm{d}t],$$

is the probability of surviving the ith failure mode and P_j is the probability of failure due to failure mode interaction.

For hardware that has survived infant mortality Ps_i can be estimated using probabilistic design techniques that account for wear and cumulative damage effects. The last term in the above equation (P_j) may or may not be significant depending on the class of equipment being analyzed. The reason the term is included is that the inter-dependences of failure modes could lead to a reliability that is less than the product of the individual survival probabilities. For example, a part has a probability of surviving the effects of corrosion or fatigue over the useful life, but still may have an additional probability of failure due to effects of both corrosion and fatigue acting together.

The constant failure rate (random failure) assumption is valid when making reliability predictions for major helicopter subsystems (e.g., power plant, transmission, avionics, etc.) during early design evaluation. It is generally valid when a large number of failure mechanisms contribute to the failure of a component. Standard MIL-HDBK-217 prediction techniques, assuming a constant failure rate, will provide sufficient accuracy.

During the detailed design phase more detailed reliability analyses are conducted. Cumulative damage theory, wear theory and fatigue theory all assume an end-of-life characteristic for the individual parts, and usually an increasing hazard rate better approximates the actual reliability characteristics. Mechanical reliability prediction techniques (stress–strength-interference and probabilistic fatigue analysis) have been developed to estimate part failure probability and, hence, reliability. These techniques are probabilistic extensions of standard mechanical design analysis techniques; they provide a great deal of insight into factors that

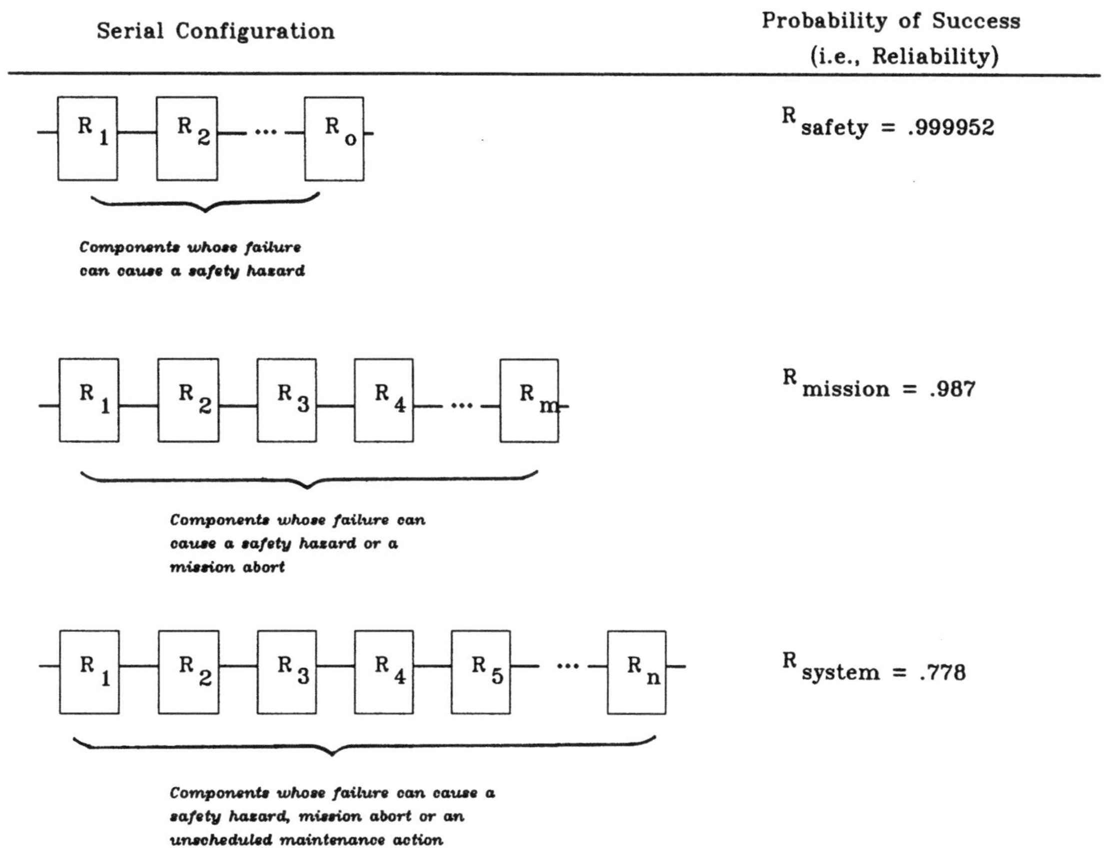

Figure 3–8 Reliability Block Diagrams for Safety, Mission Abort and Unscheduled Maintenance Actions

lead to unreliability and would lead to early problem identification and correction.

Both a high level of analytical competence and sufficient time are required to implement a probabilistic (reliability) design analysis program. Fortunately, many mechanical components can be easily designed to include a large safety margin, and there are many parts whose failure will not contribute to system failure; thus these parts can be eliminated from the probabilistic analysis without extensive investigation. A failure mode, effects and criticality analysis would help identify critical parts for more extensive probabilistic design analysis. Parts whose failure can cause a safety hazard require the highest level of attention and, therefore, should be subjected to the most rigorous and thorough design analysis.

It should be noted that, generally, parts capable of causing safety hazards are but a fraction of a total system part count. The various categories of reliability are illustrated in block diagram form in Figure 3–8. The reliability requirement and level of analytical and test effort are dependent on the criticality of the part failure. The expense of a probabilistic design analysis program to identify and correct inherent design reliability problems early, would be compensated by the more costly hardware redesign programs that could potentially be eliminated.

Maintainability models are developed in a similar fashion to that described for reliability. The evaluation of maintainability requires measurement of the factors which would tend to delay or speed up any maintenance action. The maintainability model must account for these factors. The most direct approach to developing the model is one which focuses on an accurate appraisal of system downtime. Basically, system downtime can be broken into two categories, preventive maintenance downtime and corrective maintenance downtime.

Preventive maintenance downtime is that time during which a system is shutdown so that maintenance can be performed in order to prevent any anticipated failures. However, a preventive maintenance action does not always contribute to system downtime. Many tasks, such as adjustments, lubrications and light cleanings, do not require a system shutdown. Thus, when considering preventive maintenance downtime, estimates must be made as to that time (e.g., remove/replace time) which requires an actual system shutdown.

Many factors enter into the estimation of preventive maintenance downtime, such as the number of maintenance persons and their skill levels. For simplicity, personnel availability often is not considered due to the assumption that a preventive maintenance task is being carried out by a person who can handle all portions of the task and, if the person causes a failure, the failure can be repaired immediately.

The number of persons performing maintenance is obviously a factor in that increases in manpower will almost always reduce the amount of time required to effect a maintenance action. Likewise, the skill level also needs to be considered since it is usually the case that an above average maintenance person will need less time per maintenance task than a below average person.

Preventive maintenance downtime may be allocated into time intervals based on consideration of the number of maintenance persons and their skill level. Unlike preventive maintenance, corrective maintenance downtime consists exclusively of system downtime, the notable exception being when redundant or backup systems are used.

Corrective maintenance downtime is that time which includes increments of preparation time, fault location time, fix time, alignment time and checkout time. The allocation of quantities to each of these increments is strongly influenced by design factors such as modularity, accessibility, interchangeability and, particularly, the degree of built-in test and fault isolation capabilities. These design factors must be carefully assessed when estimating downtime.

Corrective maintenance downtime must also take into consideration the skill level and number of personnel. However, estimates of corrective maintenance downtime must also consider a factor to account for the availability of a repair person, since it is highly unlikely that there will be a repair person present and available to service every random failure. The availability of a repair person is dependent on the facility's maintenance philosophy. The maintenance philosophy establishes the location of the maintenance pool relative to the item to be serviced and, consequently, determines the amount of travel time.

The models and the resulting MTBF and MTTR data can be used to quantitatively evaluate the impact of the elements that drive the operating R&M of the system including those design, production, operation and maintenance factors depicted in Figure 3–9. This R&M improvement process involves first assessing the MTBF and MTTR numerics and the observed preventive maintenance downtime data with respect to the defined objectives and operating requirements for the system. The intent is to determine quantitatively the extent of improvement considered necessary. Close coordination with cognizant design personnel is considered essential at this point in order to assure that improvement goals established are consistent with the overall objectives for the

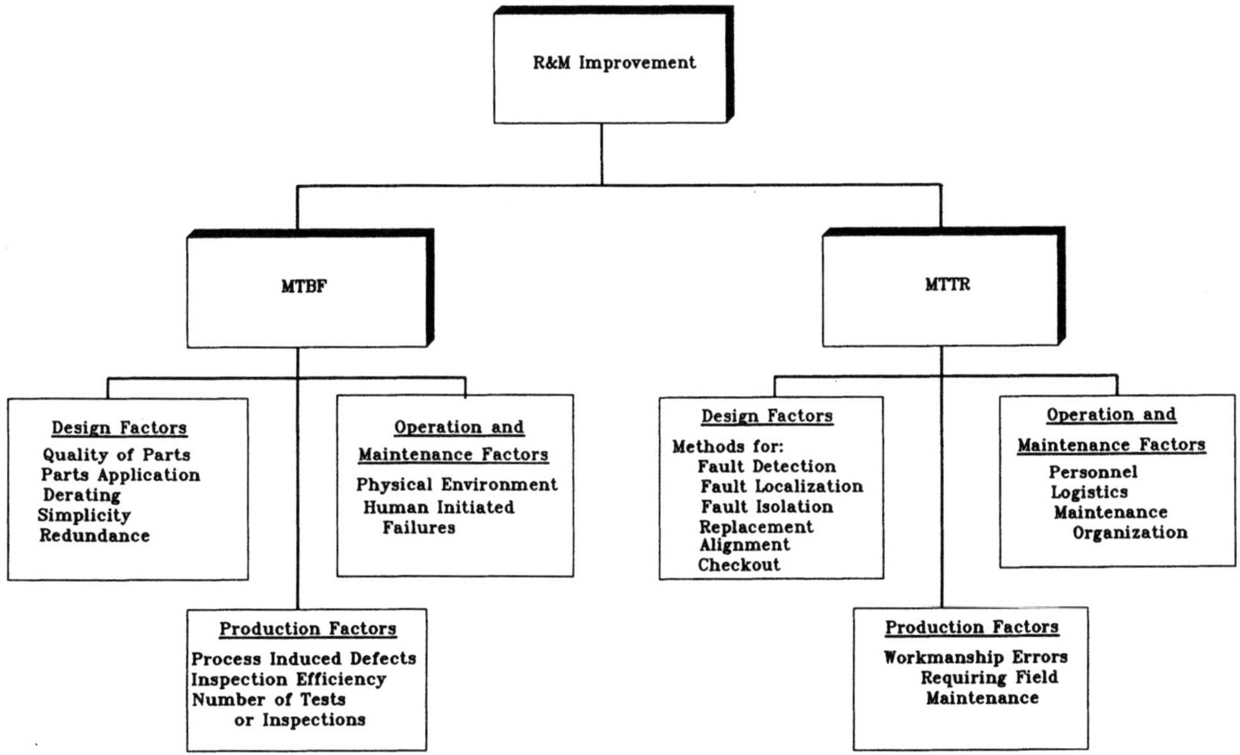

Figure 3–9 R&M Improvement Factors

particular system or component. Once quantitative improvement goals are established, the next step is to review the R&M models and their backup data to identify areas and criteria for improvement and to formulate recommendations that would meet the criteria.

Improving system reliability (MTBF) involves a systematic review of several concepts, among which are the following:

- The reduction of failure rates by operating components at reduced (derated) stress levels. This is accomplished by selecting components which have ratings well in excess of those required for their system application.
- The use of special components for which reliability has been significantly increased through special manufacturing techniques, quality control procedures and testing methods.
- Design simplification to eliminate parts or components.
- The substitution of functionally equivalent items with higher reliability.
- The overall reduction of failure rate through increased control of the internal system environment—e.g., through reduction of ambient temperature, isolation from handling effects and protection from dust.
- The provision of design features which enable prediction of incipient failures, and permit remedial action to be taken before an operational failure occurs.
- The provision of design features which reduce the probability of human initiated errors.
- The provision of multiple, identical parts, paths or higher functional levels (redundancy) in order to prevent a system failure in the event that one element fails.
- The reduction of failure rate through increased control of the environment external to the equipment—as through reduction of ambient temperature, isolation from handling effects, isolation of operator from ambient noise and protection of equipment from dust.
- The use of production screen tests for the purpose of significantly reducing incipient failures due to undetected defects in workmanship or components.

Improving system maintainability (MTTR) involves reducing

(1) Localization time through increased use of special built-in circuits for fault detection, error warning lights, etc.

(2) Isolation time by:
- Designing for replacement at higher levels.
- Utilizing test indications which are less time consuming and/or less difficult to interpret.
- Designing for minimum diagnostic strategies.
- Making accessible and obvious both the purpose of test points and their relationship to the item tested.
- Improving the quality of technical manuals or maintenance aids.
- Using technicians with a higher skill level.
- Increasing depth of penetration of localization features.

(3) Disassembly and reassembly time by:
- Designing accesses for ease of entry.
- Reducing number of access barriers.
- Increasing ruggedness of equipment elements.
- Reducing need for isolation access by bringing test point, controls and displays out to accessible locations.

(4) Interchange time through an evaluation of its major factors including the following:
- Functional level of replacement (part, component, assembly, unit, etc.).
- Type of replaceable element (e.g., plug-in subassemblies, quick-disconnect units, etc.).
- Number of interconnections per replaceable item.
- Type of connections (hydraulic, fuel, electrical, etc.), insertion and removal forces, and special tool requirements.
- Orientation and location of replaceable elements.
- Skill level of maintenance personnel.
- Packaging density.

(5) Alignment time through an evaluation of its major factors including:
- Functional level at which alignment is accomplished.
- Difference between functional levels of replacement and alignment.
- Type and extent of alignment required.
- Number of alignment parameters.
- Accessibility of alignment features and spacing in relation to surrounding elements.
- Skill level of maintenance personnel.
- Alignment sensitivity.

- Difficulty of adjustment criteria.
- Requirement for external test equipment and tools.
- Accuracy, completeness and ease of use of instructions and data.

(6) Checkout time through an evaluation of the tasks required to verify that the system has been fully restored to operational capability. Normally, if localization time elements are minimized then the time required for checkout is also minimized.

Computing the impact of the recommendations, which appear most useful for cost trade-off consideration, on MTBF, MTTR, overall downtime and system availability using the methods and techniques previously described is the next step in the improvement process.

Critical to the analysis process is the ability to assess quantitatively the cost of the R&M improvements. The cost of each recommended change must take into account total cost throughout the life-cycle of the system and accordingly must include cost elements associated with design, manufacture, procurement and field use (i.e., operation, maintenance and logistics).

The final activity is to compute cost–benefit factors, i.e., develop a numeric for each R&M recommendation which reflects the total cost of the change and its impact on system performance. This will allow the determination of those change recommendations which have maximum cost-effectiveness. The recommended changes can then be ranked in decreasing order of cost-effectiveness as defined by the computed cost–benefit factors.

3.7 LIFE-CYCLE ACTIVITIES

The RCM driven ILS process has a life-cycle perspective. The driving force is reduction of the maintenance burden and support cost while maintaining the designed-in R&M characteristics and necessary operational readiness state. A life-cycle program must be planned and implemented to integrate R&M engineering practice into the process and to effectively interface with ILS activity.

The complete RCM & ILS process requires an extensive base of reliability and maintainability (R&M) data. This R&M data, to be useful for a specific analysis, must be continually refined and updated to include the most recent information. Early R&M and RCM logic analyses are based primarily on data derived from existing or similar fielded equipment. Analyses performed later in development are based on detailed design engineering data. Assessments performed after deployment during field operation are based on statistical analysis of actual experience data. The major data items necessary for input to the RCM/ILS process are mean time between failures (MTBF), mean time to repair (MTTR), and failure mode, effects and criticality analysis (FMECA) data.

An essential aspect of the process is the identification and analysis of potential failure modes to establish the initial basis for formulating corrective maintenance requirements. The objective is to systematically identify the likely modes of failure, the possible effects of each failure, and the criticality of each failure on function, safety and maintenance.

R&M prediction and failure mode analysis are performed, as part of a complete life-cycle program, in support of the system development, production, maintenance and logistic process. As indicated in Chapter 2, the R&M tasks must be coordinated with the ILS tasks and the results available as essential input. This coordination addresses the timing of the tasks, the level of detail and the documentation requirements. Chapter 2 described the interface of RCM with the ILS process.

Fundamental to an effective RCM&ILS process is the establishment of the system's R&M characteristics. A complete, well planned life-cycle R&M program must be established that embodies the following basic considerations:

(1) R&M are quantitative characteristics that are predictable in design, measurable in test, controllable in production and sustainable in the field.
(2) The actual in-service reliability of a hardware system is a function of its design, as well as subsequent life-cycle activities, where:
   (a) *Design* establishes the 'inherent' R&M potential of a system and is defined by its engineering documentation, and
   (b) *Subsequent life-cycle activities* can only degrade R&M below this inherent design level. For example, the transition of a system from a 'paper' design to initial production hardware results in reliability below the inherent level. Consequently, analysis of operational reliability must be approached first via its design characteristics (which establish an upper limit of reliability) and then in conjunction with a series of modification factors that account for production and operation and maintenance (O&M) degradation and improvement.

(3) The improvement and growth of reliability are best accomplished in the early stages of design and development by implementing highly disciplined and systematic engineering analysis and test activities which enhance inherent reliability by forcing the design to be iterated, prevent defects through process improvements, and minimize production and operational and maintenance (O&M) degradation by eliminating potential failures and manufacturing flaws prior to production.

(4) In order to restore and retain the designed-in R&M levels, deliberate and positive engineering action must be taken continuously throughout a system's life-cycle. This action must include the development of optimum RCM derived maintenance plans and support requirements.

Figure 3–10 shows that the achievement and retention of R&M requires a total reliability and quality management process. R&M and quality engineering tasks and controls are applied beginning at system concept and extending through all life-cycle phases. R&M and quality engineering considerations are an integral part of system acquisition and operation during which the required levels of effectiveness are planned, achieved and maintained at minimum total cost. The *Conceptual Phase* involves planning for R&M, performing trade-off studies and identifying areas of high technical risk. The *Validation Phase* involves preparing R&M inputs to requests for proposal (RFP), work statements and specifications for R&M. Also, proposals are evaluated for compliance with the R&M design and program specifications. The *Development Phase* involves implementing the applicable R&M program elements including management and monitoring; R&M analysis, testing, failure analysis and data reporting; and RCM decision analysis and maintenance support planning. During this phase the inherent R&M of the design is established to comply with the specified requirements; test and controls are implemented to prevent defects and to ensure that the R&M levels are not appreciably degraded during production and are achieved in the field; and the RCM logic analysis is applied, ILS requirements are defined and the maintenance support plan is prepared. The *Production Phase* involves planning and implementing factory screening; statistical process controls and continuous process improvements: performing failure analysis and data collection activities; evaluating change proposals, the maintenance tasks and ILS requirements, and updating the maintenance support plan, as required. The *Operational Phase* involves collecting and analyzing field

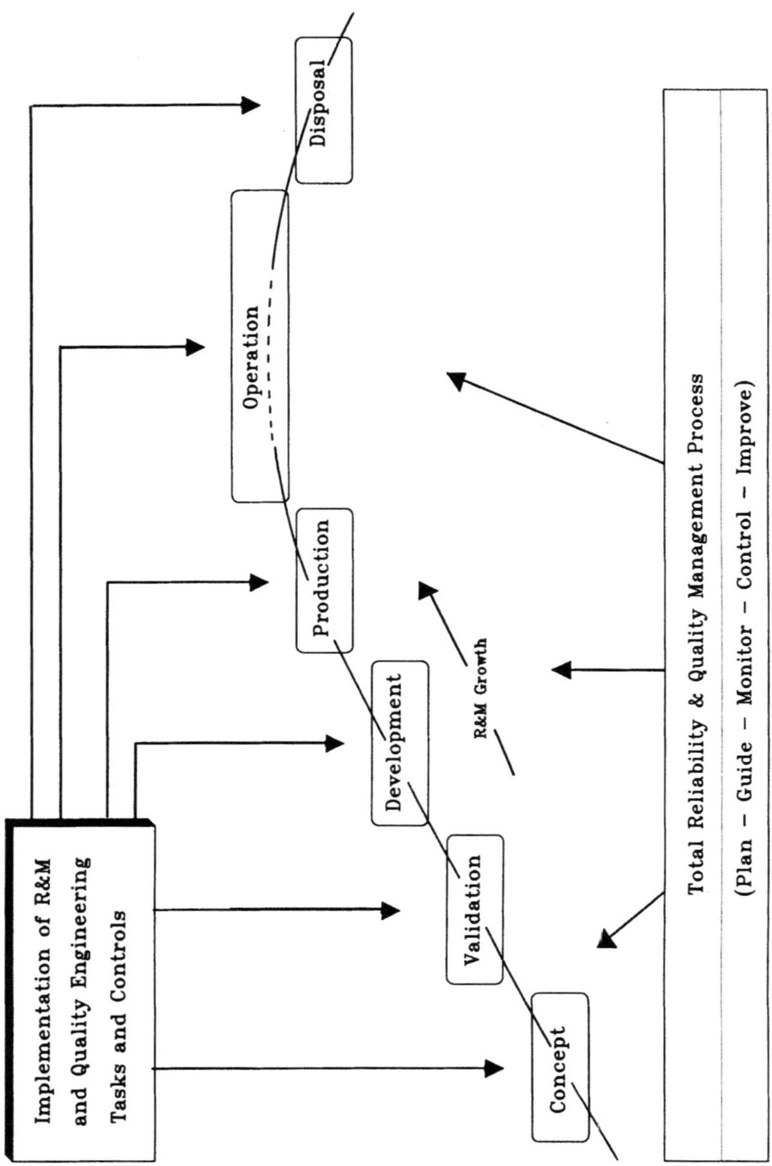

Figure 3–10 R&M Activities During System Life-Cycle

experience data with respect to R&M, performing a sustaining engineering activity to re-establish the RCM/ILS requirements; assuring the R&M of spare parts, and evaluating the R&M characteristics of proposed product improvements. The *Disposal Phase* involves compiling and retaining essential R&M experience data once the system is removed from the inventory.

Essential to the overall life-cycle R&M process are the following three major activities:

(1) The performance of trade-off and cost-effectiveness studies to formulate detailed requirements for the statement of work in terms of program elements and numerical specifications which are cost-effective and have a high probability of achievement.
(2) The evaluation of proposal documents, control provisions, program planning and technical expertise to select the most qualified contractor with respect to R&M.
(3) The implementation of the R&M program during development and production, including the preparation and review of deliverable data items, conducting independent R&M assessments and participation in design reviews.

The *R&M and cost trade-off studies* (*Item* 1) are performed during design concept and validation to aid in the achievement of a balanced design. Table 3–1 defines activities and work effort involved in performing R&M and cost trade-off studies. The steps involved in performing trade-off studies are broadly described as follows:

Step 1. Perform Preliminary Analysis: (a) define trade-off criteria for R&M, cost schedule, etc., (b) define the level of effort to be applied to the trade-off process consistent with the level of system definition available.
Step 2. Perform Design Analysis. Further define the constraints associated with specific system or hardware items and support system characteristics. Define the limitations between which increments in R&M may vary.
Step 3. Define Parameters. Establish the parameters of a standard or 'baseline' design which just meets all requirements and which establishes a starting point for all parameters of interest during trade-offs.
Step 4. Gather Data. Collect, sort and validate system data and, to a lesser extent, component and part level data from available source.
Step 5. Perform Trade-Off Studies. Generate and evaluate design approaches for R&M which satisfy the trade-off criteria. Generate

TABLE 3-1
R & M and Cost Trade-off Study Activities

- Standardize the data base in order to provide a baseline for comparing competing design configurations
- Provide standardized definitions for failure consistent with safety, mission, and unscheduled maintenance reliability requirements
- Establish credibility of early predictions and/or assessments

This is established through:
  Reliability Models—detailed enough for trade-off studies
  Assumptions–environment (ground, flight, etc.), component technology, etc.
  Application Factors—temperature, stress, etc.
  Other Ground Rules–component quality levels
- Provide R & M estimates for cost trade-off studies
- Define R & M cost sensitive factors (from cost comparisons and studies)
- Monitor, evaluate and control special R & M projects which include trade-off studies with respect to cost sensitive factors such as:
  - Mechanical, electrical and thermal stress (derating)
  - Component substitution
  - Component quality
  - Environmental stress screening and burn-in
  - Redundancy
  - Packaging (environmental resistance)
  - Built-in test equipment
  - Modularization
  - Accessibility, inspectability
  - Maintenance skill levels
  - Maintenance support equipment

sensitivity curves which show the break-points for R&M with respect to cost, and for given performance inputs.

Step 6. Refine Studies. Apply design details, as they become available, for refinement of the trade-off studies so that the optimum design approach becomes apparent.

Requirements for R&M parameters and program elements are then established and specifically tailored to the system procurement approach for system development and production. There are three basic procurement approach options that exist to meet the specified acquisition need:

> Option 1 provides for the procurement of existing commercial systems in order to obtain a low cost, quick-response capability for certain requirements. Advantages of this option include use of a proven design, reduced lead-times and minimal development expense.

Possible disadvantages include inability to meet R&M requirements, limited performance, parts availability, reduced control of model changes and increased logistic support requirements.

Option 2 provides for the procurement of modified versions of existing commercial systems. This option provides for use of the basic commercial configuration with modifications to meet certain required specifications. Possible advantages of this form of procurement are quicker availability and lower development cost than for a new design. Possible disadvantages include the loss of integrity of the commercial product, the addition of unproven components and the compromise of mission capability.

Option 3 provides for the procurement of systems to full customer specifications. Within this option are two procurement categories: (a) *existing development*, which is characterized by an existing technical data package (TDP); engineering change orders which do not significantly impact schedule, require extensive requalification or involve substantial redesign; a smooth transition to production which involves existing production facilities; (b) *new development*, which involves a complete new design or changes to major components and major redesign of existing system. New development is characterized by the establishment of a program office and preparation or restructuring of a TDP. The possible advantages of procuring a newly designed item are that the item can fully meet customer requirements, that the design and configuration can be controlled, and that the logistic support can be assured. Possible advantages of the procurement of an existing design are the shorter lead-times involved, the use of less costly production improvements to reach required performance objectives and the utilization of existing technology.

Table 3–2 identifies the essential elements which would comprise an effective R&M program for a new system that requires a full development program [in accordance with option 3(b)]. Methods for accomplishing the tasks listed in the table are described in the R&M program plan. Further discussion of a few of the program tasks, particularly those applicable to early development, are given in the following paragraphs.

R&M Management involves defining and assuring the effectiveness of the various R&M program elements planned for use during system development and production. This, in general, involves reviewing the SOW and preparing an R&M program plan. The plan must detail the approach, criteria and procedures to be followed to meet the objectives of the development program as specified in the SOW, and, generally, MIL-STD-785 and MIL-STD-470. The program plan must recognize that, in order to achieve an actual field reliability that approaches the predicted reliability, the thrust of the R&M program must be: (1) to emphasize early R&M analysis and prediction and the use of accurate and detailed models that account for design and field application factors; (2) to force out defects through an aggressive reliability growth process; and (3) to measure reliability under environmental conditions which duplicate field conditions.

The R&M organization must have direct access to program management and must have an effective relationship to design, manufacturing, procurement, quality control, cost and other organizational functions with separate responsibilities and authorities. It must be comprised of a highly effective team of specialists with experience in all R&M areas. Key personnel must be committed to the program. The R&M engineering tasks, analyses and control elements must be adequately described, scheduled and properly timed such that they coincide with major project design points.

R&M Apportionment involves the subdivision of equipment reliability requirements (or goals) into the various major items that comprise the system. These apportionments become, in turn, design requirements for the individual equipment items. An apportionment study, to serve the needs of the impending design effort, must be completed shortly after contract award. The apportionment study allocates failure rates or repair rates quantitatively to the lowest practical functional level. Ideally, the apportionment should show an increment in reliability (MTBF) or maintainability (MTTR) over and above a strict subdivision of numerical requirements at the given level which serve as design goals. In addition, the apportionment should be based on system criticality, complexity of design and function, operational use environment, previous experience with similar equipment and relation to the state-of-the-art.

R&M Prediction involves identifying the equipments' reliability and maintainability parameters quantitatively (e.g., MTBF and MTTR) through preparation of system models, identification of applicable distribution functions and use of appropriate failure and repair rates. The models, distributions and input data must account for the mission profile, environmental application stresses, maintenance conditions and other operating/non-operating use factors. The methodology must be defined, including the modeling assump-

TABLE 3-2
R & M Program Tasks During Development and Production

Management—the organization, planning, controls documentation, and definition necessary to carry out the R & M program

R & M Apportionment—the process of subdividing system level reliability requirements down to component and parts levels of assembly

R & M Prediction—the effort to estimate system R & M based on the design characteristics; the modeling details procedures and data base are documented

Maintenance Concept—the scope of the maintenance activities at field sites, intermediate facilities and depot organizations. Fault isolation, support equipment and skill requirements and maintenance task criteria are defined

Failure Mode Analysis—the effort to perform an analytical part-by-part evaluation of the system to determine the consequences of potential failure on system operation

Component Control and Standardization—the effort to select, specify and control all 'critical' mechanical, electrical and electromechanical parts and components

Design Review—the methods and results of performing formal evaluations of design activities and their status at key development milestones

Reliability Growth Testing—the planning and implementation of tests specifically designed for the purpose of identifying defects so that they can be analyzed and corrected thus forcing reliability to grow

R & M Qualification Test—planning and implementation of a test to show that the system complies with specified R & M requirements

Failure Analysis—the engineering effort to determine the cause of reported failures and to report these findings for subsequent corrective action

Environmental Stress Screening—the process of applying non-destructive screens to critical assemblies in order to force-out latent defects during manufacturing prior to field operations

Reliability Assessments—the effort to determine the actual reliability of the system based on testing or field use data

Data Collection and Feedback—the collection of test and operational experience data for feedback to the development process

tions, configuration basis, the data sources and the level of detail applicable to the status of the design.

To be fully cognizant of the impact of actual field use conditions, and to reflect non-ideal maintenance/supply conditions, prediction efforts must recognize the failure and repair cycle. The prediction effort must be performed with the ultimate field use in mind.

Design Reviews involve evaluations of performance, reliability, maintainability and various other characteristics of the system at major design and testing milestones. The design review program must be geared to ascertain that methods have been established which provide for review of all system elements down to the component level; that the program includes subcontractor's design review activities; that it adequately defines the participants and their responsibilities; and that it describes the deficiency-follow-up control procedure. In addition, design review procedures must include a detailed and comprehensive check list and criteria against which the design can be evaluated. The check list must be relevant to the design phase under review.

Design reviews provide the means for assessment and monitoring of the contractor's design effort and generally coincide with major program milestones.

Failure Mode Analysis involves a part-by-part analysis to determine system and/or component effects when considering all significant failure modes. The analysis is intended to:

- Relate parts, assemblies, or functions to their failure effect
- Determine quantitatively the occurrence probability of the failure effects
- Determine failure mode criticality
- Provide a basis for corrective action well in advance of equipment fabrication
- Aid in the generation of test plans and procedures
- Aid in the analysis of failures

To be effective, these analyses must detect, analyze and evaluate all significant failure modes and must convey their findings in time to be used at appropriate project decision points. Plans for corrective action, resulting from identification of modes which reduce system capability below acceptable minimum standards, must be defined.

The individual failure modes of each safety critical part, identified by failure mode analysis, are used for RCM decision logic analysis and ultimately to determine how impending failures can be detected and corrected in order to preserve, to the degree possible, the inherent levels of reliability and safety designed in the item. The RCM logic analysis identifies specific preventive maintenance tasks and requirements for:

- Detecting and correcting incipient failures either before they occur or before they develop into major defects
- Reducing the probability of failure
- Detecting hidden failures that have occurred
- Increasing the cost-effectiveness of the maintenance program

Component Control and Standardization involves effort directed to select, specify and control all critical or primary mechanical, electrical or electromechanical components used or planned for use in the system. This effort includes methods to secure necessary approval for components for system use. Consequently, detailed effort to justify, test and assure quality forms an extensive part of the control process.

Procedures must be prepared to provide adequate definition of all facets of critical component specification and qualification, including special controls covering source and incoming inspection.

Some of the key considerations that must govern the evaluation of a proposed R&M program (item 2) are as follows:

- The relationship of the R&M program to other project requirements must be defined.
- A formally organized program with central management, a documented program plan, and separate accountability for program resources must be structured and implemented. The R&M program must be negotiated together with the negotiation of the overall project contract (rather than after contract execution). A realistic program that delineates scope and cost of all R&M efforts must be established.
- Periodic reviews of the program are required which provide for revisions of the program plan, if necessary, depending on the results of the reviews.
- The prime contractor must maintain control of his own R&M effort as well as that of subcontractor and supplier R&M programs and must determine their effect on reliability of the overall system.
- All project data must be accessible. In order to provide for the most convenient accessibility, a central file or data center for documentation must be established.
- The project must be covered by one integrated test program (including reliability growth and qualification and factory acceptance) instead of separately managed testing programs. This requirement prevents both duplications and omissions in testing, and provides a single test baseline in parallel with a closely interrelated program of reliability assessment. This approach emphasizes the intimate tie-in of the reliability assessment effort with the requirements of the project and underscores its role as an input to the various project decision points.

The evaluation process starts with the review of contractor proposals where not only the most qualified contractor with respect to R&M is selected, but also the course for subsequent R&M management activities during development and production are established.

Proposals are evaluated for:

Compliance with Requirements

- Must show compliance with the specified R&M design and program requirements defined in the RFP.
- Must comply with the data item requirements.
- The R&M of the proposed design must be capable of demonstration without minimizing performance capability or incurring excessive cost.

Understanding of the Problem

- Must demonstrate understanding of the scope or range of tasks which make up the R&M effort.
- Must show an understanding of R&M technology: mathematical/statistical modeling, hardware engineering (stress factors), physics of failure, etc.
- Must show a knowledge of advanced, yet proven methods for R&M programs.
- Must show an understanding of the interaction between various R&M elements and the system design and development process, including the interface aspects of R&M which are development milestones.

Soundness of Approach

- Must indicate that the manpower, facilities and other resources are adequate to implement the approach described.
- Must show that the approach to R&M possesses sufficient flexibility to accommodate design changes, program delays or extension of R&M elements.
- Must indicate that the contractor can meet the objectives of the R&M program within the scheduled time period.
- Should contain any suggested extensions or exceptions beneficial to the Government.

Technical Expertise

- Must contain sufficient background or prior experience in R&M and related areas.

Management

- Must show how the contractor's R&M management structure for the proposed program functions within the overall corporate and program management. This includes personnel assigned, their technical expertise, management techniques, and lines of communications.

The R&M program is carried out by the contractor during development and production (item 3). Effective, systematic and timely management activities, engineering tasks and controlled tests are implemented which, in general, are performed in accordance with the requirements of US MIL-STDs-785 and -470. An effective management and control system is defined and implemented that will directly enable R&M personnel to influence design, provide timely outputs consistent with major decision points, and, in general, provide the means to develop and build a hardware system that meets cost-effective objectives and requirements.

During early development R&M estimates and cost analyses are made to support design trade-off studies. As development progresses, detailed engineering analyses and tasks are performed to identify and help correct problems and to force the design to be iterated, as necessary, prior to the build-up of hardware and to provide initial data for maintenance support planning.

The program includes early procurement, build-up, and reliability growth testing of critical components. Also, formal reliability growth testing is performed later in development with emphasis on failure analysis and corrective action and a test cycle that reflects the application environments, including mechanical stresses and climatic extremes. A production reliability assurance activity is planned which provides the necessary controls and procedures to allow a smooth transition from design and development to production without degrading reliability and which emphasizes the application of statistical process control (SPC) and environmental stress screening (ESS).

Sound R&M practice must also be carried out through the production and operation phases, as well, or any benefits gained during development could be lost. During production, efforts focus on quality control (QC) including SPC, modification and change control, and ESS. Quality controls are applied to assure that the parts are not defective and that the workmanship and manufacturing techniques employed in assembly are consistent and of high standard, to assure that the system is built to meet the designed-in R&M levels. Proposed changes to the system are evaluated for compliance with the R&M requirements set in the procurement specification or for lower assemblies established by the allocations. Controlled ESSs are applied to components and assemblies during manufacturing to force out latent defects prior to fielding. During operation, efforts involve compiling, reducing and analyzing field experience data, performing age-exploration and determining the impact on R&M of product improvements.

CHAPTER 4

R&M Engineering

Implementation of a comprehensive life-cycle RCM program requires detailed knowledge of several very specialized R&M engineering techniques. These techniques include:

- Reliability allocation and prediction
- Maintainability allocation and prediction
- Failure mode analyses including the fault tree analysis (FTA) and the failure mode, effects and criticality analysis (FMECA) procedures
- Reliability testing
- Production, storage, and operations and maintenance (O&M) control
- Environmental stress screening (ESS)

Each of these techniques is described in the following sections of this chapter.

4.1 RELIABILITY ALLOCATION AND PREDICTION

Reliability allocation and prediction are important techniques to be applied during the design of a complex system. Reliability allocation is the process of apportioning the overall system reliability requirement down to the subsystems and lower levels of assembly. Reliability prediction is the process of estimating system reliability starting at the lower level assemblies and proceeding up to the higher levels of assembly. In actual application there is considerable overlap between allocation and prediction. Early allocations help formulate alternative design concepts while predictions are used to assess the impact of the alternative designs on system reliability and to help establish the design approach. As the design progresses, predictions are performed in more detail in support of the development process, to help assure that the evolving system will meet its reliability requirement.

The objectives of reliability allocation are to:

- Apportion the system reliability requirement among the subsystems and components before a commitment is made to a particular design approach.
- Focus attention on the relationship between the various subdivisions of the system, and the contribution of each, to overall system reliability early during design when changes can be made easily and economically.
- Set realistic reliability design targets for subsystems and lower subdivisions based on pertinent design factors.
- Determine the need for incorporation of specific reliability design features, e.g., the need for redundancy can be established during the conceptual phase and then reflected in the specification for system development.

There are several basic techniques available for allocating system reliability to the subsystems and the components within. The particular technique to be applied in a given situation would depend on many factors, such as the amount and type of data available and the overall configuration of the system. Some of the techniques available are listed below in order of increasing complexity:

- Equivalent subsystems in series configuration
  Allocated subsystem reliability is defined by:

$$R_s = R^{1/n}$$

where n is the number of subsystems and R_s is system reliability.

- Non-equivalent subsystems in series configuration
  Weights are applied to each subsystem using the following formula:

$$W = \frac{C_i}{C_1 + C_2 \ldots C_n}$$

where C_i is the complexity of the ith subsystem and allocated subsystem reliability is defined by

$$R_i = R_s^{w_i}$$

- Consideration of subsystem importance and complexity
  Weights are applied to each subsystem using the following formula

$$\text{MTBF}_i = \frac{k_i \cdot t_i}{\left(\frac{n_i}{N}\right)(-\ln R_s)}$$

where k_i is the probability that the system fails if i fails, n_i is the number of components in ith subsystem, i is the total number of subsystems, and allocated subsystem reliability is defined by

$$R_i = 1 - k_i[1 - \exp(-t_i/n_i)]$$

It should be noted that an allocation is valid only to the extent that the final figures are achievable by the components to which they are assigned. If the reliability allocations are not achievable, redundancy may be required to meet the overall system requirement. However, the application of redundancy requires complex iterative analysis that trades-off reliability with cost, weight and other system effectiveness parameters.

The objectives of reliability prediction are to:

- Establish the inherent reliability of the design
- Aid design trade-off decisions
- Provide reliability estimates for assessing compliance to the specified requirement during design
- Provide criteria for reliability growth and demonstration testing
- Identify and help eliminate potential design failure modes
- Provide quantitative input for early maintenance support planning
- Support economic analyses and life-cycle cost studies

Reliability prediction involves evaluating each part and its failure modes, designed-in safety margins, stresses, loads and material strength factors early in the design process when corrective changes can be made easily on paper—before the build-up of prototype or production hardware.

Some of the procedures for predicting system reliability are described in the following paragraphs.

Similar Equipment

Most system developers maintain a data base consisting of past hardware failure information. The failure data are, generally, accumulated to the subsystem and part level and are collected through various data reporting systems. Internal data reporting of test information and data from operational experience are included in the file. The data base can be large and include a variety of system types and classes operating in a full spectrum of environmental conditions. Consequently, failure rate data may be available on a system with similiar technology and complexity.

If similar equipment data exist, they can be used to provide a gross estimate of reliability that is useful during the early system definition or concept phase. However, care must be taken, particularly if the data base relies on maintenance information only. The data may include operator and maintenance errors and reflect scavenging and environmental damage. To obtain a meaningful estimate of system reliability, adjustments must be made to the data to eliminate these factors.

Semi-empirical Component Reliability Prediction

This technique is applicable to hardware that is in a mature state of development. New helicopter systems, for example, may be required to carry larger payloads and have larger diameter rotors. Engines will develop higher shaft horse-power and higher turbine temperatures with higher horsepower-to-weight ratios. Performance improvements generally increase stresses and loads, which the designer compensates for by increasing the strength of the components. For the existing materials technology and design environment, agreements could be expected between MTBF and helicopter performance parameters.

Empirical (MTBF) data on fielded helicopter components and subsystems form the basis for deriving formulae that allow interpolation and prediction of the reliability for the new performance improved components and subsystems. The formulae can be as simple as a single performance parameter related to MTBF or a nonlinear regression model consisting of several key performance variables. For example, the MTBF for an airframe may be expressed in terms of weight and/or shaft horsepower. The engine MTBF may be expressed in terms of horsepower, operating hours, turbine temperatures and fuel type, and the rotor blade MTBF may be expressed in terms of rotor diameter, payload and horsepower. In any case, once appropriate correlation coefficients are determined, the parametric formulae could then be used to estimate the MTBF of the improved subsystems or components, keeping in mind that the results are valid only if the design features and material characteristics of the improvements are similar to those of the existing fielded systems.

Part Count

The part count reliability prediction method provides an estimate of reliability based on a count by part type. This method is applicable during early development where the degree of design detail is limited. It involves counting the number of parts of each type and multiplying this number by a generic failure rate for each part within a functional component and/or block depicted in the system block diagram. Generic part failure rates can be derived from the developer's specific data system or from industry-wide data sources including:

(1) The Government–Industry Data Exchange Program (GIDEP)
(2) US MIL-HDBK-217, 'Reliability Prediction of Electronic Equipment'
(3) The Nonelectronic Parts Reliability Data Notebook (NPRD-3)[9]
(4) The IEEE Reliability Data Manual (IEEE-STD-500)[10]

The advantage of the part count method is that it allows rapid estimation of reliability in order to quickly determine the feasibility (from a reliability standpoint) of a given design approach. The technique uses information derived from available engineering information and does not require part-by-part stress and design data.

US MIL-HDBK-217 Reliability Prediction

US MIL-HDBK-217, 'Reliability Prediction of Electronic Equipment', establishes uniform methods for predicting system reliability. It presents failure rate models for nearly all electronic components along with qualifying factors which design or reliability engineers can use to perform reliability analysis on electronic systems in specific applications. It provides a common basis for reliability predictions and a basis for comparing and evaluating reliability predictions of related or competing designs.

The basic concept which underlies the prediction of reliability is that system failure is a reflection of part failure. Consequently, individual part failure rates are applicable within a series reliability model such that the system failure rate is described by the sum of the individual part failure rates.

US MIL-HDBK-217 presents part failure rate prediction models which have been derived from large scale data collection and analysis activities and failure mode and physics of failure studies. These models, in general, incorporate basic stress dependent generic part failure rate data which are modified by suitable adjustment factors derived specifically for the item under study. The basic modal failure rates, data and adjustment functions are derived from established sources.

These models vary with part types; however, their general form is:

$$\lambda_p = \lambda_b \pi_E \pi_A \pi_Q \ldots \pi_n$$

where:

λ_p is the total part failure rate. λ_b is the base failure rate. The value is obtained from reduced part test data for each generic part category, where the data are generally presented in the form of failure rate vs normalized stress and temperature. It takes into account the part's primary load stress conditions, i.e., the part's strength compared with its applied or operating stress (safety margin). The stress ratio is generally a normalized operating stress with respect to part strength at a reference temperature, e.g., 25°C ambient. Rated strength data are compiled from part drawings and other design information. Operating stress data are evaluated through a stress analysis and other design and stress information, in conjunction with actual measurements.

π_E is the environmental adjustment factor, which accounts for the influence of environment other than temperature (it is related to operating conditions such as vibration, humidity, etc).

π_A is the application adjustment factor (it depends on how the part is used and takes into account secondary stress and application factors that are considered to be reliability significant).

π_Q is the quality adjustment factor (it is used to account for the degree of manufacturing control under which the part was fabricated and tested prior to its shipment).

π_n represents adjustment factors which are used to account for cycling effects, construction type, and other design and application characteristics.

These factors represent the kinds of data required to perform a part-by-part prediction. The implementation of these concepts is illustrated in Figure 4–1.

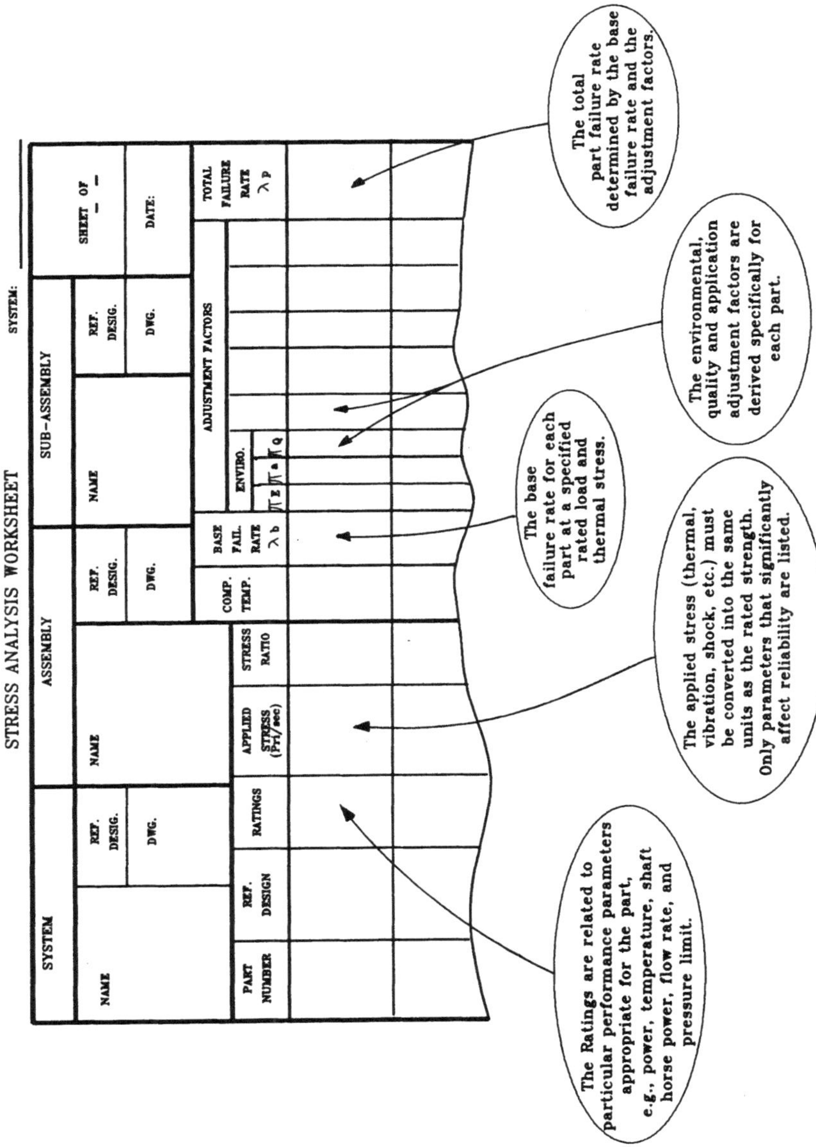

Figure 4–1 MIL-HDBK-217 Reliability Prediction Worksheet Procedure

Stress – Strength Analysis (Safety Factor and Interference)

Reliability theory is based on the premise that a given part has certain physical strength properties which, if exceeded, will result in failure. Further, this property, as with all properties of nonhomogeneous material, varies from specimen to specimen. Thus, for a particular part or material an estimate of the mean value and of the dispersion of the strength property may be found by testing.

The operating stress imposed on a part also varies. These stresses vary from time to time in a particular part, from part to part in a particular design, and from environment to environment. An estimate of the mean value and the dispersion value of the operating stress must be determined by test, analysis or experiment.

In most cases, both the applied stress and strength of a part may be described with sufficient accuracy for reliability prediction purposes by the normal distribution curve. Figure 4–2 illustrates the stress–strength interaction based on the normal distribution. Figure 4–2(a) illustrates the distribution of a typical stress–strength density curve for an item having low reliability and/or inadequate design margin. The shaded area indicates that stress exceeds strength for a certain percentage of time, with resultant failure. In contrast, Figure 4–2(b) shows the separation of the stress–strength distribution indicative of a high design safety factor (adequate design margin) and high reliability.

A factor of safety, F.S., can be defined by the ratio of the means of an item's strength and stress distributions:

$$\text{F.S.} = \frac{\mu_t}{\mu_s}$$

The failure probability P_F can then be related to the F.S. for the assumed normal distributions of stress and strength.[11] The parametric relations of P_F to F.S. are shown in Figure 4–3. If the standard deviation, σ, to mean value ratios are known and connected by a straight line, the points of intersection yield the relationship between the probability of failure, P_F, and the safety factor, F.S. For example, if

$$\sigma_s \text{ is } 20\% \text{ of } \mu_s; \ \frac{\sigma_s}{\mu_s} = 0.20 \text{ (Point A)}$$

$$\sigma_t \text{ is } 10\% \text{ of } \mu_t; \ \frac{\sigma_t}{\mu_t} = 0.10 \text{ (Point B)}$$

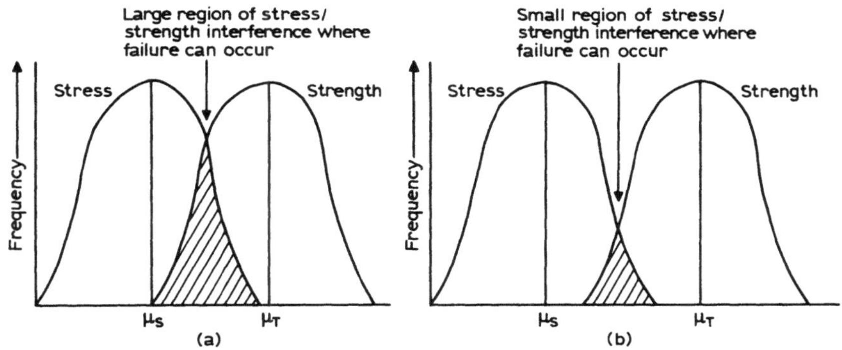

Figure 4–2 Typical Stress–Strength Interaction Diagram

A line is constructed from A to B. If F.S. is 1.6 (the intersection of the A–B line at C) then $P_F = 10^{-2} = 0.01$ or $R = 0.99$.

Thus a rough estimate of reliability can be obtained using the above relationship if design analysis data on the mean and deviation of the loads and strength are available.

Bayesian Techniques

Bayesian statistics provides a methodology for allowing prior information concerning a random process to be integrated with more current information including available test or field experience data, thus yielding a result which utilizes the widest possible range of available information or knowledge. For example, listed below are the MTBF's of components of the hot section module of a helicopter engine before and after testing.

Components	*Prior MTBF (hr)*	*Operating hours*	*Experience failure*	*Posterior MTBF (hr)*
Turbine rotor assembly	24,510	1487·5	1	19,493
Combustion liner	13,889	1370·6	1	11,820
Stage 1 turbine nozzle	15,432	1335·2	0	16,234
Stage 2 turbine nozzle	35,971	1335·2	0	37,453
Total module	4,869	1382·1	2	4,460

The prior MTBF could have been analytically computed using reliability prediction techniques. The posterior MTBF is a new estimate of reliability

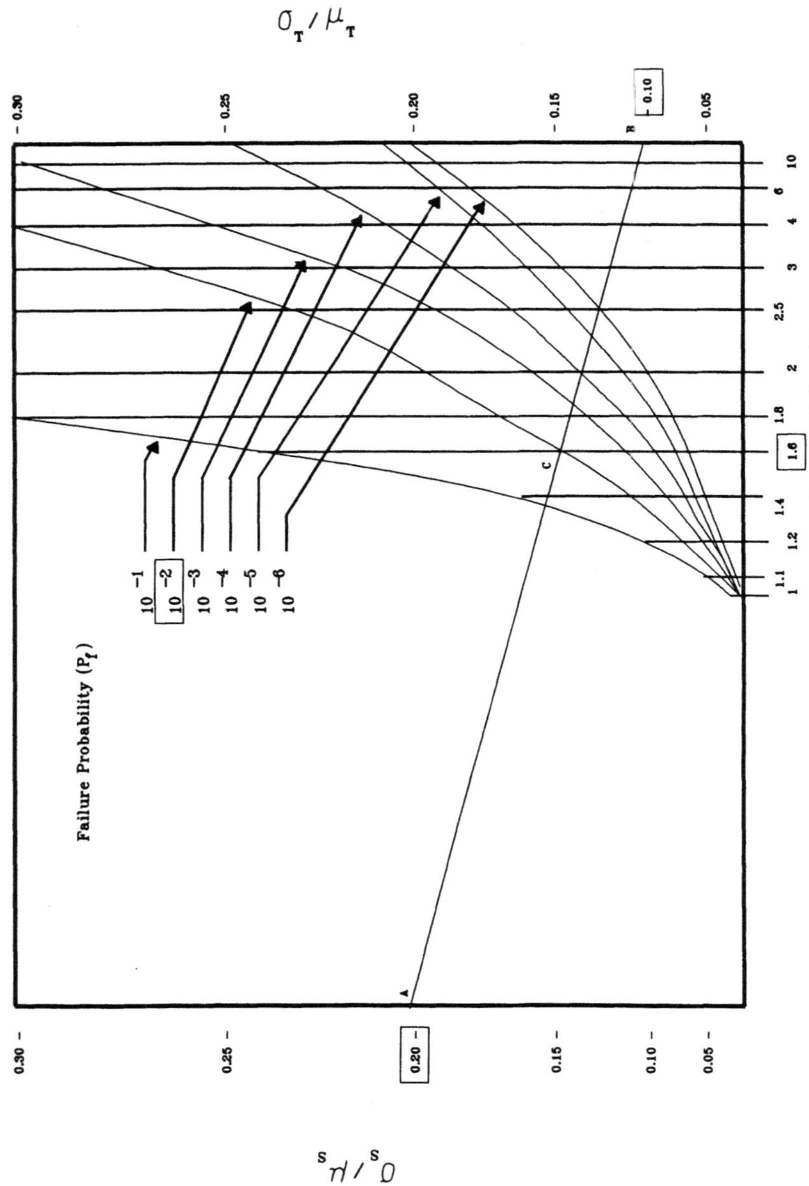

Figure 4–3 Safety Factor vs Probability of Failure

$$P(A/B) = \frac{P(A)P(B/A)}{\sum_A P(A)P(B/A)}$$

A	*P(A)*	*B* Test result *T*	F	P(B/A)	P(A)P(B/A)	P(A/B)
0·5	0·25			0·5	0·1250	0·6173
0·8	0·25	1	1	0·2	0·0500	0·2469
0·9	0·25			0·1	0·0250	0·1235
0·99	0·25			0·01	0·0025	0·0123
					Σ0·2025	

A is a hypothesis or statement of belief. ('The reliability of this component is 0·50 or 0·90.')

B is a piece of evidence, such as a reliability test result, which has bearing upon the truth or credibility of the hypothesis. ('The component failed on a single mission trial attempt; *T* is the number of tests and *F* is the number of failures.')

P(A) is the *prior* probability, or the probability, we assign to the hypothesis *A* before evidence *B* becomes available.

P(B/A) is the *likelihood*, or the probability of the evidence *assuming given* the truth of the hypothesis. ('The probability of the observed failure, given that the true component reliability is indeed 0·90, is obviously 0·10.')

P(B) is the probability of the evidence *B*, evaluated over the entire weighed ensemble of hypothesis *A*.

P(A/B) is the *posterior* probability of *A given* the evidence *B*.

Figure 4–4 Illustrative Example of the Bayesian Formula

that reflects the experience but still gives credit to the analytical reliability prediction.

The Bayesian formula is used to update probability prediction with new information. The formula is normally written in discrete probability terms as

$$P(A/B) = \frac{P(A)\,P(B/A)}{\sum_A P(A)\,P(B/A)}$$

Figure 4–4 provides an illustrative example using the Bayesian formula with the probability terminology defined. Note that the posterior probability represents a discrete distribution. What is desired is a reliability point estimate. It is possible to obtain a reliability estimate by defining a loss function which is associated with an incorrect estimate. Also, prior and present information could be weighted to reflect their importance. For standard continuous distributions both loss functions and weighting factors were applied, and the following formulae were obtained.

1. *Normal Distribution*—Suppose the mean (μ) is to be estimated and the prior distribution of μ is said to come from a normal distribution also. If the loss function is $(\mu-\mu^*)^2$, then the Bayes estimate of μ is

$$\mu^* = \frac{n\bar{X}-n\mu_0}{2n}$$

where n is the sample size, $\bar{X}$ the mean from the sample, and μ_0 the prior estimate of μ.

2. *Poisson Distribution* — Suppose the mean (λ) is to be estimated, and the prior distribution of λ is estimated to come from a gamma distribution. If the loss function is $(\lambda-\lambda^*)^2$, the Bayes estimate of λ is

$$\lambda^* = \frac{2x}{(x/\lambda_0)+T}$$

where T is the time interval of the test, x the number of occurrences in time T, and λ_0 the prior estimate of λ.

3. *Exponential Distribution* — Suppose the mean (θ) is to be estimated and the prior distribution of θ is said to come from a gamma distribution. If the loss function is $(\theta-\theta^*)^2$, then the Bayes estimate of θ is

$$\theta^* = \frac{2}{(1/\theta_0)+\bar{t}}$$

where $\bar{t}$ is the observed average time between occurrences and θ_0 is the prior estimate of θ.

θ is actually the failure rate of the component, and MTBF $= 1/\theta$.

Further utility of the Bayesian technique may be demonstrated for the case where current information may be in the form of an updated reliability estimate based upon a higher level of more selected information, rather than being in the form of test data *per se*. To illustrate, the Bayesian technique may be helpful in combining reliability estimates from a similiar equipment or empirical technique early in the system development, with a later reliability estimate based upon a detailed stress analysis.

4.2 MAINTAINABILITY ALLOCATION AND PREDICTION

Maintainability allocation and prediction are also important techniques to be applied during the design of a complex system. An allocation performed early in development helps to define the optimum maintainability design approach, i.e., the combination of ease of maintenance features (such as built-in automatic detection, location and failure diagnostic equipment, and the incorporation of easily accessible and interchangeable modules and subassemblies) that best meets the system maintainability requirement. Predictions are performed as the design progresses to assess compliance of the design to the specified requirement.

Maintainability allocation is the process of apportioning the system maintainability requirement, generally in terms of a mean time to repair (MTTR), down to the subsystems and lower levels of assembly. It is accomplished in such a manner that the statistical mean of all the subsystems' MTTR's will be less than or equal to the MTTR required of the total system.

A maintainability prediction is performed after the basic system has been defined. It is at this stage that sufficient engineering data are available to perform a meaningful quantitative evaluation of the system's design characteristics in terms of performance, serviceability and support. The objectives of maintainability predictions are to:

- Establish the inherent maintainability of the design
- Aid design trade-off decisions
- Provide maintainability estimates for assessing compliance with the specified requirement during design
- Provide quantitative input for early maintenance support planning
- Support cost of ownership and acquisition cost studies

Maintainability prediction data are used as input to early maintenance planning and logistics support analysis to determine, for example, the personnel (number and skill level) required to maintain a given number of systems within a specified time period. Maintainability prediction facilitates making decisions regarding difficulty of maintenance (which translates into personnel skill levels), tools and equipment required, consumable items used while performing maintenance and facilities required.

Maintainability prediction involves estimating repair times, maintenance frequency per operating hour, preventive maintenance time and

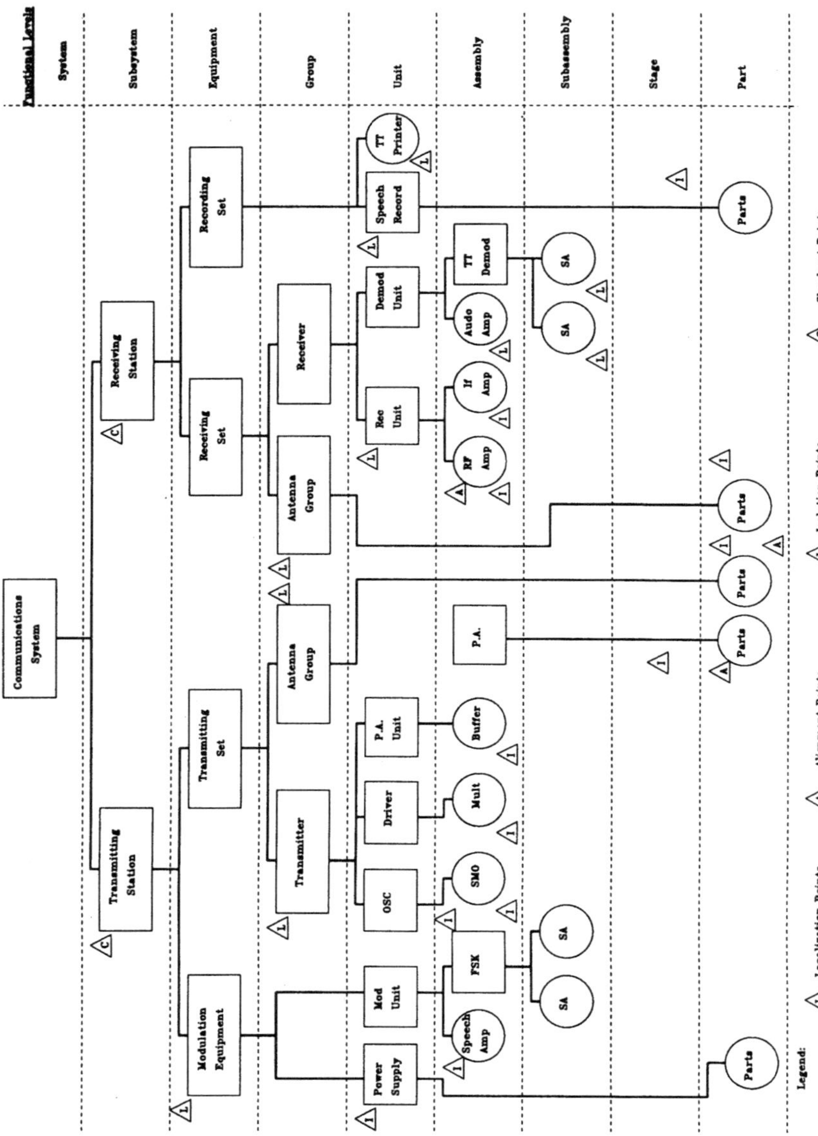

Figure 4–5 Functional Level Diagram of a Typical Communications System (taken from MIL-HDBK-472)

other factors. It provides a measure of the ease and speed with which a system can be restored to operational status following a failure. The basic approach to maintainability prediction is to estimate the total time-to-repair of each replaceable part in the system, including both failure diagnosis and repair time, and then expressing this as a probability that a repair will be completed by a designated time when corrective maintenance is performed in accordance with prescribed procedures and resources.

Techniques for predicting maintainability during system development involve, in general, the determination of a weighted average MTTR based on failure rates obtained from reliability prediction studies and maintenance time factors derived from a review of the system design characteristics. Conceptually, the repair of a hardware item after the occurrence of a failure necessitates the initiation of a corrective maintenance task which ultimately results in the interchange of a replaceable part or assembly. In order to achieve a complete 'repair', various activities both before and after the actual interchange are necessary. This includes activities for localization, isolation, disassembly, interchange, reassembly, alignment and checkout.

The prediction process generally involves preparing a functional-level diagram (Figure 4–5) for the system to facilitate determining the repair time for each replaceable item (which is indicated by a circle at the termination of a branch). The functional-level diagram reflects the overall maintenance concept and the complete replacement breakdown for all items that comprise the system.

US MIL-HDBK-472, 'Maintainability Prediction', provides procedures for maintainability prediction including the determination of the appropriate maintenance time factors. The handbook describes five maintainability procedures. Table 4–1 provides a comparative summary of the significant attributes of each of the maintainability procedures. In general, Procedures I and III are applicable solely to electronic systems and equipments. Procedures II and IV can be used for all systems and equipment. In applying Procedure II to non-electronic equipment the appropriate task times must be estimated. Procedure V can be used to predict maintainability parameters of avionics, ground and shipboard electronics at the organizational, intermediate and depot levels of maintenance.

MIL-HDBK-472 Procedure I

This procedure is built around 'elemental activity' times, which are fundamental elements of downtime from which other more comprehensive

TABLE 4-1
Comparison Matrix of Maintainability Predication Procedures (from MIL-HDBK-472)

Procedure	*Applicability*	*Point of application*	*Basic parameters of measure*	*Information required*
I	To predict flight-line maintenance of airborne electronic and electromechanical systems involving modular replacement	After establishment of the design concept, provided that data as listed in the column entitled 'Information Required' are available	Distribution of downtime for various elemental activities, maintenance categories, repair times and system downtime	(a) Location and failure rate of components (b) Number of: 1. Replaceable components 2. Spares 3. Test points
II	To predict the maintainability of shipboard and shore electronic equipment and systems. It can also be used to predict the maintainability of mechanical systems provided that required task times and function levels can be established	Applicable during the final design stage	Part A of procedure: Corrective maintenance expressed as an arithmetic or geometric mean time to repair in hours. Part B of procedure: Active maintenance in terms of: (a) Mean corrective maintenance time in man hours (b) Mean preventive maintenance time in man hours (c) Mean active maintenance time	For corrective maintenance (Part A): (a) Packing: to the extent that detailed hardware configurations can be established (b) Diagnostic procedure (c) Repair methods (d) Parts listing (e) Operating stresses (f) Mounting methods (g) Functional levels at which alignment and checkout occur.

			in terms of mean man hours per maintenance action	For active maintenance (Part B): The respective maintenance task times for corrective and preventive maintenance must have been determined
III	To predict the mean and maximum active corrective maintenance downtime for Air Force ground electronic systems and equipment. It may also be used to predict preventive maintenance downtime	Applied during the design development and control stages	(a) Mean and maximum active corrective downtime (95th percentile) (b) Mean and maximum preventive downtime (c) Mean downtime	(a) Schematic diagrams (b) Physical layouts (c) Functional operation (d) Tools and test equipment (e) Maintenance aids (f) Operational and maintenance environment
IV	To predict the mean and/or total corrective and preventive maintenance downtime of systems and equipments	Applicable throughout the design, development cycle with various degrees of detail	(a) Mean system maintenance downtime (b) Mean corrective maintenance downtime per operational period (c) Total corrective maintenance downtime per operational period	Complete system documentation portraying: (a) Functional diagrams (b) Physical layouts (c) Front panel layouts

TABLE 4-1—*contd.*
Comparison Matrix of Maintainability Predication Procedures (from MIL-HDBK-472)

Procedure	*Applicability*	*Point of application*	*Basic parameters of measure*	*Information required*
			(d) Total preventive maintenance downtime per operational period	
V	To predict maintainability parameters of avionics, ground and shipboard electronics at the organizational, intermediate and depot levels of maintenance	Applied at any equipment or system level, at any level of maintenance pertinent to avionics, ground electronics, and shipboard electronics	(a) Mean time to repair (MTTR) (b) Maximum corrective maintenance time [$M_{max}(0)$] (c) Mean maintenance man hours per repair (MMH/repair) (d) Mean maintenance man hours per operating hour (MMH/OH) (e) Mean maintenance man hours per flight hour (MMH/FH)	Early prediction: (a) Primary replaceable items (b) Failure rates (c) Fault isolation strategy (d) Replacement concept (e) Packaging philosophy (f) Fault isolation resolution Detailed prediction: (a) Replacement concept (b) Fault detection and isolation outputs (c) Failure rate (d) Maintenance

measures of downtime are developed. An elemental activity is defined as a simple maintenance action of short duration and relatively small variance which does not change much from one system to another. Typical examples would be the opening of an equipment compartment or checking of maintenance records. An extensive listing of these elemental activities along with associated time distributions is given in MIL-HDBK-472. The activities in this listing have been broken into five categories. They are:

1. Preparation Time
2. Malfunction Verification Time
3. Fault Location Time
4. Part Procurement Time
5. Repair Time

The times in these various categories along with information concerning 'Final Malfunction Test Time' are combined step-by-step and used to build up a 'Total System Downtime'. The progression of this building-up process through several steps is graphically illustrated in Figure 4–6.

A Monte Carlo procedure is outlined in MIL-HDBK-472 to accomplish the build-up of elemental times into a distribution of system downtime. Successful application of this Monte Carlo procedure depends on (1) an accurate description of the distribution of time required for the performance of an elemental activity and (2) the probability of occurrence of an elemental activity. The distribution of time has been found to be independent of the type or design of the system involved and is specified for each elemental activity in the procedure. However, the probability of occurrence is related to various design parameters and a method is given for acknowledging this relationship and calculating these probabilities.

Both the distributions of activity times and their probabilities of occurrence are used by the Monte Carlo procedure to produce a predicted distribution of system downtime.

MIL-HDBK-472 PROCEDURE II

This procedure consists of prediction methods which can be used during the final design stage of a product to predict corrective, preventive and active maintenance procedures. (Corrective and preventive maintenance predictions include only actual repair time when the equipment under repair is shutdown. Active maintenance predictions combine both

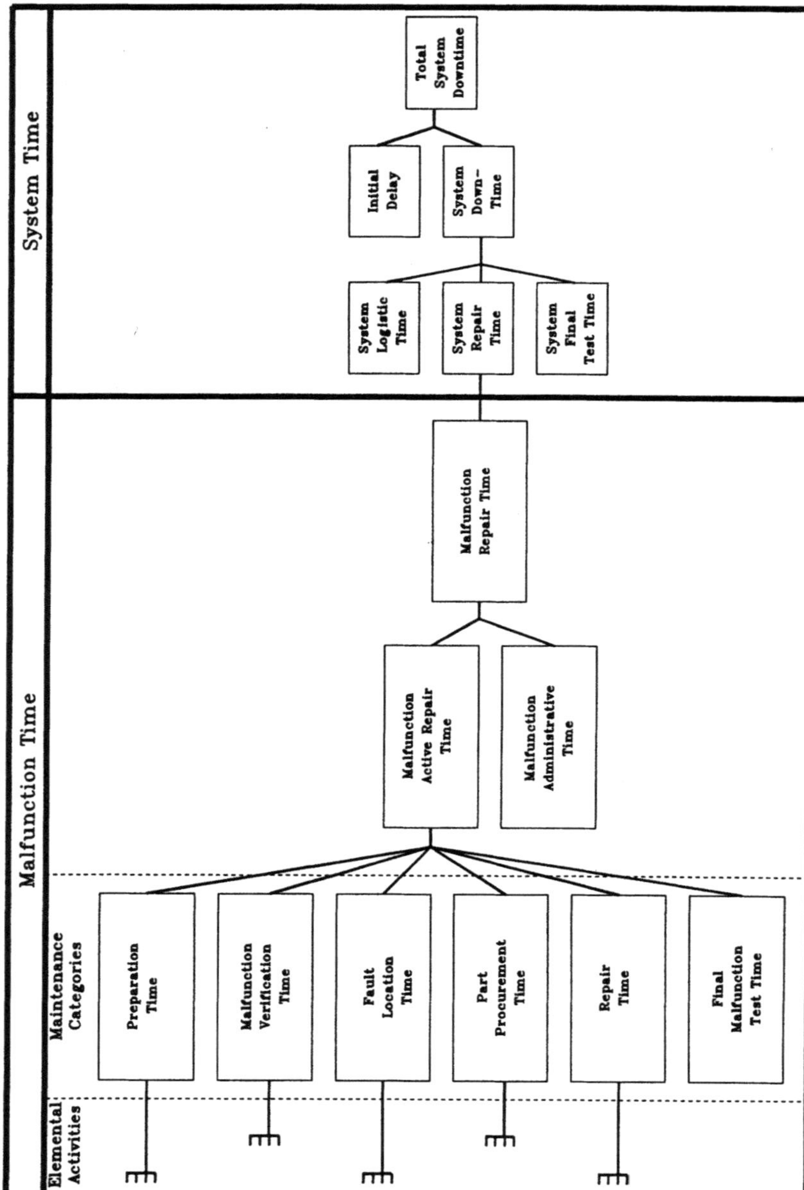

Figure 4–6 Build-Up of Time Elements (taken from MIL-HDBK-472)

corrective and preventive maintenance.) In addition, there are two distinct and different approaches within Procedure II that can be used for predicting maintenance times. Both approaches require the analyst to develop a description of the maintenance tasks under study. Times are then applied to various elements of the descriptions, summed up and combined with failure rate information to arrive at a predicted maintenance time.

The maintenance task descriptions are in terms of the functional level at which repair takes place and in terms of generalized maintenance tasks. The different functional levels are a recognition that repair time is dependent on the level at which the repair is accomplished.

The first approach relies on the use of predetermined maintenance times for assigning times to each maintenance task including localization, isolation, disassembly, interchange, reassembly, alignment and checkout. These predetermined times apply to corrective maintenance only and come from two sources. The primary source consists of tabulated data compiled as a result of over 300 observations of maintenance activity in the US Navy fleet. These data are supplemented with predetermined time standards developed by using a synthetic basic motion time system.

The second approach uses estimated times as determined by the analyst for assigning times to elements of the described maintenance tasks. This approach is subjective and requires a thorough understanding of equipment groupings, diagnostic and repair methods, etc., on the part of the analyst. Times are estimated for both corrective and preventive maintenance and are eventually combined to determine the mean man hours of active maintenance time.

MIL-HDBK-472 PROCEDURE III

This procedure is for predicting the mean and maximum corrective maintenance downtime of ground electronic systems. The basic assumptions upon which it is based are:

1. System downtime is principally due to the failure of replaceable items.
2. Length of system downtime is a function of specific design parameters which govern replacement time. These parameters are:
   (a) Physical configuration of the system
   (b) Facilities provided for maintenance

(c) Degree of maintenance skills required to replace the failed item.

3. Similar classes of equipment require a similar type of maintenance activity when repair by replacement is used.
4. Uniformity of design within a class of equipment will permit the use of a random sample of replaceable items to establish repair times for the entire class.

The basic procedure consists of the following steps:

1. Select a random sample of replaceable items.
2. Conduct a maintainability analysis for every item in the sample.
3. Assign a 'score' to each maintenance task associated with a sample item. This is done with the aid of design check lists that provide scoring criteria.
4. Convert scores to downtime through the use of an equation given in the procedure.

US MIL-HDBK-472 describes each step and provides sample size data for generic part classes and examples of the factors affecting maintenance time and criteria for scoring each factor.

MIL-HDBK-472 PROCEDURE IV

This procedure provides estimates of:

(a) The elapsed time to perform preventive maintenance action, assuming that no detectable malfunctions exist in the system.
(b) The elapsed time to correct malfunctioning end items detected during each preventive maintenance action of an operational function.
(c) The distribution of corrective maintenance times for detectable malfunctioning end items for each preventive maintenance action of an operational function.
(d) The mean corrective downtime (MCDT) for detectable malfunctioning end items for each preventive maintenance action of an operational function.
(e) The distribution of corrective maintenance task times for the system and subsystems.
(f) The preventive downtime (PDT) for the system and subsystems for a specified calendar time.
(g) The MCDT for the system and subsystem for a specified calendar time.

(h) The total mean downtime for integrated preventive and corrective maintenance for the system and subsystems for a specified calendar time.

The procedure itself focuses on end-item maintenance task analysis. It requires detailed system and subsystem block and flow diagrams, failure rates, a description of the operational resources (facilities, personnel, support equipment, etc.) and a description of each maintenance task to be performed.

End items are identified down to the smallest piece of equipment on which a specific maintenance action will be accomplished. The failure rate for each end item is then identified along with the preventive and corrective maintenance actions to be performed on the item. A task analysis is conducted for all maintenance actions to determine the troubleshooting, repair and verification time for each end item. The resulting preventive and corrective maintenance times along with their associated frequency of occurrence are then integrated over a previously specified calendar time to derive the total preventive downtime, total mean corrective downtime and the total mean downtime.

MIL-HDBK-272 PROCEDURE V

This procedure can be used for any type of system in any operational environment to predict the parameters of maintainability applicable to organizational, intermediate and depot levels of maintenance. It is based on the following assumptions and ground rules:

1. Failure rates experienced are all in the same proportion to those predicted.
2. Only one failure at a time is considered.
3. Maintenance is performed in accordance with established maintenance procedures.
4. Maintenance is performed by persons possessing the appropriate skills and training.
5. Only active maintenance time is addressed; administrative and logistic delays and clean-up are excluded.

The application of the procedure enables the user to monitor the overall system maintainability throughout the design and development of a system. The user can identify whether or not the specified maintainability design requirements will be met before the system is complete. Thus, if it appears the maintainability requirements will not be met, the designers can be informed and the necessary changes can be made before they become prohibitively expensive.

Procedure V consists of two distinct approaches. The first approach is an early prediction method that makes use of estimated design data. The following information must be provided to support this approach:

- The number and contents of the primary replaceable items (either actual or estimated).
- The failure rates, either predicted or estimated, associated with each replaceable item.
- The basic fault isolation test strategy of each replaceable item.
- The replacement concept, if fault isolation is to a group of replaceable items.
- The packaging philosophy.
- The fault isolation resolution, either estimated or required (i.e., percent of faults isolated to one replaceable item or the average replaceable item group size).

The second method uses actual detailed design data to predict the maintainability parameters. The following information must be provided to support this approach:

- The replacement concept for each replaceable item or group of items.
- The fault detection and isolation outputs associated with each replaceable item.
- The failure rate of each replaceable item.
- The maintenance procedure that is followed to remove and replace each replaceable item.

The two approaches of this maintainability prediction procedure are both time synthesis model techniques and employ the same general mean time to repair (MTTR) prediction model:

$$\text{MTTR} = \frac{\sum_{n-1}^{N} \lambda_n R_n}{\sum_{n-1}^{N} \lambda_n}$$

where N is the number of replaceable items (RI), λ_n is the failure rate of the nth RI and R_n is the mean repair time of the nth RI.

MTTR is the primary maintainability parameter that can be predicted using this procedure. The other maintainability parameters that can be predicted using this procedure are: maximum corrective maintenance time, percent of faults isolatable to a single replaceable item, percent of faults isolatable to *N* replaceable items, mean maintenance man hours per repair, mean maintenance man hours per operating hour, and mean maintenance man hours per flight hour.

The prediction procedures described above take into account the maintenance features, i.e., the condition monitoring and diagnostics equipment, modular packaging, etc., incorporated into the design and the failure rates of the replaceable parts and components. The built-in ease of maintenance features reduce or eliminate the human response time involved in fault location, isolation and checkout. If complete 'location' and 'isolation' are achieved through condition monitoring and built-in test equipment, the maintenance task is reduced to 'removal' and 'replacement'. If 'checkout' is accomplished by the built-in test equipment, the MTTR then approximates the mean time to remove and replace (MTTR/R). However, it is difficult to design built-in monitoring and diagnostic equipment that is completely effective. First, since the number of possible faults which may occur in a system is directly related to the number of parts, and the failure modes within, it is often impossible to associate a fault signal with each element and failure mode. Consequently, since in most cases it is not practical to locate and isolate all conceivable failures, the level of location and isolation must be traded-off with cost, reliability, weight and other system effectiveness parameters.

Also, the location and isolation of a particular fault requires detection of an error signal unique to that fault and to the element in which that fault occurs. In many instances, compromises must be accepted in the application of built-in fault isolation techniques. As described above, complexity of the system may prohibit a one-to-one correspondence between elements and fault indicators, but unlike the problem of diagnostic inefficiency a fault indication may be observed for one of several elements. In this circumstance, ambiguity exists and complete isolation to a particular element is not achievable. Ambiguity also exists because:

(1) a complex system is composed of a variety of elements, several of which may involve similar responses when at fault and
(2) an element may have several failure modes, some of which exhibit responses identical to failure modes of dissimilar elements.

The prediction of maintainability must account for the repair times associated with complete fault location and isolation when ambiguity exists. The time to identify the actual fault from a subset of conceivable faults identified by the built-in test equipment must be included. Also, the realization of increased maintainability requires that the built-in test equipment has high reliability. Experience with complex systems has shown that as high as 30% of operational failures are due to the built-in test equipment alone. Two modes of failure are conceivable: the first where the built-in test malfunctions and does not detect a system fault, and the second where the built-in test malfunctions, indicating a system fault that has not occurred.

In the first mode of failure, a system malfunction would be realized and thought to be due to a fault identifiable by built-in test, leading the repairman to checkout system elements not at fault. Since no failure could be found, the failure would eventually be localized to the built-in test equipment and from there would be isolated. In the second mode of failure, no system malfunction would be evident, but maintenance time would likely be expended to assure the indicated fault did not exist. No verification of the indicated fault would localize the malfunction to the built-in test equipment. Additional repair time would then be expended in isolating the fault. Either mode of failure leads to significant losses in maintainability since considerable repair time is spent in locating and isolating false faults in addition to the time required to repair the built-in test equipment.

Maintenance practices in many complex systems use a time-based overhaul maintenance concept for some of its components to reduce operational failures. The determination of an optimum time between overhaul (TBO) is based on leaving the component in operation as long as possible without experiencing an in-service failure. This maintenance approach is applicable only to components exhibiting a significant wearout characteristic and should not be used for parts and components having a long useful life where the failure rates are constant. To be economically feasible, the cost of an in-service failure (or loss of system function) must be greater than the cost of the scheduled overhaul. Therefore, estimation of TBO is also based on an evaluation of the total cost of the in-service failure. In general, the higher the in-service failure cost the shorter the TBO. Similarly, for a given failure cost, the more rapidly the failure rate increases with time, the shorter the TBO requirement.

Thus, factors having a predominant influence on the specification of an optimum TBO are the times during which the system is out of service (awaiting maintenance or having maintenance performed) and the resources required to perform the overhaul activity. The applicability of a TBO to a particular component is determined as part of the RCM analysis process previously described.

One of the advantages of the TBO concept is that scheduled overhaul is planned in advance and waiting time is kept to a minimum, as opposed to unexpected in-service failures where time will be lost due to being unprepared for the maintenance activity. However, this advantage is limited by a lack of techniques for accurately estimating the cost of in-service failures. If the cost is overestimated, the TBO interval may be too short. Thus, the actual cost will increase by not taking advantage of the full system life. If the cost is underestimated, the TBO interval may be too long and actual costs again increase due to more in-service failures.

Condition monitoring equipment is used with the goal of providing continuous status of subsystems and the ability to isolate failures to a replaceable component. Implementation of condition monitoring requires an on-board system capable of interrogating the operational status of all subsystems through sensors and transducers installed in the subsystems for operational purposes. The system provides, on command, fault indications and locations in the form of printed readouts or displays.

An objective of condition monitoring is to reduce the inventory requirement for test equipment, spares, manpower, time involved in getting to and from the equipment, etc. Logistically, it eliminates the need to know what equipment is at what operational location and whether a particular test equipment item is configured for the next system it will be required to test.

Among the many advantages of condition monitoring are:

- Immediate failure indication
- Increased flight safety
- Increased mission effectiveness
- Increased availability
- Reduced test time
- Reduced maintenance man hours per flight hours
- Insight into scheduling preventive maintenance
- Reduced incorrect fault diagnosis
- Reduced skill level requirements

Condition monitoring will significantly reduce mean time to repair by decreasing fault isolation time and providing a ready source to requalify subsystems after their repair. Also, fewer and less skilled maintenance men are required, no time is required for test equipment set-up and disassembly and less time is required for checkout. Decreased delay times normally associated with logistics considerations further reduce mean time to repair. Since all units removed are known to be defective, demand on shop test equipment is reduced and time is not wasted in checking out good equipment and recertifying it after test. Overall this leads to fewer units in the pipeline, a decrease in the number of units which must be stocked and less time wasted in setting up and dismantling units.

However, condition monitoring is not without drawbacks. A system capable of performing the sophisticated functions described above is of sufficient complexity to significantly reduce overall reliability. In a helicopter system, for example, the penalty paid in decreased reliability must be offset by the increase in the probability of mission success achieved through the ability to detect failures in flight and to provide knowledge needed to evaluate the ability of the helicopter to continue its mission. Other disadvantages include weight, volume and power requirements encountered when adding any hardware to the system. Effective utilization of the condition monitoring concept requires an analysis of the advantages and disadvantages inherent in its application.

4.3 FAILURE MODE ANALYSIS

Failure mode analysis involves determining what parts in a system or component can fail, the modes of failure that are possible, and the effect of each mode of failure on the complete system. The more complex the system the greater the interaction between its constituent components, and the greater the need for a formal and systematic process to identify and classify effects.

The results of a failure mode analysis are tabulated in such a manner that enables the design engineer to:

(1) Identify reliability and safety critical areas and single failure points for design change or improvement, design review and configuration control
(2) Determine the need for redundancy, fail-safe design, further derating, design simplification, and more reliable materials and parts

(3) Assure that the test program is responsive to identified potential failure modes and safety hazards
(4) Pinpoint key areas for concentrating quality control and inspection, environmental stress screening and manufacturing process controls
(5) Identify data recording requirements

In addition, failure mode analysis provides necessary data for RCM decision logic analysis and maintenance support planning. It provides information for selection of preventive and corrective maintenance points and development of trouble-shooting guides. It provides information to facilitate the establishment of on-condition maintenance inspections, condition monitoring requirements and hard time limits for the allowable use time (or number of stress cycles) for short-life components where wearout or aging is the dominant failure mechanism. Failure mode analysis can also be used to facilitate the investigation of actual field failures and the determination of their impact on mission success and overall reliability.

As mentioned, the initial step in the RCM logic process is to perform a failure mode analysis to identify the safety critical items and to develop part failure mode criticality data. For Army aircraft a critical item is defined as a part, assembly or installation procedure with one or more critical characteristics that, if not conforming to the design data or quality requirements, could result in the loss of, or serious damage to the aircraft and/or serious injury or death of crew members.

There are two basic approaches to failure mode analysis: (1) fault tree analysis (FTA) and (2) failure mode, effects and criticality analysis (FMECA). These are described in the following subsections.

4.3.1 Fault Tree Analysis

The fault tree analysis (FTA) process is a tool that lends itself well to analyzing potential failure modes. Its objectives are:

- To assess the magnitude of potential failures, particularly those affecting safety at an early stage of system development, and
- To identify and prioritize all possible failure modes and hazardous conditions so that effective corrective measures can be formulated and instituted prior to system deployment.

FTA is a top-down technique. The analysis starts with a specific failure condition (e.g., loss of power), and proceeds downward to define possible system and subsystem faults, conditions and user actions whose occurrence singly or in combination can cause this event. Logic diagrams are used to portray these basic faults, conditions and events.

The FTA procedure can be characterized as an iterative documented process of a systematic nature performed to identify basic faults, determine their causes and effects, and establish their probabilities of occurrence. The approach involves, first, the structuring of a highly detailed logic diagram that depicts basic faults and conditions that can lead to system failure and/or user hazard; next, the use of computational techniques to analyze the basic faults and determine failure mode probabilities; and, finally, the formulation of corrective suggestions that when implemented would eliminate (or minimize) those faults considered critical.

FTA can be applied at any time during a product's life-cycle. However, it is most effective when applied:

(a) During early development, based on preliminary design information.
(b) After final design, prior to full scale production, based on manufacturing drawings.

The first analysis is performed to identify failure modes and formulate corrective measures (primarily in the design area), with priority given to those faults with the highest criticalities. The second analysis is performed to show that the system, as manufactured, is acceptable with respect to reliability and safety. Corrective actions or measures, if any, resulting from the second analysis would emphasize controls and procedural actions that can be implemented with respect to the 'as-manufactured' design configuration.

The outputs of the analysis include:

(a) A detailed logic diagram that depicts all basic faults and conditions that must occur to result in the hazardous condition(s) under study.
(b) A probability of occurrence numeric for each hazardous condition under study.
(c) A detailed fault matrix that provides a tabulation of all basic faults, their occurrence probabilities and criticalities, and the suggested change or corrective measures involving circuit design, component part selection, inspection, quality control, etc., which, if implemented, would eliminate or minimize the hazardous effect of each basic fault.

The steps, and some of the factors associated with each step that must be considered during the analysis, are shown in Figure 4–7. The following paragraphs discuss each of the steps in further detail.

Step 1 Diagram Fault Tree

The first step in the fault tree analysis is to develop a detailed logic diagram that portrays the combination of events that may lead to the condition under study. All events (i.e., component faults, human errors, operating conditions, etc.) that must occur to result in the defined fault condition are interconnected systematically through basic logic elements ('and' gate, 'or' gate, etc.) to form the fault tree. The fault tree symbols and a representative logic configuration are shown in Figure 4–8.

It is necessary to have a knowledge of the system design, its functional operation and maintenance requirements, and how it is used. Then the fault tree is developed, beginning with the defined failure condition and proceeding downward with a series of engineering judgments to define the basic input events. This logic structuring process continues until each input event chain has been terminated in terms of a basic fault. When the fault tree structure is complete, the undesired event is completely defined in terms of:

(a) Basic faults (hardware and human) whose occurrence alone or in combination can result in the defined hazard regardless of their apparent frequency of occurrence.
(b) Independent input events.
(c) Basic faults (e.g., component failure modes) for which failure rate data are available or may be estimated.

Step 2 Collect Basic Fault Data

After the fault tree has been structured, the next step in the process is to collect failure rate data for each basic fault that comprises the fault tree. Failure rate data are necessary inputs for determining occurrence probabilities and assessing criticality. These data consist of two general classes:

(a) Component failure rate data.
(b) Human error rate data.

In general, the component failure rates are determined through a review of component items identified as faults on the fault tree. This involves reviewing the failure modes of each basic element which comprises the

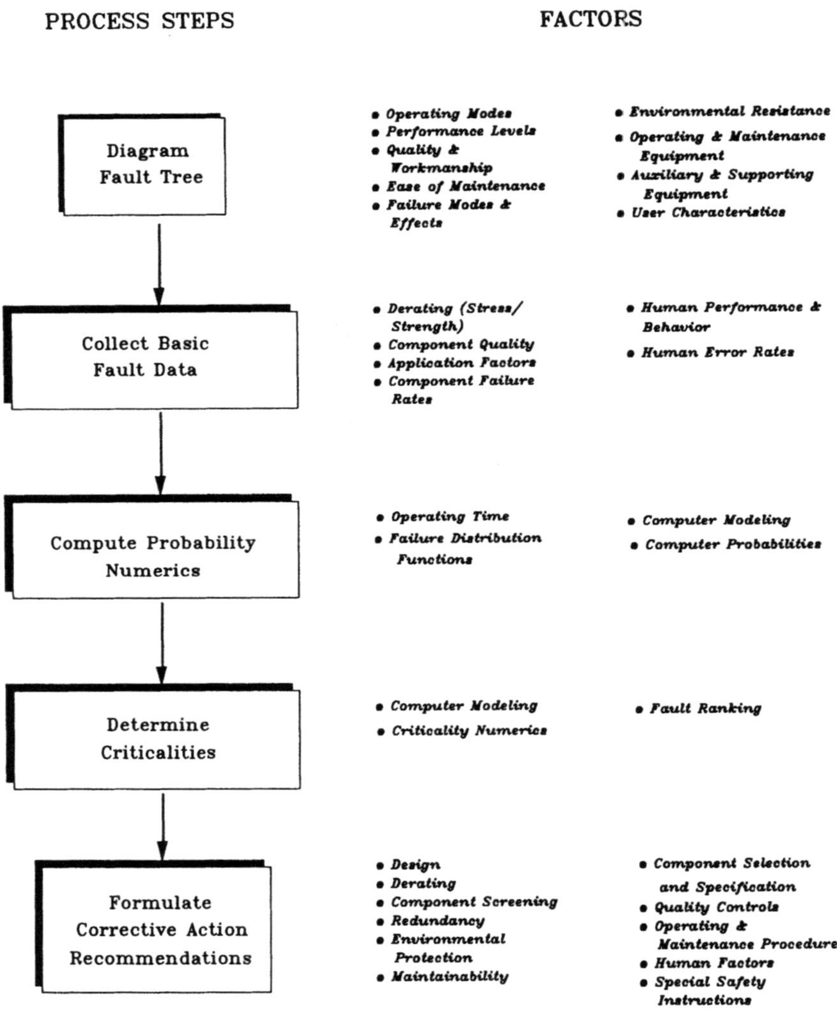

Figure 4–7 Steps and Factors Involved in the Application of Fault Tree Analysis

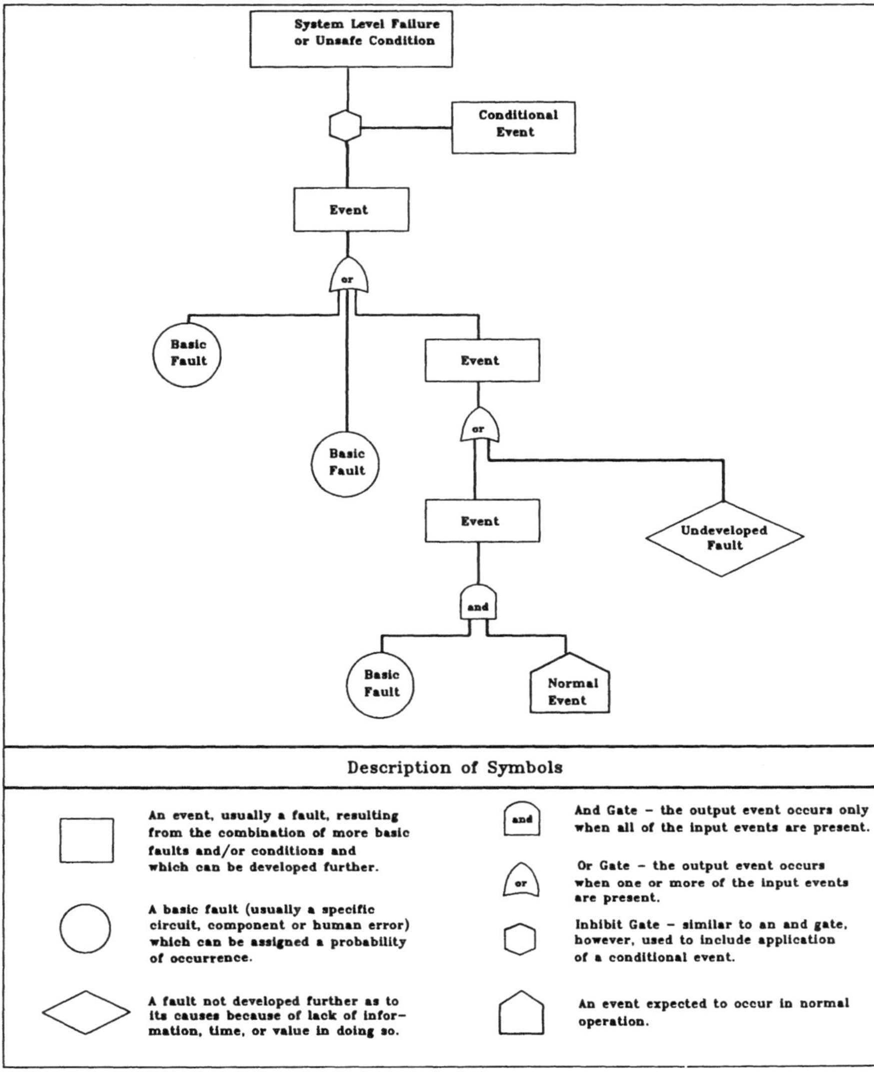

Figure 4–8 Fault Tree Symbols

identified fault and establishing a modal failure rate, based on historical generic part data and available design application information. Standard reliability prediction techniques can be used to estimate these failure rate numerics.

Human error rates mean the expected rate at which a failure caused by operating or maintenance personnel takes place, whether intentionally or unintentionally. It is very difficult to obtain an error rate since very few data exist regarding this area. Since a large scale data base is lacking, human error rates can be developed through subjective techniques based on discussions with personnel familiar with the system operation and maintenance environment. These techniques involve detailing each human error depicted on the fault tree into basic task elements. The intent is to define small segments of human performance—where an error rate can be more easily assessed. Assessing the error rate for these individual elements would involve a literature survey, including a review of currently available human error data and/or prior estimation information from personnel familiar with the operational elements. The final error rate numerics must account for the nature of human performance and its sensitivity to learning, fatigue and other behavioral factors.

Step 3 Compute Probability Numerics

After the fault tree is structured and all fault data collected, the next step in the analysis process is to compute probability numerics. This involves computing the occurrence probabilistics for all basic faults, events and hazardous conditions (top faults) based on the combinatorial properties of the logic elements in the fault tree. The analysis involves repeated application of the following basic probability expressions for the fault tree logic gates:

'And' gate $$P(A) = \prod_{i=1}^{n} P(X_i)$$

'Or' gate $$P(A) = 1 - \prod_{i=1}^{n} [1 - P(X_i)]$$

where $P(A)$ is the output probability, $P(X_i)$ is the probability of the ith input and n is the number of inputs.

Given a fault tree diagram whose basic faults and output events are properly interconnected, the output event probabilities are computed, starting with the lowest levels and continuing to the highest levels in the tree.

Step 4 Determine Criticalities

After the occurrence probabilities have been computed, the next step in the analysis process is to determine the criticality of each basic fault. Criticality is a measure of the relative seriousness of the effects of each fault. It involves both qualitative engineering evaluation and quantitative analysis and serves to provide a basis for ranking the faults for corrective action priorities. The object is to assign a criticality numeric to each fault based on its occurrence probability and its contribution to the overall probability for the fault condition under study.

Criticality is defined quantitatively by the following expression:

$$\mathrm{CR} = P(X_i)\,P(H|X_i)$$

where $P(H|X_i)$ is the conditional probability of the overall hazardous condition given that the basic fault (X_i) has occurred.

Step 5 Formulate Corrective Action Recommendations

Finally, after all probabilities and criticalities are computed, all data are reviewed and evaluated in order to formulate general corrective suggestions. These suggestions can be related quantitatively to the fault elements and failure modes identified by the fault tree analysis. These suggestions, in general, would involve:

- Areas for redesign
- Component part selection
- Design and procurement criteria
- Maintenance procedures
- Inspection procedures
- Quality controls
- Special safety instructions

The scope and extent of the suggested corrective measures would depend on the faults identified and their criticality and should be considered in relation to their effectiveness, practicality and cost.

A fault matrix is then prepared to aid in the evaluation and the formulation of the specific recommendations. The fault matrix provides a tabulation of the following information for each basic fault:

- Basic fault identification number

- Basic fault description
- The occurrence probability $P(X_t)$
- The criticality numerics
- The recommended corrective action(s) for those faults considered critical involving design, controls, tests, procedures, inspection, etc., that can be implemented in order to eliminate or reduce the hazardous effect

FTA Example

Figure 4–9 presents a very simplified fault tree diagram where loss of hydraulic power of a helicopter is the top event; some of the potential faults and conditions which can lead to this event are identified. Note that the figure is given to illustrate use of the technique; it is not intended to fully define the hydraulic reliability problem as it can exist with helicopters in their operating environment. Actual diagrams for specific helicopter systems would be much more complex and can only be developed from a detailed review and evaluation of individual design characteristics and application considerations.

The fault tree indicates that loss of hydraulic power can be caused by:

(a) a hydraulic pump failure
(b) a hydraulic by-pass solenoid failure
(c) a line or fitting failure
(d) a servo actuator failure
(e) a filter system failure
(f) a hydraulic relief valve failure

The fault tree further indicates that two conditions are necessary to cause the filter system failure; i.e., as shown by the 'and' gate, both a clogged filter and a by-pass switch failure must occur for the filter system to fail.

Figure 4–9 also presents a fault matrix prepared for this failure event. The matrix lists the criticality numerics computed for each fault, based on the sample fault probabilities, $P(X_t)$ data used in this example and possible corrective measures.

4.3.2 Failure Mode, Effects and Criticality Analysis

The failure mode, effects and criticality analysis (FMECA) technique can be characterized as a systematic method of cataloging failure modes starting at the lower level of assembly and assessing the consequences at

Fault Tree Diagram

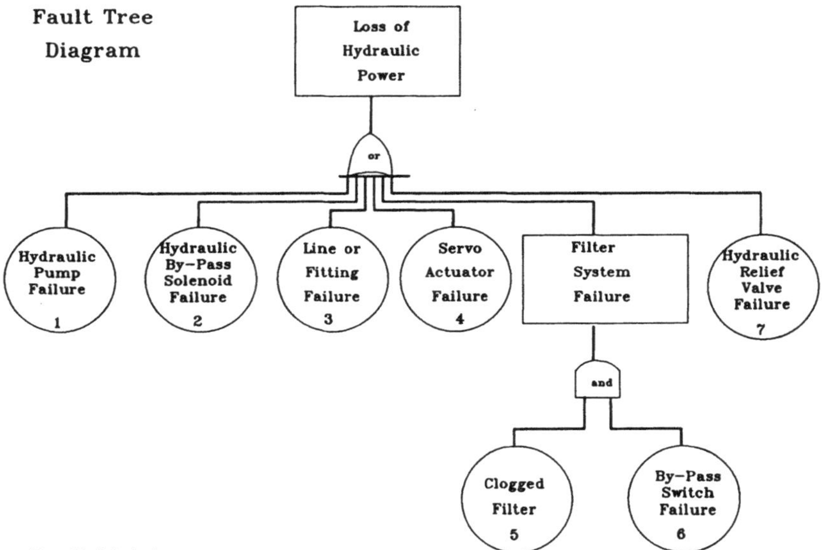

Fault Matrix

Identification Number	Basic Fault	Failure Effect	P(x's)	Criticality	Corrective Action Recommendations
1	Hydraulic Pump Failure	Loss of Hydraulic Power	0.8×10^{-6}	0.8×10^{-6}	Install Improved Pump
2	Hydraulic By-Pass Solenoid Failure	"	0.2×10^{-6}	0.2×10^{-6}	Install Redesigned Solenoid
3	Line or Fitting Failure	"	0.8×10^{-6}	0.8×10^{-6}	Inspect Periodically for Fretting or Contact Abrasion
4	Servo Actuator Failure	"	0.8×10^{-6}	0.8×10^{-6}	Incorporate Safety Wire or Jam-Nut Locking Tab
5	Clogged Filter	Filter System Failure	$< 1 \times 10^{-6}$	1×10^{-10}	Increase Frequency of Inspection/Cleaning
6	By-Pass Switch Failure	"	$< 1 \times 10^{-6}$	1×10^{-10}	Install Redesigned Switch
7	Hydraulic Relief Valve Failure	Loss of Hydraulic Power	0.2×10^{-6}	0.2×10^{-6}	Incorporate Improved Relief Valve

Figure 4–9 Fault Tree Sample Analysis

higher levels of assembly. As with FTA, the FMECA can be performed utilizing either actual failure modes from field data or hypothesized failure modes derived from design analyses, reliability prediction activities and experiences relative to the manner in which components fail. The FMECA provides insight into failure cause and effect relationships. It provides a disciplined method to proceed part-by-part through the system to assess failure consequences. In its most complete form failure modes are identified at the part level. Each identified part failure mode is analytically induced into the system, and its failure effects are evaluated and noted, including severity and frequency (or probability) of occurrence.

The FMECA is generally performed in accordance with US MIL-STD-1629A, 'Procedures for Performing a Failure Mode, Effects and Criticality Analysis'.[12] That standard provides a procedure for performing an FMECA which involves systematically evaluating and documenting, by failure mode analysis, the potential impact of the subject indenture level failure on system output. Data and information used as input to the procedure include the design specification, schematics and other configuration information obtained from the cognizant design engineer.

The FMECA procedure involves several steps. The first step is to list all failure modes at the lowest practical level of assembly. For each failure mode listed, the corresponding effect on performance at the next higher level of assembly is determined. The resulting failure effect becomes, in essence, the failure mode that impacts the next higher level. Iteration of this process results in establishing the ultimate effect at the system level. The analysis is complete once all identified failure modes have been evaluated and their relationships to the end effects defined.

Probabilities for the occurrence of the system effect can be calculated, based on the probability of occurrence of the lower level failure modes (i.e., the product of modal failure rate and time). A criticality number is then calculated from these probabilities and a severity factor assigned to the various system effects. Criticality numerics provide a method of ranking the system level effects for corrective action priorities, engineering change proposals or field retrofit actions.

Performing an FMECA involves the tabulation of information on a standard worksheet, such as that given in Figure 4–10.

The numbers on the worksheet correspond to those given below, which describe each entry:

FMECA WORKSHEET

SYSTEM ________ COMPONENT ________ PAGE ___ OF ___ DATE ___ ENGR ___

PART	FAILURE MODE	COMPONENT EFFECT	SYSTEM EFFECT/SEVERITY	DETECT-ABILITY	FAILURE FREQUENCY	CRITICALITY
Hydraulic Pump	Leaky Pump	Loss of Hyd. Power	Flight Abort. (Precautionary)	—	—	—
Hydraulic By-Pass Solenoid	—	—	—	—	—	—
—	—	—	—	—	—	—
(1)	(2)	(3)	(4)	(5)	(6)	(7)

Figure 4–10 Sample FMECA Worksheet

(1) & (2) Identification of the part and failure mode(s) associated with that part.
(3) Identification of the component effect resulting from that failure mode.
(4) Identification of a system effects and severity factor based on the following:

Code no.	*Severity factor*	*Description*
I	10	Most severe—crew safety jeopardized and mission aborted
II	8	Mission aborted—safety hazard avoided with pilot skill
III	5	Mission aborted—no safety hazard
IV	5	Undesirable condition—flight can continue
V	1	Operates normally, no effect

(5) Detectability of the failure mode by final test/inspection—a check mark (/) indicates that failure mode is *not* detectable.
(6) Frequency of the failure mode (or failure probability) based on:

$$P(x) = \lambda M t$$

where $P(x)$ is the failure mode probability, λ is the part failure rate (from MTBF studies), M is the mode distribution, and t is the time (e.g., 2 hour mission). Note: the above parameters when combined are intended to express relative probabilities. Actual probability would also include terms to reflect induced reliability and quality defects. This analysis assumes that these defects will not significantly vary on a part-by-part basis.

(7) Criticality — the criticality ranking is derived from $C = P(x) \times S$, where C is criticality, $P(x)$ is probability of occurrence of the failure mode and S is the severity factor.

4.4 RELIABILITY TESTING

The attainment of a specified reliability level, within allocated resources, is largely dependent on the extent to which reliability testing is applied during development. Since, at the beginning of the development process, the reliability of a new system would be much lower than the required level, reliability tests and other improvement techniques must be applied to grow reliability to the specified level.

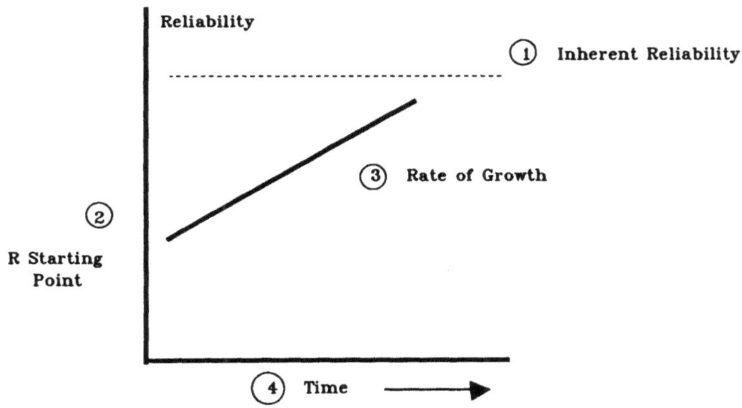

Figure 4–11 Reliability Growth Plot — Log–Log Scale

Reliability growth can be defined as the positive improvement of reliability through the systematic and permanent removal of failure mechanisms. A well planned growth program, that includes reliability testing, provides a means by which reliability can be measured and managed during system development. It provides a means for extrapolating a current reliability design level to some future level, thus allowing trade-off decisions to be made early in the development process.

Prior to conducting a reliability test, a detailed test plan must first be prepared. The plan must describe the growth process and show how it will be applied to the system or component under test. It must identify the specified and predicted (inherent) reliabilities and provide a description of the method used to predict reliability (model, data base, etc.). It must define the starting point for the reliability growth test and provide the criteria for estimating the reliability of the initial test hardware. It must also define the test, fix, retest conditions, requirements and criteria, as they relate to and impact the reliability growth rate. Finally, the plan must include a timeline showing the total length of the test with appropriate allowances provided for corrective action and repair downtimes.

Figure 4–11 illustrates the relationships of these factors when plotted on a log–log scale. The circled numbers refer to the above factors. Each of these factors affects reliability growth significantly as discussed in the following paragraphs.

Inherent reliability (1) represents the value of reliability established by the design and may correspond to the specified value. Ordinarily, the specified value of reliability is somewhat less than the inherent value. The relationship of the inherent (or specified) reliability to the starting point greatly influences the total test time.

Starting point (2) represents an initial value of reliability usually based on a percentage of the estimated inherent designed-in reliability. Starting points must take into account system technology, complexity and the results of the R&M efforts applied during design. Higher starting points minimize test time.

Rate of growth (3) represented by the slope of the growth curve is influenced by the rigor and efficiency with which failures are discovered, analyzed and corrected. Rigorous test programs which foster the discovery of failures, coupled with management supported analysis and timely corrective action, will result in a faster growth rate and shorter total test time.

The length of the test (4) is dependent on the planned test time as well as how efficiently repairs are made, failures analyzed and corrective actions implemented. Estimates of repair time and operating/non-operating time as they relate to calendar time must be made. Lengthy delays for failure analysis and implementation of corrective action will extend the overall growth test period.

Each of the factors discussed above impacts the total time (or resources) scheduled to grow reliability to the specified value. A reliability growth model can be used to help plan a growth test program and to allocate resources. The model can provide estimates of the total test time needed to grow to a given reliability value under various levels of corrective action and to provide insight into cost, schedule and the number and type of test units needed to grow reliability to a desired value during development. A reliability growth model is used to help plan a program based on these factors.

The reliability growth models are based on a mathematical formula (or curve) that reflects the reliability of the system as a function of test and development time. It is commonly assumed that these curves are non-decreasing. That is, once the system's reliability has reached a certain level, it will not drop below this level during the remainder of the development program. It is important to note that this is equivalent to assuming that any design or engineering changes made will not decrease system reliability.

If the shape of the reliability growth curve is known, then the model is a deterministic one. In this case, the time and effort needed to meet the reliability requirement can easily be determined. In those cases where the shape of the curve is not known, it is generally assumed that it belongs to a particular class of parametric curves. This is analogous to life testing when it is assumed that the life distribution of the items is a member of some parametric class such as the exponential, gamma or Weibull families. The analysis then reduces to a statistical problem of estimating the unknown parameters from the experimental data. These estimates may be revised as more data are obtained during the progress of the development program. Using these estimates, reliability can then be monitored and projected and the necessary trade-off decisions can be made.

Other models are nonparametric and allow for the estimation of the present system reliability from experimental data without attempting to fit a particular parametric curve. The estimates are usually conservative, and projections of future system reliability are generally not possible.

A brief description of some of the more common reliability growth models is given in the following paragraphs:

Model 1. This approach considers a reliability growth model in which the mean time to failure of a system with exponential life distribution is increased by removing the observed failure modes. In particular, it shows that, when certain conditions hold, the increase of mean time to failure is approximately at a constant percent per trial. That is, if $\theta(i)$ is the mean time to failure of the system at trial i, then $\theta(i)$ may be approximated under certain conditions by

$$\theta(i) = A \exp(Ci)$$

where A and C are parameters. Note that

$$\theta(i+1) = (\exp C)\ \theta(i)$$

The maximum likelihood estimates of A and C are given.

Model 2. This model considers a situation where the system failures are classified according to two types. The first type is termed 'inherent cause' and the second type is termed 'assignable cause'. Inherent cause failures reflect the state-of-the-art and may occur on any trial, while assignable cause failures are known and, whenever one of these modes contributes a failure, the mode is removed permanently from the system. This approach uses a Markov-chain approach to derive the reliability of the system at the nth trial when the failure probabilities are known.

Model 3. This model was derived from the analysis of data available on several systems to determine if systematic changes in reliability improvement occurred during the development effort. The analysis showed, for those systems, that the cumulative failure rate vs cumulative operating hours approximated a straight line when plotted on a log–log scale.

Mathematically, the failure rate equation may be expressed by

$$\lambda_\Sigma = KT^{-\alpha}$$

$K < 0, 0 \leqslant \alpha \leqslant 1$, where T is the cumulative failure rate of the system at operating time T, and K and α are parameters.

Model 4. This model assumes that the system is being modified at successive stages of development. At stage i, the system reliability (probability of success) is P_i. The model of reliability growth under which one obtains the maximum likelihood estimates of P_1, P_2 ..., P_i assumes that

$$P_1 \leqslant P_2 \leqslant \ldots \leqslant P_K$$

That is, it is required that system reliability is not degraded from stage to stage of development. No particular mathematical form of growth is imposed on the reliability. In order to obtain a conservative lower confidence bound on P_K, it suffices to require only that

$$P_K \geqslant \max_{i<K} P_i$$

That is, it is only necessary that the reliability in the latest stage of development be at least as high as that achieved earlier in the development program.

Data consist of x_i, successes in n_i trials in stage i, $i = 1$, ..., K.

Model 5. This model assumes that at stage i of development the distribution of system life length is F_i. The model of reliability growth, for the maximum likelihood estimates of $F_1(t), F_2(t), \ldots, F_K(t)$ is:

$$\bar{F}_1(t) \leqslant \bar{F}_2(t) \leqslant \ldots \leqslant \bar{F}_K(t)$$

where

$$\bar{F}_i(t) = 1 - F_i(t)$$

for a fixed $t \geqslant 0$. In order to obtain a conservative upper confidence curve on $F_K(t)$ and, thereby, a conservative lower confidence curve on $F_K(t)$ for all non-negative values of t, it suffices only to require that

$$\bar{F}_K(t) \geqslant \max_{i<K} \bar{F}_i(t)$$

for all $t \geqslant 0$. That is, the probability of system survival beyond any time t in the latest stage of development is at least as high as that achieved earlier in the development program.

4.4.1 The Design, Development and Production Reliability Growth and Qualification Process

Figure 4–12 suggests a typical growth and qualification process based on the factors discussed above.

The figure depicts the MTBF of a hardware item as it progresses through its design, development and production stages. Seven discrete stages are shown:

(1) Represents the value of MTBF of the design as estimated by reliability prediction. This value is the upper limit of reliability as established by the system design configuration, its technology and the designed-in reliability attributes. Maximizing this inherent reliability level requires the selection of proven, high quality, well derated long-life parts, with emphasis placed on the use of adequate design margins and ample means for heat dissipation and shielding. Well established, systematic reliability engineering techniques, i.e., reliability prediction and failure mode, effects and criticality analysis, are applied to enhance reliability growth through design iteration.

(2) Represents the value of MTBF for the initial assembled hardware. This value usually falls within the range of 10–20% of the inherent (or predicted) reliability. Estimates of the starting point can be derived from prior test experience.

(3) Represents the MTBF growth during environmental qualification testing on the engineering test units. These tests, generally performed in accordance with US MIL-STD-810, 'Environmental Test Methods', are to assure that the hardware meets its end-item application environments before initiation of a formal reliability growth test.[13] Although not specifically designed for reliability growth, some growth will occur, from design improvements, as depicted by the slope of the curve.

(4) Represents the MTBF growth during testing on early test units. The rate of growth depicted by the slope of the curve is governed by the amount of control, rigor and efficiency with which failures are discovered, analyzed, and, most importantly, corrected through design and manufacturing action. Rigorous test programs which foster the discovery of failure mechanisms, coupled with management supported analysis and timely corrective action, will result in a faster growth rate and consequently less total test time and expense.

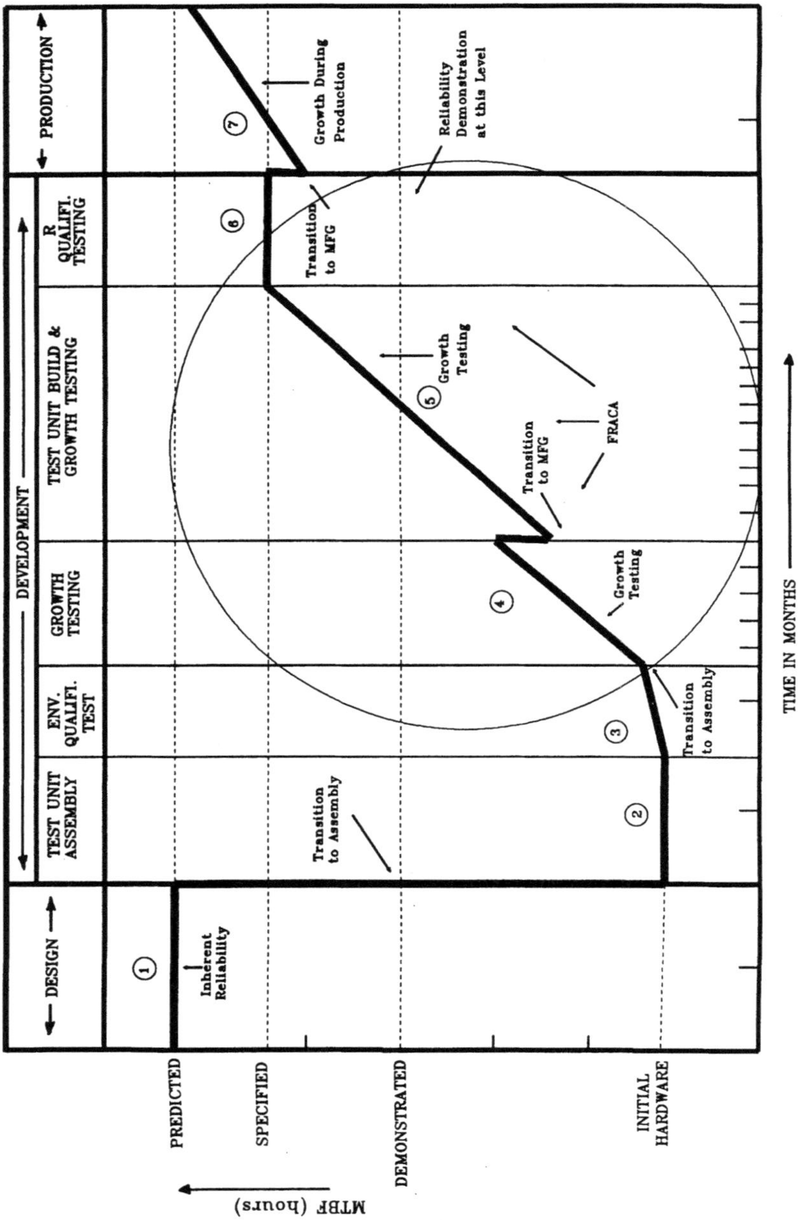

Figure 4–12 Reliability Growth Process

(5) Represents the value of MTBF during assembly of advance test units built by manufacturing. An initial MTBF degradation is shown to reflect transition of the design configuration to manufacturing. The extent of this degradation is dependent on the applied manufacturing and quality control methods as well as on the effectiveness of environmental stress screening (ESS) applied to incoming components, fabricated boards and modules, and finished assemblies and subsystems to remove defects. This stage also shows the MTBF growth during growth testing. The rate of growth, like the earlier rate, is governed by the amount of control, rigor, and efficiency with which failures are discovered, analyzed, and corrected through design and manufacturing action.

(6) Represents the value of MTBF during reliability qualification testing of pilot units. Reliability qualification tests, performed in accordance with US MIL-STD-781 'Reliability Qualification and Production Acceptance Test—Exponential Distribution',[14] are designed for the purpose of proving, with statistical confidence, a specific reliability requirement, not specifically to detect problems or to grow reliability. Note that reliability growth testing differs from reliability qualification testing in that growth testing is intended to disclose failures, while qualification testing is not.

(7) Shows the reliability degradation when the final qualified design configuration is moved to production and the subsequent growth that can be expected due to manufacturing experience and to the application of environmental stress screening and corrective actions. Screening, which is central to the manufacturing inspection and acceptance process, consists of a family of techniques in which electrical, thermal and mechanical stresses are applied to accelerate the occurrence of potential failures. By this means, latent failure-producing defects, which are not usually caught during normal quality inspection and testing, are removed from the production stream. Burn-in is a specific subclass of screen which employs stress cycling for a specified period of time. Note that, when the production hardware is accepted for operation, it should have achieved the specified MTBF level or, under ideal conditions, the inherent or predicted level.

It should be emphasized that design establishes the inherent reliability potential of a system or equipment item, and the transition from the paper design to actual hardware and ultimately to operation often results in an actual reliability that is far below the inherent level. The degree of degradation from the inherent level is directly related to the reliability, testability and maintainability features designed and built into the system as well as to the effectiveness of the test program applied during development to eliminate potential failures and deterioration factors. Lack of a well planned, carefully executed growth test program can result in an actual system reliability perhaps as low as 10% of its inherent reliability potential.

As shown in Figure 4–12, the reliability of the initial hardware starts at some level that might be considered the state-of-the-art at the beginning of development. During development, through application of a formal test program, reliability grows to the level established and then qualified prior to release for production. Every failure is analyzed as part of the program to determine its root cause, develop sound corrective action and verify by continued testing that the detected failure has been eliminated.

The actual growth test is performed after successful completion of environmental qualification testing and prior to reliability qualification (demonstration) testing. Although all testing should be viewed and planned as contributing to reliability growth, the formal reliability growth test is deferred until after environmental qualification, when the design of the hardware which is to be used in the test reflects the anticipated configuration and manufacturing processes. The hardware to be tested should have all significant changes required as a result of environmental qualification testing incorporated before initiating the reliability growth test. Note also that, after the growth test is successfully concluded, all significant fixes are incorporated into the test hardware prior to initiating the reliability qualification test.

It should also be emphasized that failure modes are detected through testing. However, the reliability achieved as a result of the growth process becomes meaningful only when the necessary changes developed and proven to achieve that reliability are properly and fully incorporated in configuration control documentation for production hardware. As a consequence, the reliability growth process becomes familiarly known as one of test, analyze and fix (TAAF).

As indicated, a reliability qualification test is also conducted as part of the overall test program. Reliability qualification testing is intended to provide reasonable assurance that minimum acceptable reliability requirements have been met before items are committed to production. The test must be operationally realistic and must provide estimates of demonstrated reliability. It is a pre-production test that must be completed in time to provide management information as input to a production release decision.

Reliability qualification tests are normally required for items that are newly designed, have undergone major modification, and/or have not met their allocated reliability requirements for the new system under equal (or more severe) environmental stress. Off-the-shelf items may be considered qualified provided they have met their allocated reliability requirements for the new system under equal (or more severe) environmental stress.

4.4.2 The Planning and Implementation of Reliability Testing

The achievement of a cost-effective test program requires careful planning followed by well executed tests with engineering activities that start early during development and continue through full scale development and initial production. The test activities must be planned and conducted as an integral part of the overall reliability program, evaluated along with other engineering activity and optimized relative to the specific needs of the system under development. This requires establishing adequate test requirements, assigning responsibilities and providing necessary resources (test units, facilities, personnel, etc.).

Planning an optimum test program is a complex procedure involving the evaluation of a large number of interactive factors and the use of trade-off analysis to arrive at an optimum combination. Some of the major factors are: the technology and state-of-the-art of the hardware, its complexity, criticality, physical size, and program cost constraints and other limitations. The planning process is driven by the objective of meeting a specified level of system reliability at minimum cost. It involves setting interim reliability goals to be met during development and establishing the necessary resources to attain these goals. The program, once designed and optimized, will then reflect the desired balance between reliability and test cost (and time).

Reliability test planning addresses program schedules, amount of testing, resources available and the realism of the test program in achieving the requirements. An optimum program requires careful consideration and trade-off of numerous factors and the execution of well timed, properly sequenced engineering analysis and test activities. A reliability program growth curve is constructed which establishes interim reliability goals throughout the program. Periodic assessments of reliability are made during the test program (e.g., at the end of a test phase) and compared with the planned reliability growth values. These assessments provide visibility of achievements and focus on deficiencies in time to affect the system design. By making appropriate decisions, particularly in regard to the timely incorporation of effective corrective changes, commensurate with attaining the milestones and requirements, management can control the growth process.

For complex systems, the model used most often for planning the overall reliability growth process, and in particular reliability growth testing, is one originally published by J. T. Duane[15] (previously identified as Model 3). As described, the following mathematical expression would hold as long as the reliability improvement effort continues:

$$\lambda_\Sigma = \frac{F}{T} = \frac{1}{K}T^{-\alpha}$$

where λ_Σ is the cumulative failure rate, T is the total test hours, F is the failures, during T, K is the constant determined by circumstances and α is the growth rate.

Cumulative MTBF (M_c) is that value determined at any time during the test by dividing cumulative test time by cumulative chargeable failures and is represented by:

$$M_c = \frac{1}{\lambda_\Sigma} = KT^{\alpha}$$

Differentiating with respect to time:

$$\lambda(t) = \frac{\partial F}{\partial T} = \frac{K}{T}(1-\alpha)T^{-\alpha} = (1-\alpha)\lambda_\Sigma$$

so that the 'instantaneous' or current failure rate is (1-α) times the cumulative failure rate, or the 'instaneous MTBF' is 1/(1-α) times the cumulative MTBF. An adequate interpretation of 'instantaneous MTBF' is:

> 'The MTBF that the equipment currently on test would exhibit if we stopped the reliability growth and continued testing. Thus the instantaneous or current-status curves are straight lines displaced from the cumulative plot by a factor $(1-\alpha)$, which shows up as a fixed distance on a logarithmic plot, as shown in Figure 4–13.'

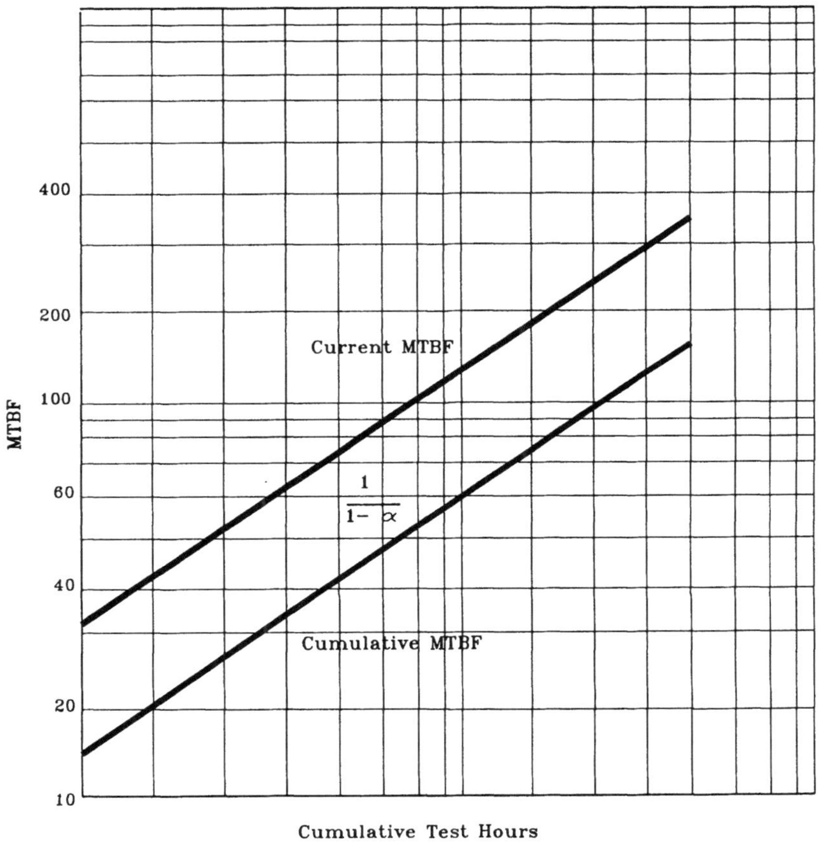

Figure 4–13 Duane Plot

Normally, the cumulative MTBF (M_c) is measured in test and converted to instantaneous (or current) MTBF (M_I) by dividing by 1 - α, that is,

$$M_I = \frac{M_c}{1-\alpha}$$

The cumulative MTBF is plotted vs cumulative test time, a straight line is fitted to the data and its slope, α, is measured. The current MTBF line is then drawn parallel to the cumulative line but displaced upward by an offset equal to $1/(1-\alpha)$. The corresponding test time at which this line reaches the required MTBF is the expected duration of the growth test.

As discussed earlier, the test plan is to be prepared such that it defines the cumulative test time required to grow to the specified or targeted MTBF, the number of test units subjected to growth and qualification tests, and the anticipated test time per unit. The plan should provide:

(1) Values for specified and predicted (inherent) MTBF. Methods for predicting reliability (model, data base, etc.) must also be described. Ordinarily, the contract specified value of MTBF is somewhat less than the inherent value. The relationship of the inherent (or specified) MTBF to the starting point greatly influences the total test time.
(2) Criteria for reliability starting points, i.e., criteria for estimating the MTBF of initially fabricated hardware.
(3) Reliability growth rate (or rates). To support the selected growth rate, the rigor with which the test–analyze–fix conditions are structured must be completely defined.
(4) Anticipate schedule. This section should also relate test time, corrective action time and repair time to each other.

Each of the factors listed above impacts the total time (or resources) which must be scheduled to grow reliability to the specified value. A reliability growth model graphically depicts the relationships of these factors and is used to estimate the appropriate test time as well as to structure the overall program. Figure 4–14 shows the relationship of these factors based on the Duane model and shows several growth lines having different slopes, depending upon the emphasis given to the growth test program.

The plan should reflect an MTBF starting value between 10 and 20% of its predicted MTBF. This would be for projects with no previous growth testing. A conservative value should be used whenever possible to assure adequate funding and time to complete the effort. Higher starting points may be indicated based on whether or not previous reliability or burn-in tests were performed with aggressive corrective action support. Furthermore, a conservative growth rate of 0.3–0.4 should be planned in all cases. Although rates of 0.5 and 0.6 are theoretically possible, they are seldom achieved. It also should be kept in mind that inadequately burned-in units can seriously bias test results, yielding greatly inflated growth rates that cannot be sustained.

Figure 4–14 shows that the value of the parameter α can vary from 0.1 up to 0.6 as a theoretical upper limit. A growth rate of 0.1 can be expected

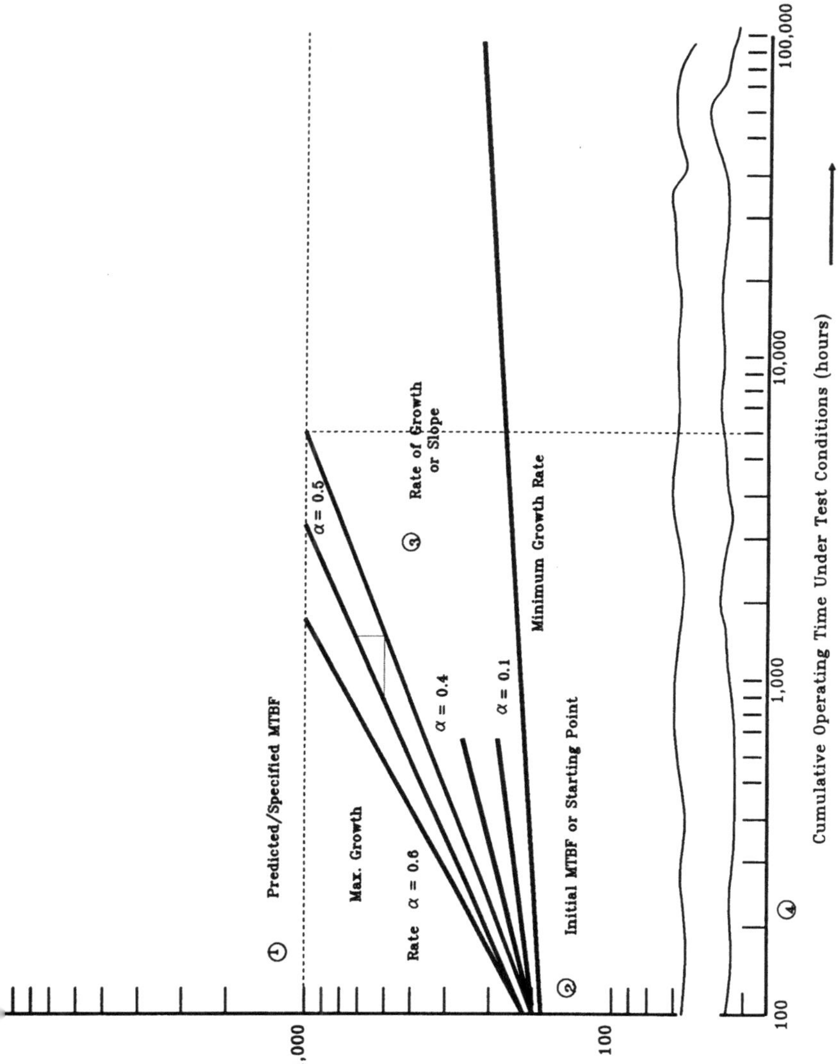

Figure 4–14 Reliability Growth Plot

in those programs where no specific consideration is given to reliability. In those cases, growth is largely due to solution of problems impacting production and corrective action taken as a result of user experience. The higher growth rates can be realized if an aggressive reliability program with management support is implemented. This latter type of program must include a formal stress-oriented test program designed to aggravate and force defects and vigorous corrective action.

Figure 4–14 shows the requisite hours of operating and/or test time and continuous effort required for reliability growth. It shows the dramatic effect that the rate of growth has on the cumulative operating time required to achieve a predetermined reliability level. For example, the figure shows, for an item whose MTBF potential is 1000 hours, that 100,000 hours of cumulative operating time is required to achieve an MTBF of 400 hours when the growth rate is 0.1, and, as previously stated, a 0.1 rate is expected when no specific attention is given to reliability growth. However, if the growth rate can be accelerated to 0.4 (by formal failure analysis and corrective action activities) then the specified 1000 hour MTBF can be achieved in approximately 6000 hours.

Trade-off studies and analyses are a critical part of an equipment development process and, in particular, the determination of an optimum test program. Decisions regarding resource allocation, test type and total test time must be made relative to the specific requirements and constraints of the particular equipment under development, before the test plan is prepared. For example, a formal growth test program may require 10,000 hours of test time. The calendar time required for such a test can be a year or more. In such cases a phased program dividing the test program between early engineering test units and advance manufacturing test units may be an acceptable trade-off.

Also, trade-offs relating directly to the need for the specified or desired MTBF may be made. Engineering experience suggests that the achievement of a reliability (in MTBF) equal to 80% of the design's potential is possible. Thus, the question becomes one of examining the system MTBF need. Experience also shows that most specified MTBF values relate to a program goal at the next higher system level or an individual project goal based on improvement over the superseded equipment or system. Also, most of these goals are based on the need to lower maintenance costs and not on improving mission reliability. Thus, this trade-off reduces to one of considering the cost of improving MTBF vs life-cycle cost–benefits of that improvement. Experience with life-cycle cost studies has shown that incremental improvements in MTBF, especially those greater than 10%, will nearly always produce life-cycle cost savings 10–100 times that of the cost to realize that improvement and sometimes much greater. These savings figures are much higher for projects that eventually go to competitive production than for one- or two-of-a-kind projects. Any trade-off accepting a lower MTBF should be backed up with a life-cycle cost analysis substantiating an unacceptable ratio of extra test costs vs life-cycle savings.

The extent of reliability growth and qualification testing is a key trade-off. Reliability qualification testing is used to certify that the reliability level has been achieved. It generally requires a test time of nearly three times the specified MTBF thus always requires a significant time period and a significant expense. Reliability qualification is important on many development systems after growth testing but qualification testing should never replace growth testing. Inasmuch as the physical properties of the tests themselves are virtually identical, the costs differ only in the length of test time. Test time required for systems with longer MTBF requirements and the costs can be the same. The major difference is in the way test data are analyzed and the implementation of corrective actions. This difference can be minimized in the data analysis; credible qualification accessments can be made from growth test data. From the opposite point of view, credible growth test results can be achieved from qualification test efforts when effective corrective actions are considered individually and later implemented as deemed necessary. In summary, qualification tests should not be performed at the expense of growth tests unless the growth test has been completed, but the growth test discipline could continue during qualification testing to realize as much growth as possible.

Reliability growth is monitored throughout the test program using a graphic plot of the achieved MTBF expressed as a point estimate. The point estimate is the cumulative MTBF and is calculated by dividing the cumulated test time by the total of the failures that have occurred up to that time. The plot, identified as 'achieved reliability', is made on the same graph as the reliability growth test model and is not adjusted to reflect corrected failures. A second plot may be made to reflect the level at which the achieved reliability would be if those failures were discounted for which acceptable corrective action has resolved the failure to the satisfaction of the sponsor activity. This second plot is identified as 'Adjusted Reliability'.

The best-fit (the method of least squares may be used) straight line through the first few (three to six) plotted MTBF points will establish the growth rate and can be compared directly with the planned growth line. The slope (α) of the best-fit straight line drawn through the plotted points represents the growth rate. As long as the achieved reliability compares favorably with the planned growth, as presented in the test plan, satisfactory performance may be assumed. If the growth is significantly less than planned (after enough data have been collected to establish a growth rate), a careful analysis must be made to determine the reasons for the poor performance and to develop a corrective action plan.

The reliability growth test and its associated failure analysis and corrective action activity can be considered satisfactory if any of the following conditions exist:

(1) The plotted MTBF values remain on or above the planned growth line.
(2) The best-fit straight line is congruent with or above the planned line.
(3) The best-fit straight line is below the planned line, but its slope is such that a projection of the line crosses the horizontal required MTBF line by the time that the planned growth line reaches the same point.

If none of the above conditions exists, it can be assumed that the planned reliability growth cannot be achieved with the current level of activity. This situation requires that a corrective action plan be generated and, after approval, implemented. Before the corrective action plan can be determined, a careful analysis of the equipment design and related failures must be accomplished to ascertain the problem areas and possible design modifications. As the reliability growth test continues, a moving average of achieved reliability may be constructed by arranging the failure times (accumulated test times between failures) for the equipment on test in chronological order of occurrence. The moving average for any specific number of failures is computed as the arithmetic mean of the failure times selected sequentially and in reverse order. For example, the moving average for two failures is obtained by adding the last two failure times and dividing by two; for three failures, by summing the last three failure times and dividing by three; and so forth. The number of failures to be used in the computation is arbitrary but is restricted to ten or fewer. This curve, if used, will be identified as the moving average and the number of failures used for computation will be noted.

A current MTBF growth line may also be constructed and progress toward the predicted MTBF noted. The current MTBF is the cumulative

MTBF divided by 1-α. This line will parallel the best-fit cumulative MTBF line and indicate MTBF's at a factor of $1/(1-\alpha)$ above the cumulative MTBF's. This will provide an indication of what the smoothed present reliability is at the current cumulative test time and will represent the probable value of MTBF if no further corrective action were implemented. Therefore, when the current MTBF reaches the predicted MTBF value, the reliability growth necessary to demonstrate the specified reliability at a reasonable confidence level, during a formal reliability qualification test, has been achieved.

4.4.3 Failure Reporting, Analysis and Corrective Action

Failures that occur during reliability growth and qualification tests, as well as during production tests and field operation and maintenance activities, must be accurately reported, thoroughly analyzed and corrected on a timely basis to prevent recurrence. A failure recurrence control program must be in place to assure that this takes place and that the sequence of events that occurs upon detection of a failure is documented, including the analysis, corrective action, personnel responsibilities, scheduling and applicability of the analysis to reliability growth.

Implementation of an extensive failure recurrence control program requires the availability of reliability analysts, physics of failure specialists, chemists and metallurgists who have years of experience in analysis of systems and components. The program requires that failure analyses are performed on failed components and material to determine the root causes and underlying mechanisms of failure and that the results of all failure analysis activity are documented on forms designed for this purpose. These forms provide entries for part identification data, conditions under which failures occurred, operating parameters indicating degradation, reference to applicable plans or procedures, and complete details leading up to or surrounding the failure incident. The analysis methods used, including test, X-ray, dissection, SEM, chemical analysis, etc., to determine failure causes are also described.

A simplified flow chart showing the sequence of the events that must take place when a failure occurs is given in Figure 4–15. A failure is defined as any deviation from acceptable limits called out in the applicable system specification. Also, any operating discrepancy that requires an unscheduled adjustment or calibration to be made (except normal operating adjustments or scheduled maintenance actions) is defined as a failure for reporting purposes.

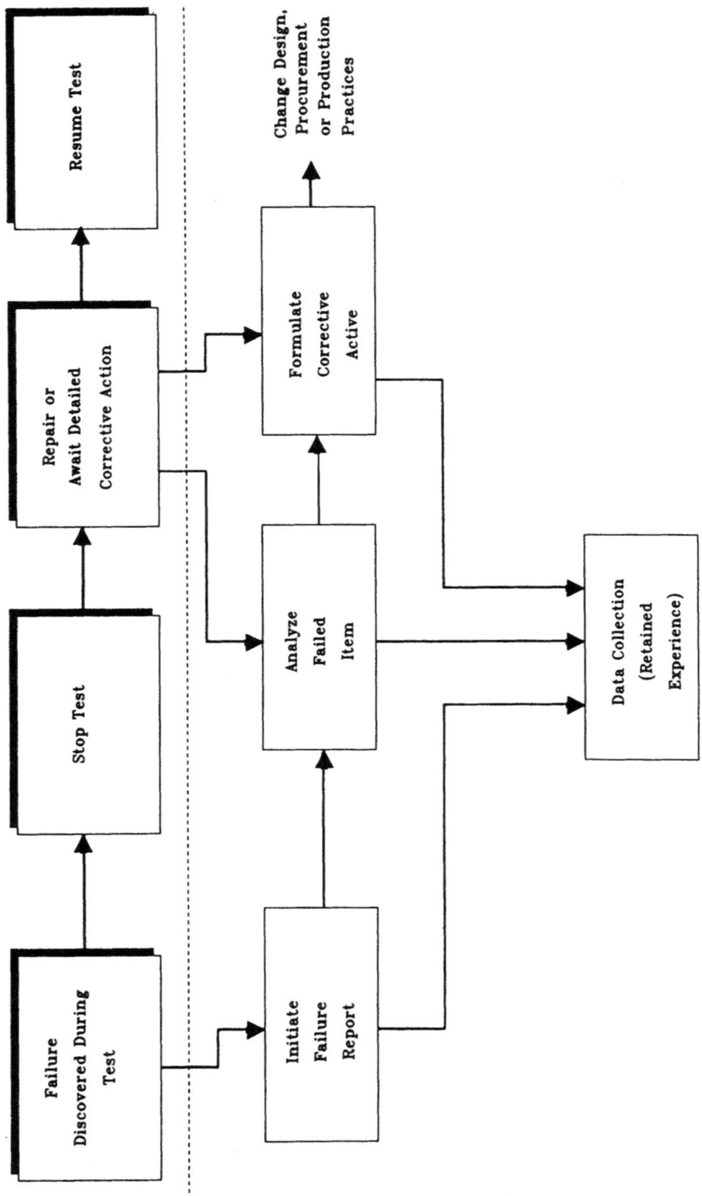

Figure 4–15 Simplified Flow Chart of Failure Analysis Activities

When a failure occurs, the test is stopped, a failure report is initiated, the discrepancy is entered into the log and the appropriate organizations are notified. If the failure is critical, i.e., if it impacts safety or would cause significant damage if the test was allowed to continue, the repair action is held pending the results of the subsequent failure analysis investigation and corrective action implementation plan. If the failure is not critical, repair action takes place immediately and, after the unit is repaired and in a serviceable condition, it is returned to test.

After a failure report is prepared, it is determined if a complete failure analysis investigation is to be conducted in order to identify the root cause of the failure. In general, a failure analysis is conducted for all failures that occur during reliability growth and qualification testing. In the case of failures that occur during production testing, a failure analysis investigation is, generally, conducted only when a definite trend or pattern has been established. To facilitate the establishment of trends or patterns, and subsequent failure activities, previous failure reports (generally contained in an R&M data bank) are searched to determine if previous reports were processed with identical or similiar failures.

A failure analysis investigation, if required, is performed to determine the root cause of failure and to provide a basis to establish the necessary corrective action to prevent recurrence. The failure analysis may involve:

- Component/material/assembly
  — Microscopic examination
  — Characterization
  — Empirical (thermal/electrical application measurement)
- Design
  — Derating
  — Circuit (drift/tolerance build-up)
- Test
  — Material compatibility
  — Accelerated life
  — Transient/over-voltage, etc.

Based on the results of the failure analysis the most appropriate corrective action to eliminate the failure is determined. Corrective action measures could involve:

- System redesign
- Part selection criteria
- Part derating criteria

- The application of tests to weed out specific failure mechanisms
- Special in-process fabrication inspections and tests
- Special reliability and safety assurance provisions

The appropriate corrective actions, e.g., engineering change, process change or maintenance change, are then determined and an implementation plan is prepared. The corrective action implementation plan describes the specific action to be taken on the units in test as well as those in-production or in-service. It also provides criteria including effective time periods for verifying that the corrective actions are effective.

To facilitate failure analysis and the development of appropriate corrective action plans, as well as to provide a basis for assessing hardware reliability based on actual experience, an R&M data bank should be maintained. The data bank would include all failure analysis and corrective action reports from prior reliability and production tests and field operation and maintenance activities. These prior data are then used to correlate reported failures and to establish failure trends. The data bank would also facilitate subsequent failure analysis corrective action investigations.

The data can also be used as direct input to reliability and safety engineering analyses as well as to provide a basis to track, measure and report reliability results so that corrective action in the design, manufacturing processes and tests, or operational maintenance procedures, can be taken and adjustments made as necessary to minimize cost and maximize effectiveness. In addition, this experience base can be used for comparative analyses and assessments and applied to the design of subsequent system development programs.

The data bank would provide the following information:

- Field return rates (at start-up and during use)
- Failure rates/modes (and responsible failure mechanisms)
- Manufacturing inspection/test and screening reject rates/efficiency factors
- System reliability plots vs time
- Development/manufacturing/field reliability correlation factors
- Problem areas priority list
- Accomplished corrective actions
- Corrective action tracking/effectiveness

The data bank can be programmed to facilitate trend analyses, failure analyses, failure histories and development of component/part failure rate data. It can also have the capability of automatically computing and flagging trends from all data recorded in the bank.

4.5 PRODUCTION, STORAGE AND IN-SERVICE DEGRADATION CONTROL

Historically, the actual in-service reliability of many systems has been much lower than the levels predicted during design and demonstrated during reliability and growth testing. This discrepancy may be, in part, because the in-service operating and maintenance stresses for these systems were not completely anticipated or understood, or the reliability level was not realistically specified and, consequently, not fully dealt with during design, development and testing.

However, in many cases, the reliability discrepancy was due to the fact that design based reliability predictions, and the reliability growth and demonstration tests which are based on the predictions, did not take into account production, storage and in-service degradation factors. A reliability prediction, to be accurate, must reflect, in addition to the system designed-in characteristics, the impact of the production, storage and in-service degradation and growth process on inherent reliability.

The reliability design–degradation–growth cycle starts with design to establish the inherent reliability of the system and to prevent quality problems. Design efforts include: selecting, specifying and applying proven high quality, well derated, long-life parts; specifying materials that account for production process variations; incorporating adequate design margins; using carefully considered, cost-effective redundancy; and applying tests designed to identify defects or potential failures. Emphasis is placed on incorporating ease of inspection and maintenance features, including use of easily replaceable and diagnosable modules (or components) with built-in test, on-board monitoring and fault isolation capabilities. The design effort includes performing well planned reliability engineering analyses, followed by reliability growth and qualification tests that are supported by a formal system for accurately reporting, analyzing and correcting failures.

Design establishes an upper limit of reliability, and as the system is released to production its reliability will be degraded because of process induced defects: as production progresses, with resultant process improvements, the application of environmental stress screening, statistical process controls and manufacturing learning factors, reliability will grow. After the system is deployed, its reliability will again be degraded due to operating and non-operating (storage) stresses; as field operations continue, with increasing operational personnel familiarity and maintenance experience, reliability will again grow.

This section discusses reliability degradation and its control based on this life-cycle process.

Production Degradation Control

Degradation of reliability can occur as a result of defects induced during production. In order to reduce degradation and to continuously improve the quality of a system, reliability must be evaluated before and during the production process. Areas where corrective actions and process improvements can best be implemented must be identified. The process improvements evaluations must take into account the potential defects induced by production and the inspections, tests, controls and other measures applied to remove the defects. They must focus on the use of statistical process control (SPC) methods and the application of environmental stress screening (ESS) techniques.

SPC involves using statistical methods to improve the outgoing from production reliability, through the reduction of statistical variation by identifying and correcting process instabilities. SPC is an analytical technique, based on the premise that all processes exhibit variation, for distinguishing between expected variations and unexpected variations due to a malfunction. It is a tool to aid in the evaluation of the causes of any unexpected variation and for taking corrective actions to stabilize the process within acceptable limits.

ESS involves the application of a non destructive stress on a 100% basis for the purpose of converting latent (or unobservable) defects such that they may be removed by standard quality inspections and test methods. The application of ESS is described, in detail, in Section 4.6.

A key aspect of a total reliability and quality management program is to thoroughly analyze the rejects resulting from the in-process inspections and screens in order to determine their specific causes and to make corrective action and process improvement decisions. The analysis of rejects resulting from the application of ESS is of particular importance since these rejects represent the kind of defects that could cause failure during initial fielding. Taking this defect prevention approach, with emphasis on the statistical analysis of process variation, the application of

ESS and the analysis of rejects followed by effective, timely corrective actions, will improve the processes, significantly reduce reliability degradation and facilitate the growth of reliability during production prior to fielding.

As indicated, design establishes the inherent reliability potential of a system and the transition from the paper design to hardware results in an actual system reliability below this inherent level. Reliability and quality controls are applied during production to eliminate defects which can cause failure during use. These defects, whether inherent to the design or induced by the production process, can be further categorized into (1) quality and (2) latent reliability defects. The quality defect is generally apparent and detectable through standard inspection procedures. The reliability or latent defect is detectable only by application of stress. These defects, if not removed by production process inspections and screens, contribute strongly to early failures during use.

A process and inspection analysis is performed to evaluate the outgoing reliability of a hardware system as it leaves production and to assess the contributions to unreliability of the manufacturing processes, assembly techniques and quality inspections. Prior to performing a process and inspection analysis, a failure mode analysis should first be performed to identify those items that are most critical. Procedures for performing failure mode analysis were given in Section 4.3. To be both effective and practical the process and inspection analysis should focus on the critical items.

The analysis involves defining the production process in terms of a detailed flow chart and estimating the rate of defects (both quality and latent) introduced and removed at the various steps within the process. Reliability degradation is determined by comparing the defect rate leaving the process with the defect rate entering the process.

The procedure requires quantifying the process induced defects and determining the effectiveness of the quality inspections and screens applied to remove the defects. It includes determining both the latent defects attributable to purchased parts and materials and those due to faulty workmanship or assembly. Examples of some of these defects are:

- Poor welds or seals
- Poor connections
- Dirt or contamination on surfaces or in materials
- Chemical impurities in metal or insulation or protective coatings
- Incorrect positioning of parts

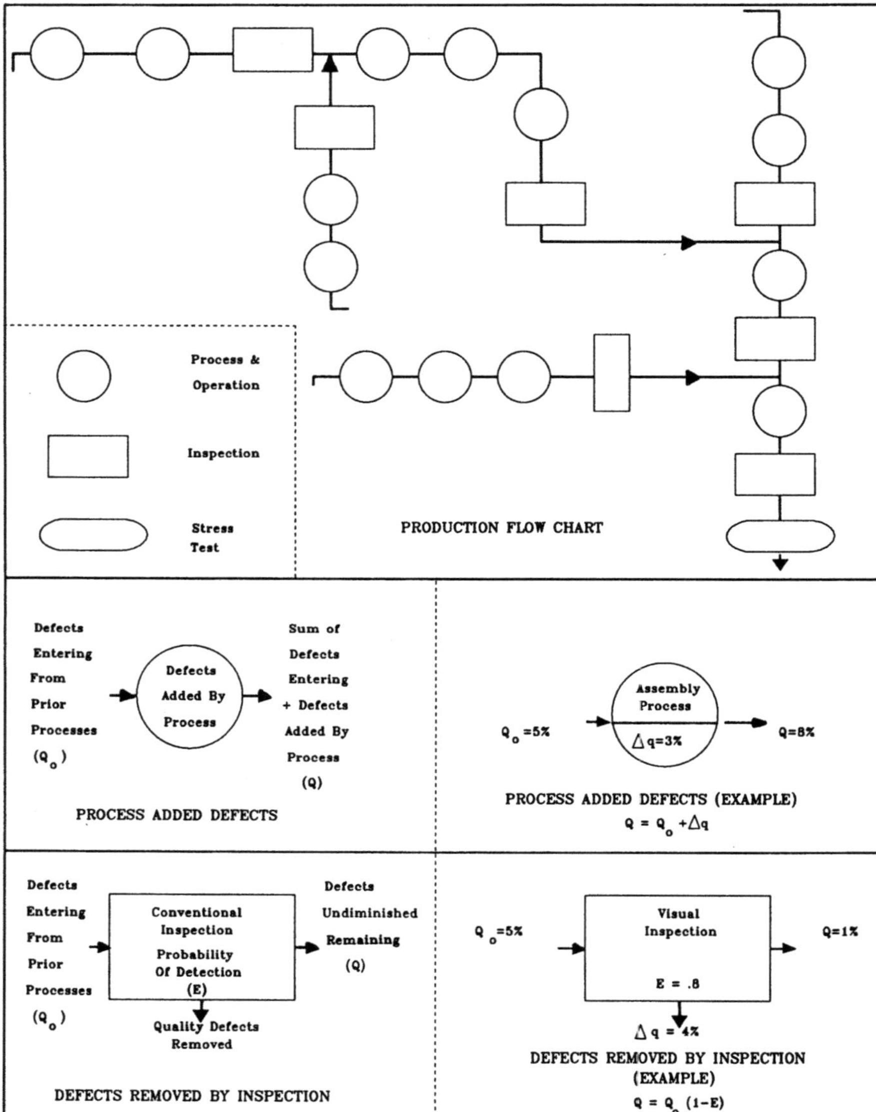

Figure 4–16 Process Flow Diagram

These workmanship and assembly errors can account for substantial degradation in reliability. They are brought about by inadequate operator learning and motivational or fatigue factors. Although inspections are applied to minimize degradation from these sources and to weed out the more obvious defects, they are not perfect. No inspection can remove all defects. A certain number of defective items will escape the process, be accepted and result in failure during early field use.

More important, these gross defects can be overshadowed by an unknown number of latent reliability defects. These weakened items, the results of latent defects or inherent flaws, will cause failure during field operation under the proper conditions of stress. An ESS is designed to apply a stress of a given magnitude over a specified duration to remove these kinds of defects, at the factory prior to fielding. As in the case of conventional quality control (QC) inspections, an ESS is not 100% effective in the removal of latent defects.

The specific steps involved in the analysis are as follows:

Step 1 — Compute the reliability of the item as it enters the production process, using one of the methods described in Section 4.1.

Step 2 — Construct a production process flow diagram (see Figure 4–16). The construction of such a flow chart involves, first, defining the various process elements, inspections and tests which take place during production, and then preparing a pictorial presentation describing how each activity flows into the next activity or inspection point. Note that the size of the diagram will vary, depending on the complexity of the process. A very simple process may require a flow diagram which contains only a few elements, whereas a complex process may involve numerous elements. Figure 4–17 shows a simplified flow chart illustrating the final assembly of a helicopter system. When fully detailed this chart, and the appropriate reject rates and inspection efficiencies established during steps 3 and 4, can be used to derive a final outgoing defect rate and reliability degradation factor (step 5). Note that a similiar flow chart can be prepared for each component in the system. Such charts are particularily useful in identifying and focusing improvement actions on those process elements which have a major impact on system unreliability.

Step 3 — Establish reject rate data associated with each inspection and test. For analysis performed on new production items, data from similiar production hardware and processes provide a basis for estimating the reject rates. The estimated reject rates should take into account historical failure modes in light of the characteristics of the test to detect that failure

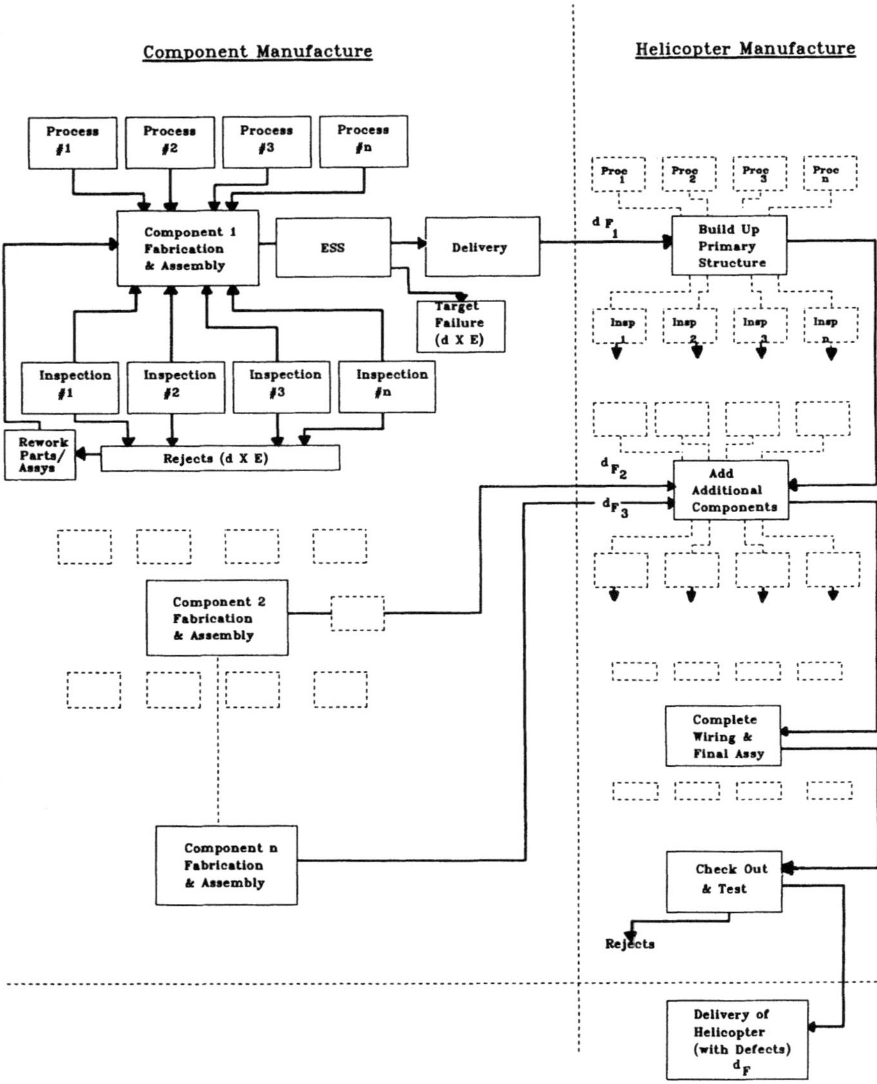

Figure 4–17 Simplified Manufacturing Flow Chart

INSPECTION STATION NUMBER XX			
Description Parameter	Characterization	Maximum Weight Factor	Assessed Weight Factor
1. Complexity of part/assembly under test	Simple part, easy access to measurement.	20	20
2. Measurement equipment	Micrometer for dimensional check, visual for surface finish.	15	10
3. Inspector experience	Highly qualified, several years in quality control.	25	20
4. Time for inspection	Product rate allows adequate time for high efficiency.	15	10
5. Sampling plan	All parts are inspected.	25	20
	Efficiency Weight Factor: W = 80/100 = 80%	= 100	= 80

Figure 4–18 Inspector Efficiency (sample data)

mode. For analysis performed on existing production processes, actual inspection and reject rate data are used.

Step 4 — Establish inspection and test efficiency factors. Efficiency factors are based on past experience for the same or a similar process when such data exist. For newly instituted or planned inspections, and tests having little or no prior history on how many defects are found, estimates of inspection and test efficiency can be made. The efficiency of an inspection depends on all factors involved in or related to the inspection. The inspections can be characterized relative to complexity, efficiency of inspectors, inspection equipment and tools, and past experience with similar inspections (see Figure 4–18).

Step 5 — Compute the output defect rate of the item based on its process defect rates and inspection efficiency factors and assess outgoing reliability.

In order to illustrate the impact that various inspections have on the outgoing from production defect rate consider the relationship between a defective component and the manufacturing inspection operations. For a defective part to escape the production process, the component failure mode must *not* be detected at receiving, in-process and final acceptance inspection stations.

The outgoing part defect rate, d_F can then be described by the following formula:

$$d_F = d_0(1-E_1)(1-E_2)(1-E_3)$$

where d_0 is the incoming part quality defect rate, E_1 is the receiving inspection efficiency, E_2 is the in-process inspection efficiency and E_3 is the final acceptance inspection efficiency.

The incoming part quality rate is dependent on the part selection criteria, the specified quality level and controls, source inspection and other experience factors with the suppliers. The inspection efficiencies, i.e., the ability of the inspections to detect the part failure modes, depend on such factors as:

- the sample size (acceptable quality level (AQL), 100% inspection, etc.)
- the adequacy of the test equipment calibration program
- the probability that all component parameters are exercised by the test procedures and test equipment
- the probability of inspection error
- the complexity of the item inspected and the inspection and measurement instructions

The formula shows the impact of any one inspection on the outgoing defect rate. For example, with perfect inspection at any one station, i.e., $E = 1$,

$$d_F = 0$$

A key facet of the process and inspection analysis is to determine the efficiencies of the inspections. Inspection efficiencies are expressed in percent ranging from 0 to 100%. Perfect error-free inspection, if possible, would be indicated by 100% efficiency. An inspection may detect (and eliminate) defects by means of either visual inspection or measurements, or combinations of the two. Regardless of the method used, an inspection technique capable of detecting the defect has an associated efficiency. The efficiency level is a measure of the ability of that inspection to detect defects.

Specifically, reliability grows during production as a result of corrective action that:

- Reduces process induced defect rates
  - — manufacturing learning
  - — improved processes
- Increases inspection efficiency
  - — inspector learning
  - — better inspection procedures
  - — incorporation of ESS

As process development and test and inspection efficiencies improve, problem areas become resolved. As corrective actions are instituted, the outgoing defect rate approximates the defect rate of the incoming parts and the outgoing reliability of the hardware system approaches the inherent (design-based) value.

Thus, the process and inspection analysis is an essential part of an effective reliability control and growth process and, as such, would allow management to exercise control, allocate resources and maintain visibility into the production and quality operations, particularly statistical process control, inspection and ESS activities — it can provide an effective and viable means to achieve a mature system prior to fielding.

A process and inspection analysis can be performed on a system or a component during the late development phase on initial production hardware to establish standardized defect rate statistics and generate inspection efficiency factors. This allows process changes, improved inspections, ESS's, and other improvements to be made and assessed prior to full scale production. Similar process analyses can then be performed during full scale production to continuously improve the processes and to assess and control actual outgoing reliability.

Storage Degradation Control

Equipment items and component parts age and deteriorate over long storage and dormant periods due to numerous failure mechanisms. These mechanisms are of the same basic kind as those found in the operating mode, though precipitated at a slower rate. Additionally, many failures which occur during the non-operating storage life are traceable to built-in (latent) production defects rather than specific aging mechanisms. These defects may escape production and become evident after non-operating periods. Protective measures must be applied to isolate hardware systems and their components from non-operating storage deteriorative influences.

Proper protection against damage and deterioration during long-term storage periods involves the evaluation of a large number of interactive factors to arrive at an optimum combination of protective controls. These factors can be grouped into three major control parameters: (1) the level of preservation, packaging and packing applied during the preparation of a material item for shipment and storage, (2) the actual storage environment and (3) the need for and frequency of periodic inspection and functional test or checkout. In addition, the application of environmental stress screening (ESS) beforehand during manufacturing will remove many of the built-in latent defects that lead to failure during storage.

Preservation, packaging and packing (PP&P) is the protection provided in the preparation of a hardware item for shipment and long-term storage. Preservation is the process of treating the corrosible surfaces of an item with an unbroken film of oil, grease or plastic to exclude moisture. Packaging provides physical protection and safeguards the preservative. In general, sealed packaging should be provided for equipment, spare parts and replacement units shipped and placed in storage. Packing is the process of using the proper exterior container to ensure safe transportation and storage. Various levels of PP&P can be applied, ranging from complete protection against direct exposure to all extremes of climatic, terrain, operational and transportation environments to protection against damage that could occur under unfavorable conditions of shipment, handling and storage.

The storage environment can vary widely in terms of protection afforded. It can include warehouse space with complete temperature and humidity control, warehouse space with no humidity and temperature control, sheds and open ground areas that are simply designated for storage.

Scheduled inspection and functional checkout are the key to assuring the actual reliability of an item during long-term storage and dormant periods. Their purpose is to provide data to enable assessment of system reliability and operational readiness, detect deterioration and other deficiencies caused by aging and improper storage methods and provide a basis for reconditioning or condition reclassification. The type, extent and frequency of in-storage inspection are dependent on the deterioration properties of the material.

An effective inspection program requires performance of a thorough and detailed visual inspection to identify defects and deterioration mechanisms due to storage stresses and dormant aging factors, followed by functional checkout to verify operability. Emphasis is placed on inspecting for evidence of deterioration, contamination and corrosion, as well as looking for loose or frozen parts, damaged parts, leakage, excessive moisture (fungus, mildew, rot), and damaged or deteriorated preservation and packaging. The functional test is defined such that it can be applied simply and quickly on storage/dormant sensitive hardware items to assure that they perform satisfactorily and are ready for use. The functional test is not intended to represent a complete and detailed inspection to determine compliance with specified requirements but is designed to verify operability fully utilizing end-item functions to indicate readiness for deployment. The tests can range from a relatively simple checkout of a critical subsystem or assembly to perhaps a full functional test of a complete system.

The US Army, in recognition of the storage deterioration/dormancy problem, has established a program which provides for assessing and controlling the quality of depot stored materiel. The program focuses on in-storage inspection, minor repair, testing, and the preservation, packaging and packing (PP&P) aspects of the stored material.

A major and most significant element within the program is the preparation and implementation of storage serviceability standards (SSS's). These standards consolidate and establish the depot quality control and reliability management procedures for assuring materiel readiness. They contain mandatory instructions for the inspection, testing and/or restoration of items in storage, encompassing storage criteria, PP&P and marking requirements, and time-phasing for inspection during the storage cycle to determine the materiel serviceability and the degree of degradation that has occurred. In the case of shelf-life items, specifically those items whose shelf-life is considered extendible, SSS's are used at the storage level to determine if such items have retained their original characteristics and are of a quality level which warrants extension of their assigned time period.

In order to assure the readiness of stored materiel, three basic inspections are employed at the depot:

(1) Inspection of materiel at receipt
(2) Inspection of materiel in storage
(3) Inspection of materiel prior to issue

These basic inspection stations are depicted in Figure 4–19, which also illustrates the receipt of materiel items from various sources along with materiel handling procedures and storage facilities.

The inspection of materiel in storage (item 2) is comprised of scheduled cyclic inspections and unscheduled special inspections. The SSS's provide instructions necessary for the performance of scheduled cyclic inspections. It should be noted that the basic assumptions of the SSS program are that all materiel when originally placed in storage is ready for issue and that all applicable preservation, packaging and packing (PP&P) requirements as defined by the appropriate technical manual have been met. Thus, the intention of the standards is not to serve as a check function for production or field repair and overhaul operations, but rather to identify, classify and ultimately eliminate failures due to long-term storage.

Scheduled cyclic inspection involves systematically inspecting the

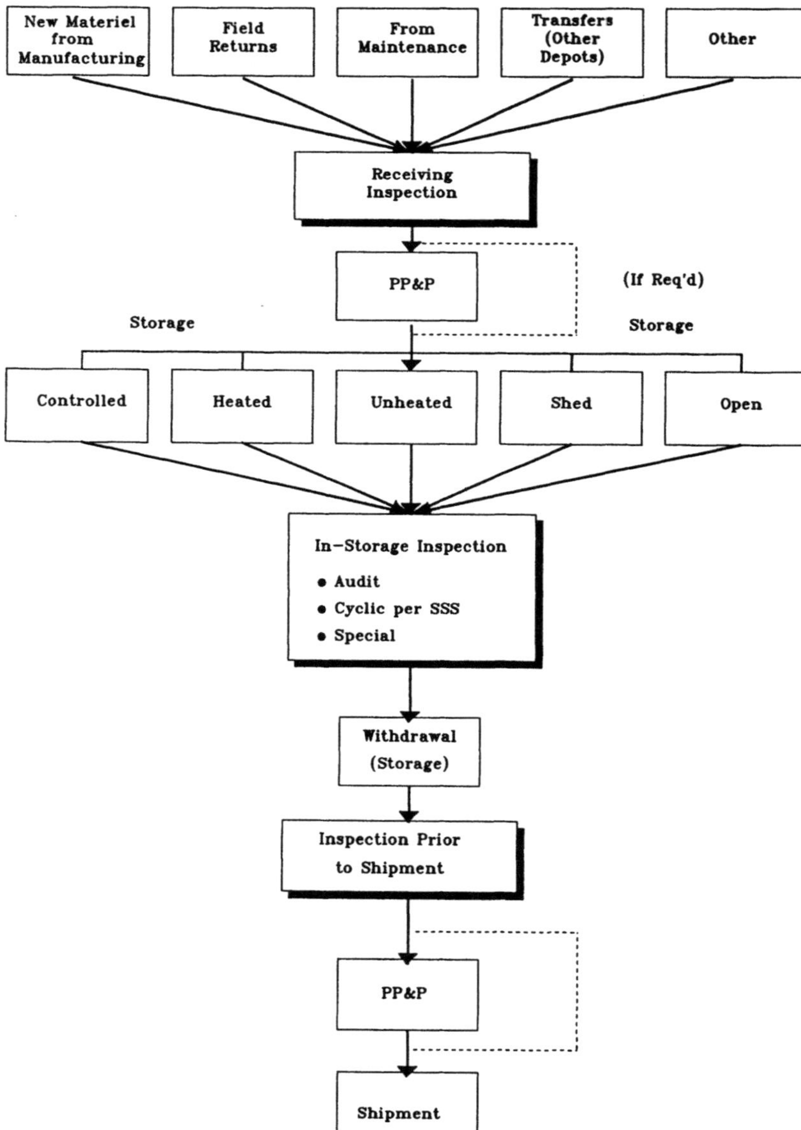

Figure 4–19 Depot Material Handling and Inspection Flow

materiel items for condition degradation, deterioration, corrosion, damage and other deficiencies as induced by improper storage methods, extended periods of storage or inherent materiel deterioration characteristics. The objective is to detect minor deficiencies before they become of major significance, thus providing for corrective actions before the materiel becomes unserviceable or unusable. In this regard, cyclic inspection identifies those items which require corrective packing and packaging or special storage control to assure that they are maintained in a serviceable condition and identifies those which require condition reclassification to a lower degree of serviceability.

Effective and efficient execution of the cyclic inspection system requirements assures that: (1) stored materiel is inspected and reclassified at intervals indicated by the assigned shelf-life code, inspection frequency code or type of storage afforded the materiel (shelf-life materiel will be controlled, regardless of other considerations); (2) quantitative data generated by the cyclic inspections will be thoroughly analyzed, summarized and furnished periodically to management to assist in the elimination of causes for deficiencies; and (3) advanced engineering and statistical techniques are used to ensure economy and cost-effectiveness of the operations.

A special inspection is performed to verify the correctness and accuracy of identity, condition, marking, packaging or other characteristics of a specific item that have become suspect. It is normally initiated as a result of customer complaints, deficiencies discovered in other depot operations (e.g., maintenance, shipping, PP&P), unexpected adverse changes in storage condition or requests from higher authority. Data resulting from these special inspections serve as a vital input to the storage serviceability standards.

To plan and specify optimum in-storage cyclic inspections requires a detailed knowledge of the materiel item, particularly its deterioration properties and risk attributes. The inspections must be practical and maintain an overall cost-effective posture that reflects readily available depot test equipment and skills.

There are, in general, two basic types of in-service cyclic inspections. The first is a subjected visual inspection where acceptance is completely described by codes covering preservation, packing, marking, storage and materiel deficiencies. Figure 4–20 indicates that this inspection is performed at three levels; i.e., at the outer package or container, the inner packing and the item itself. Critical, major or minor defects, either at the time of inspection or if expected by the next inspection, are identified as

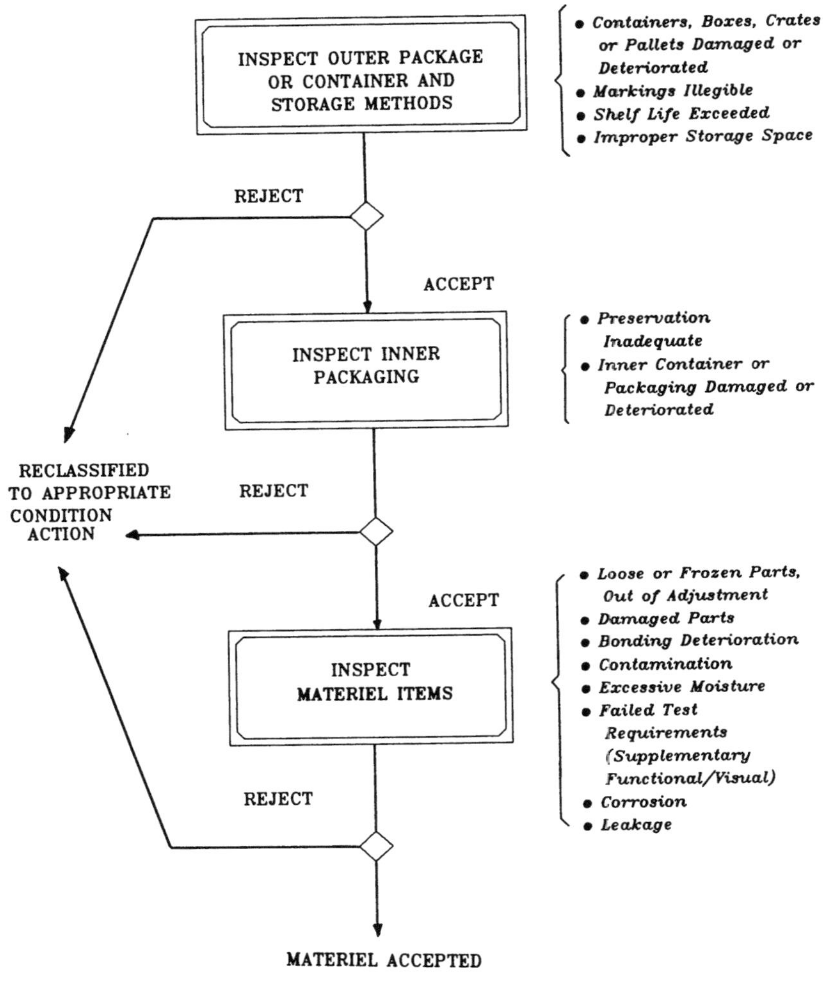

Figure 4–20 Quality Inspection Levels

such and considered as a cause for rejection. Defects of a trivial nature (e.g., nicks, dents or scratches that do not break coatings or paint films) are not considered as cause for rejection of a lot, unless some reduction in usability of function is expected prior to the next scheduled inspection.

The second type of in-storage cyclic inspection involves determining compliance with performance requirements for those items that cannot be inspected adequately by the visual coded checks. This type generally includes functional tests (derived from technical manuals) and/or special, more detailed visual inspections. Inspection procedures complete with acceptance criteria are prepared for these items and included in the standards. Emphasis is placed on defining viable test or checkout procedures based on a 'go'/'no go' concept that can be applied simply and quickly to the stored materiel items to assure that they perform satisfactorily and are ready for service and issuance with only a minimal level of evaluation, support and guidance. These supplementary tests can be applicable to parts, subsystems or complete systems, including shelf-life items as well as other items that are storage sensitive.

The functional tests are designed such that they do not require external and specialized test equipment except common and readily available equipment found at the depots and other installations (DC power supplies, volt–ohmeters, etc.). The functional tests, in general, involve first checking the operational mode of all indicators such as power lights, meters and fault lights, as applicable, and then applying a simple procedure that exercises some or all of the functions to verify operational status. Often, the item could be tested as part of a system.

The functional test procedures for a given item can be derived from a review of the maintenance and/or operating manuals. These manuals describe the operational sequence, the turn-on and shutdown procedure and the operational test and checkout procedure necessary for complete verification of operational status. Consequently, they provide a sound basis for deriving a simple, cost-effective functional test that is suitable for assessing reliability during storage.

The procedure for preparing SSS's focuses on three key parameters:

(1) The preservation, packaging and packing (PP&P) level
(2) The storage level
(3) The inspection frequency.

The procedure allows for the inspection frequencies to be adjusted at the storage facilities if the actual PP&P and storage levels differ from the preferred levels identified in the standards.

One significant factor in the preparation of the SSS's is the assignment of shelf-life codes to all materiel items for stock-keeping purposes. The shelf-life is the total period of time, beginning with the date of manufacture or restorative action, that an item may remain in the combined manufacturer and depot storage system and still remain suitable for issue and use. Shelf-life is not to be confused with service-life, which is a measurement of anticipated total in-service time.

A shelf-life item is an item of supply possessing deteriorative or unstable characteristics to the degree that a storage time period must be assigned to assure that it will perform satisfactorily in service. For the medical commodity, the definition of a shelf-life item refers only to potency expiration dated items. There are two types of shelf-life items:

(1) Type I Shelf-Life Item
An item of supply which is determined, through an evaluation of technical test data and/or actual experience, to be an item with a definite non-extendible period of shelf-life.
(2) Type II Shelf-Life Item
An item of supply having an assigned shelf-lifetime period which may be extended after the completion of prescribed inspection test and restoration action.

A code is assigned to a shelf-life item to identify the period of time beginning with the date of manufacture and terminated by a date by which the item must be used or be subjected to inspection/test/restorative or disposal action. A code is also used to signify the remaining shelf-life of an item.

Storage serviceability standards are prepared for Type II shelf-life items as well as for indefinite shelf-life items possessing storage sensitive properties. In the case of Type II shelf-life items, SSS's are used to determine:

(1) Inspection, test, and restorative requirements and criteria.
(2) Whether or not shelf-life items have retained sufficient quantities of their original characteristics and are of a quality level which warrants extension of their assigned storage time period.
(3) The authorized length of the time period extension (remaining shelf-life).

The identification and assignment of shelf-life codes to materiel items thereby becomes a significant factor in defining and controlling storage serviceability standards requirements.

An even more significant factor is the capability of an SSS to effectively increase the shelf-life of an item through the performance of specified periodic cyclic inspections and tests. This capability is illustrated in Figure

4–21. The figure shows, as an example, an item with a shelf-life of 24 months effectively being extended to a storage period of 84 months. This is accomplished by performing a number of periodic inspections and restoration actions, t_i; five in this example, each performed at a reduced interval of time between cycles. After each inspection/restoration action the shelf-life period is extended (but reduced from the initial period) and the inspection interval adjusted accordingly — the extent of the adjusted shelf-life period is dependent on the deterioration slope. The extent of deterioration recovery is dependent on the amount of deterioration and the effectiveness of the restoration action. The resultant increase (represented by the sloped line) in the upper set of curves effectively extends the 24 month shelf-life item (without inspection or restoration action) to 84 months when inspection and restoration action is applied, or by a factor of 3.5. As shown, the length of the standard shelf-life is dependent on:

(1) Material deterioration/stability properties (slope of curve)
(2) Criticality (establishes the must-issue level)
(3) Effectiveness of inspection and restoration action

The reduction in the time interval depicted between each inspection period is required to compensate for the difficulty in successive restoration actions needed to maintain the materiel at the highest level of readiness and may be regarded as the storage deterioration rate of the item.

A decision logic process has been developed to help identify what materiel items should be covered by standards and the assignment of corresponding shelf-life codes. The decision logic, illustrated in Figure 4–22, is designed to lead to the appropriate shelf-life code for the item under consideration. The logic terminates with one of the following three shelf-life code assignment decisions, previously discussed:

(A) Definite Shelf-Life Type I (non-extendible)
(B) Definite Shelf-Life Type II (extendible)
(C) Indefinite Shelf-Life 0 (non-deteriorative and not storage sensitive)
Indefinite Shelf-Life 0 (non-deteriorative, but storage sensitive)

The answers to the questions are recorded on a worksheet such as that illustrated in Figure 4–23 or entered directly into a computer system.

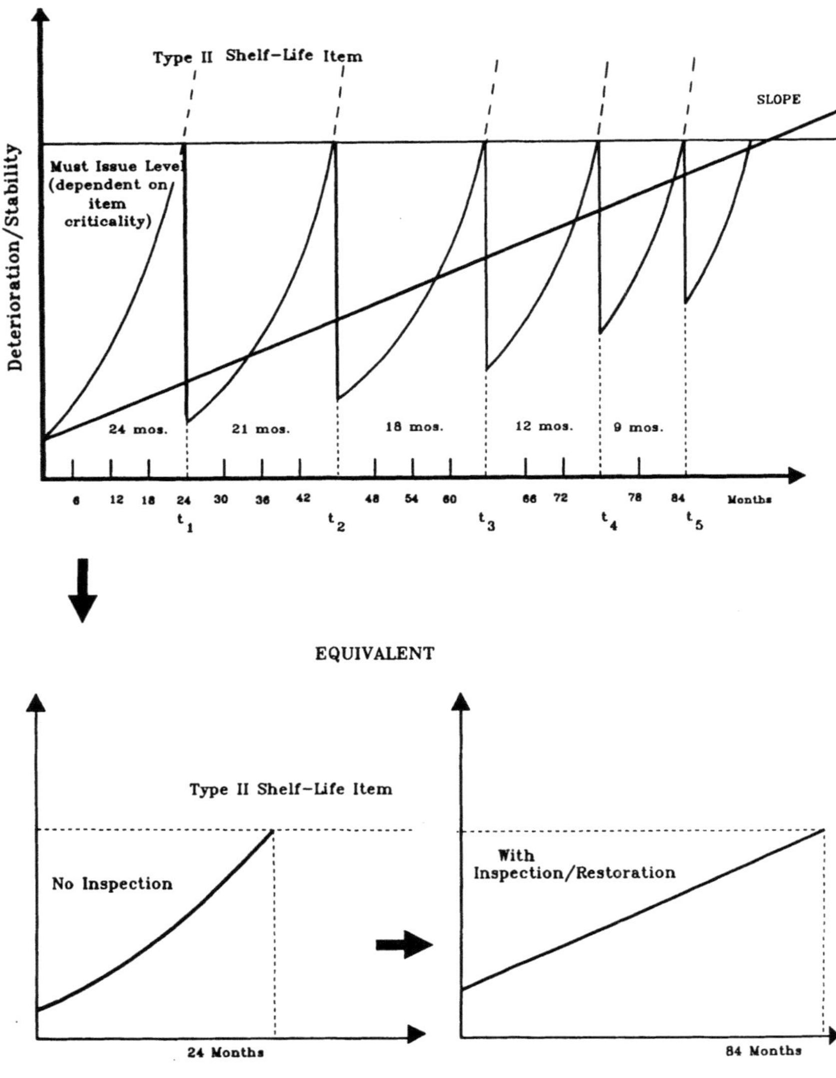

Figure 4–21 Effective Increase in Material Shelf-Life through Cyclic Inspection

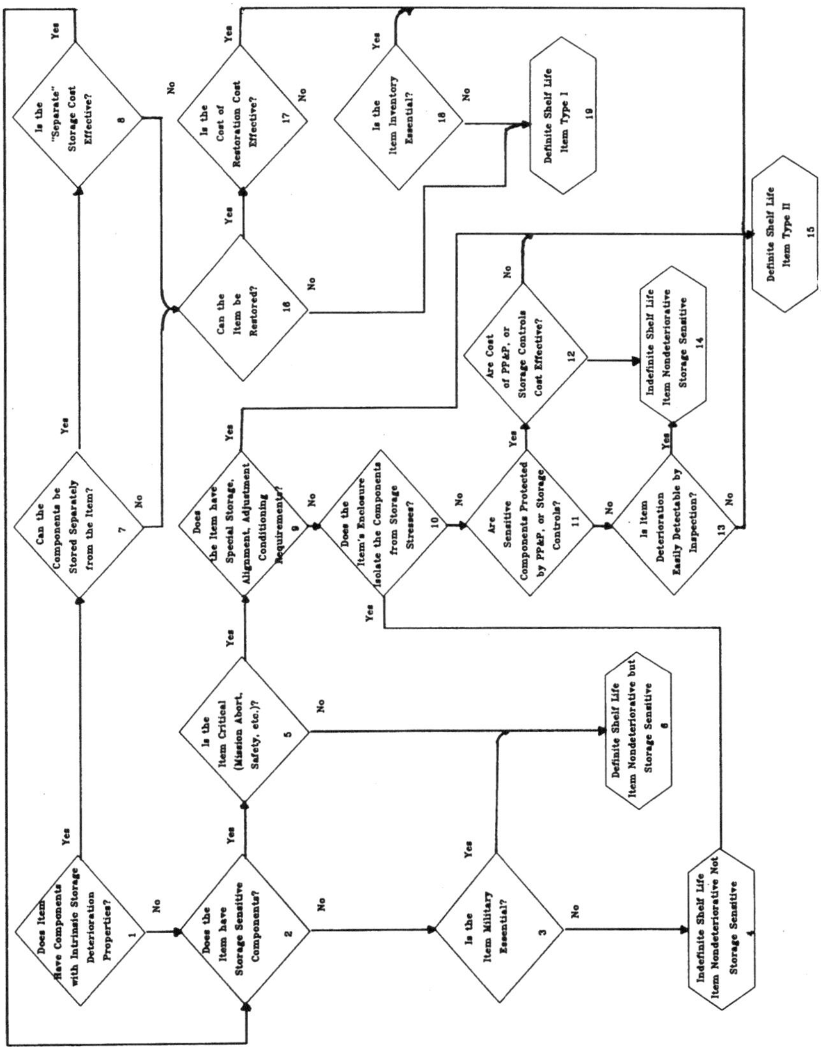

Figure 4–22 Shelf-Life Decision Logic Diagram

SHELF LIFE ASSIGNMENT PROCESS WORKSHEET			
MAJOR ITEM	MAJOR ITEM NSN	PREPARED BY	
ITEM BEING EVALUATED		ORGANIZATION	
ORIGINAL ASSIGNED SHELF LIFE CODE:	REASSIGNED SHELF LIFE CODE:	DATE	REVISION
ACTION TO BE TAKEN			

LOGIC QUESTION/DECISION	YES	NO	INFORMATION SUMMARY
1. Does Item have Components with Intrinsic Storage Deterioration Properties			
2. Does the Item have Storage Sensitive Components			
3. Is the Item Military Essential			
4. Indefinite Shelf Life Item Non-deteriorative not Storage Sensitive			
5. Is the Item Critical (mission abort, safety etc.)			
6. Definite Shelf Life Item Nondeteriorative But Storage Sensitive (code x)			
7. Can the Components be Stored Separately from the Item			
8. Is the "Separate" Storage Cost Effective			
9. Does Item have Special Storage,Alignment, Adjustment Conditioning Requirements			
10. Does the Item's Enclosure Isolate the Components from Storage Stresses			
11. Are Sensitive Components Protected by P,P&P, or Storage			
12. Are Costs of P, P&P, or Storage Controls Cost Effective			
13. Is Item Deterioration Easily Detectable by Periodic Inspection			
14. Indefinite Shelf Life Item Nondeterior-ation Storage Sensitive			
15. Definite Shelf Life Item Type II			
16. Can the Item be Restored			
17. Is the Cost of Restoration Cost Effective			
18. Is the Item Inventory Essential			
19. Definite Shelf Life Item Type I			
20. Is Storage Serviceability Standard Required			

Figure 4–23 Shelf-Life Assignment Process Worksheet

Maintenance Degradation Control

The application of the RCM decision logic translates the design characteristics of the system and its failure modes into specific hard-time replacement (HTR), on-condition maintenance (OCM) and condition monitoring (CM) tasks to be executed at the appropriate maintenance organization. However, as previously indicated, reliability can be significantly degraded by poorly executed maintenance when the system is deployed and enters the operational and maintenance phase of its life-cycle. In recognition of the degradation, recent trends in system design have been directed toward reducing the amount of human involvement by increased use of built-in test equipment and other ease of maintenance features. Also, there are techniques available (e.g., MIL-HDBK-472) which quantify, in terms of time factors, the impact of maintenance skill levels and the designed-in maintenance features in order to identify areas where improvements can be directed to reduce this degradation.

The growth of reliability toward the level designed and built into the system relies strongly on the ability to detect and identify defects inherent in the basic parts (or materials) or which may be induced during maintenance. Tests to uncover such defects need, for the most part, to be performable by average skill level maintenance personnel; not require extensive, high cost instrumentation; and, of course, not degrade the hardware being tested.

Many complex systems or components contain incipient defects which could eventually cause catastrophic failures in the field. While these incipient defects could be detected by complete disassembly and inspection, such procedures are frequently not possible under field conditions. In addition, disassembly, even by skilled servicemen, often introduces more problems than are solved. Even more difficulties are faced when personnel with the necessary skills are not available.

The sensing of effects (such as acoustical noise, mechanical vibration, visible smoke or electromagnetic noise) or phenomena not directly related to the principal functions of the system or subsystem offers opportunities which will augment, supplement or complement existing non-destructive inspection (NDI) approaches. Often, a precursor of failure can be detected without the requirement for disassembly and can be remotely detected without the need for contacting or built-in sensors. The precursors are generally sufficiently simple that they can be inspected and evaluated in the field by relatively unskilled personnel.

In the past, special equipment, such as electronic sensing and for processing of vibrations, has in some instances been successful in

TABLE 4-2
Representative List of Malfunctions

Subsystem	*Malfunction*	
Engine	Internal seizure	Defective fuel injectors
	Incorrect oil	Defective turbocharger
	Dirty oil filter	Defective turbocharger regulator
	Defective oil pump	Fan belt damage or out of adjustment
	Defective fuel pump	Defective thermostat
	Improper fuel	Low coolant level
	Water in fuel	Damaged or dirty radiator
	Dirty fuel filter	Defective water pump
	Loose or restricted fuel lines and fittings	Cracked manifold
	Restrictive air intake system	Burned or blown manifold gaskets
	Dirty air filter	Restricted exhaust system
Electrical	Low battery fluid	Faulty switches
	Bad battery	Defective rheostats
	Loose connections	Defective brushes
	Generator defective or out of limits	Faulty thermostats
	Defective regulator	Defective gyro
	Loose generator drive belt	Defective or jammed solenoid
	Defective relays	Defective ignitor
Hydraulic	Blocked or restricted lines	Incorrect gas volume in accumulator bottle
	Leaks in fittings	Clogged or defective filter
	Low oil reserve	Defective solenoid valve
	Defective pressure relief valve	Defective check valve
Transmission	Defective steer valve body	Control linkage broken or out of adjustment
	Internal and external oil leaks	Low oil level
	Low oil level	Oil collar clogged
	Faulty oil pressure regulator	Faulty oil pressure regulator
	Defective oil pump	Defective oil pump
	Incorrect grade of oil	Restricted oil lines
	Restricted oil lines	Incorrect grade of oil
	Defective oil temperature gauge	Brakes worn or out of adjustment
	Faulty steer relay valve	Faulty clutch selector valve
	Piston seal ring leakage	Excessive vent line pressure
	Output clutch seal ring leakage	Internal binding
		Reverse range clutch seal ring leakage
Tracks and suspension	Broken drive shaft	Defective idler wheel hub bearings
	Improper track adjustment	Worn sprockets
	Worn or distorted tracks	Air in lockout cylinders
	Defective road wheel bearings	Oil leak in lockout cylinders
	Defective shock absorbers	Defective pressure reducer valve
	Stuck or restricted lockout cylinder piston	
Armament	Defective firing pin	Low nitrogen pressure
	Defective sear spring	Defective retracting control valve
	Defective hammer	Defective lines of fittings
	Worn extractor	Scored cannon-mount surfaces
	Faulty gas check pad	Defective recoil cylinder
	Scored breechlock threads	Defective relief valve
	Defective counterbalance assembly	Worn rifling
	Excessive nitrogen pressure in recuperator	Jammed or defective breech drive solenoid
	Low oil in counter recoil systems	Ejector worn, defective, or out of adjustment
	Worn or defective actuating mechanism	Worn or defective recuperator cylinder seal

TABLE 4-3
Generic Failure Indicators/Mechanisms

Mechanical	
Adherence,[a] sticking, seizure	Leakage[a]
Looseners,[a] backlash	Position shift[a]
Rupture,[a] broken fracture	Creep relaxation
Fatigue,[a] brinelling	Deflection misalignment
Wear,[a] adhesion, abrasion erosion or scoring	Buckling
Vibration	Heating[a]
Timing[a]	Shock
Torque	Slippage[a]
	Pressure[a]
	Cavitation
Chemical	
Corrosion	Contamination
Deterioration	Electrolysis
Change of state, freezing, boiling	Photocatalysis
Electrical	
Deterioration,[a] insulation and contact	Timing[a]
Electrolysis	Voltage
Current	Demagnetization
Heating[a]	Photocatalysis
	Depolarization

[a]Generic failure mechanisms.

identifying potential bearing failures. Chemical analysis of wear debris in engine oil has been, with varying degrees of success, useful in identifying precursors of failure in reciprocating engines. The use of heat and infrared has also proved promising in certain cases.

A representative listing of malfunctions taken from several maintenance manuals is shown in Table 4–2. Many of these malfunctions can occur in more than one subsystem. Furthermore, these malfunctions could be classified according to the type of generic failure mechanisms involved. Table 4–3 shows a list of generic failure mechanisms that cover some of the failures that can occur. Many of these failure mechanisms can be observed directly, while some produce other effects which are indicators of failure. The US Army has developed a specific OCM program, called ACE (Airframe Condition Evaluation), to assess the condition of fielded aircraft based on indicators of failure. ACE is described in Chapter 5.

4.6 ENVIRONMENTAL STRESS SCREENING (ESS)

The purpose of environmental stress screening (ESS) is to compress the early mortality period of a new hardware item, or an item which has been overhauled or completely rebuilt, and reduce the item's failure rate to an 'acceptable' level prior to deployment. Screening involves the application of stress on a 100% basis for the purpose of revealing design, as well as workmanship and process induced, defects without weakening or destroying the hardware item or exceeding its design capabilities and, thus, decreasing its useful life. The application of selected screens serves to reveal defects which ordinarily would not be apparent during normal quality inspection and testing.

There are a large number of ESS's and screening sequences that can be applied to remove defects induced during initial hardware fabrication or during the depot repair and overhaul process. A thorough knowledge of the hardware to be screened and the effectiveness and limitations of the various available screening procedures is necessary to plan and implement an effective screening program. The tailoring of the applied screens and subsequent failure analysis and corrective action efforts determines the rate of stimulating the defects and the resulting degree of reliability improvement.

Environmental stress screening can be applied at the part, intermediate and end-item or system level. In order to detect and eliminate most intrinsic part defects, initial screening is conducted at the part level. This is the most cost-effective level at which to identify defects. Certain part defects, however, are more easily detected as part of an intermediate or an end-item or a complete system level ESS. Also, assembly defects, such as cold or missing solder joints and connector contact defects, exist and, therefore, can be detected only at higher levels of assembly. As a general rule, screens for known latent defects should be performed as early in the manufacturing or overhaul process as is possible — at the point where higher stress levels and more cost-effective screens can be safely applied.

The idealized production process starts with screened parts procured and received to a predetermined level of quality. Selected parts, such as microcircuits and semiconductors, are then rescreened as part of receiving inspection. ESS is then applied, as required, at the different levels of assembly. ESS failures are analyzed and the results used to identify appropriate modifications to the manufacturing or overhaul process and to reduce, if possible, the overall ESS burden. All ESS results, including failure rates, failure modes and time-to-failure data, are incorporated into a dynamic real-time data base from which the effectiveness of the ESS program is continuously assessed and modified as necessary.

An ESS plan is prepared that reflects the sequence of screens considered optimum based on anticipated fall-out rates, cost-effectiveness and an ESS specification process. For a new development item, ESS is generally applied first to preproduction hardware to facilitate the incorporation of changes and refinements to the plan. Once the plan has been refined a specification is prepared and incorporated into the technical data package, and the program is moved to the manufacturing. For a fielded item the specification is incorporated into its depot maintenance work requirements and used for overhaul operations. The initial planning of an ESS program takes into account the effectiveness and the economic choices between part, intermediate and end-item level screens and the parameters that must be considered.

A part level ESS is relatively economical and can be incorporated into supplier specifications. It has the potential for maximum cost avoidance, particularly when applied to complex microcircuits and other high technology devices where reliability is largely dependent on fabrication techniques and process control. Screen stress levels can be matched to requirements, which, in general, enables the safe application of higher and more effective stress levels to remove known part defects. Part screens, however, have no impact on the control of defects introduced during subsequent phases of assembly or on system level problems.

The application of part level ESS can provide cost-effective, qualified parts that meet or exceed reliability and quality targets for assembly into complex systems or components. However, screening can be inefficient and costly if the screening stresses are not carefully designed to 'attack' the specific defect present in the population. Among the risks associated with part level ESS are:

(1) Screen may damage good parts.
(2) The nature of defects may change with time.
(3) Screens in use may not attack all defects present.
(4) Screen is being used for defects no longer in the population.
(5) Population of defects may vary for different production lines.
(6) Useful life may be reduced.

An intermediate level ESS is more expensive, but can remove defects introduced at the assembly or unit level as well as those intrinsic to the parts. Because of the several part types incorporated into an assembly or unit, somewhat lower screen levels must be applied. Generally, special temperature cycling facilities are required as well as special automatic test equipment (ATE). In general, some amount of ATE is employed in virtually all large-scale screening programs. Automatic testing can not only perform functional testing rapidly after screening of complex subassemblies (or units), but is also effective during screening in the detection of evasive faults. The latter consist of marginal performance, timing problems and other defects arising from part interactions. The extent of the facilities and equipment needed is dependent on the test conditions specified.

Temperature cycling is a highly effective ESS which can reveal workmanship defects induced during assembly as well as those intrinsic part defects which escaped detection during part level screening. With electronic assemblies, temperature cycling is performed specifically to reveal assembly defects (e.g., delamination, fracture and insulation cracking), part/board bond separation, solder problems (cracking opens, etc.) and part defects. Figure 4–24 illustrates the environmental conditions and profile under which a temperature cycle can be performed. The actual number of cycles employed is dependent upon the density and technology of the parts. The number of cycles initially set represents a baseline which is adjusted during the course of the production or overhaul process to reflect the results of subsequent higher level screens or in-service performance.

A temperature-cycle ESS can be completely specified by the control parameters (circled on the figure) and the values assigned to them. The significance and impact of each of these parameters are as follows:

(1) *Temperature Range* — In general, the wider the temperature range, the more defects that will be exposed. High and low temperatures are limited by the maximum ratings of the material and the availability of environmental chambers.
(2) *High/Low Temperature Dwell Time* — The portion of the cycle that the hardware under test remains at the high or low temperature. Dwell time must include ample time for internal parts to stabilize (generally within 2°C of the specified temperature extremes) and to perform any required measurements.
(3) *Transfer Time* — The total time to transfer from the specified low temperature to the specified high temperature, or the reverse. In general, the higher the rate of change up to 15 or 20°C /minute (chamber rate), the more effective the screening.

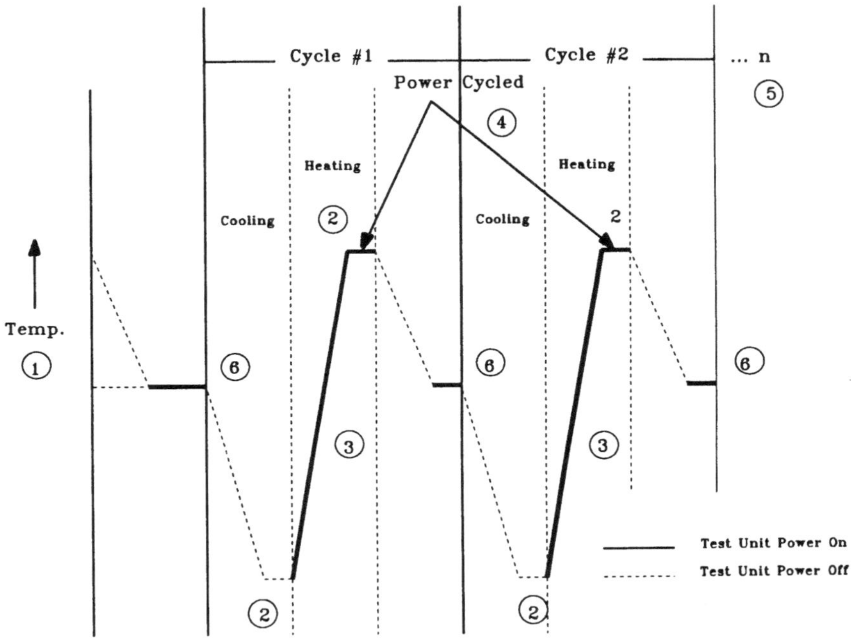

Figure 4–24 Time–Temperature Screen Test Model

(4) *Power Cycling* — The switching of the hardware's power on and off at specified intervals during the dwell time. In general, power is turned on at the completion of the low temperature period and kept on through transition to the high temperature period. During on periods the hardware should be monitored to detect intermittents.

(5) *Number of Cycles* — The number of cycles can vary from three to over 25 depending on hardware characteristics, assembly level and other considerations.

(6) *Measurement* — Measurements are conducted at points in the cycle where environmental conditions are stabilized and, therefore, definable. The two basic classes of measurements are:

- Visual inspection (V) — performed to detect physical evidence of defects or damage to the hardware.
- Functional testing (F) — performed to measure a limited number of critical parameters to assure that the hardware is

operating properly. Complete functional testing includes visual inspection and is performed prior to and after screening.

Different time–temperature ESS's can be specified through variations of these parameters. For example, a temperature-cycle screen may require a transfer time (3) of 5°C/minute whereas a thermal shock screen would require the transfer time to be less than 10 seconds with all other parameters the same. A power cycle burn-in can be specified by requiring a constant elevated temperature (parameter 1) for a specified dwell time (parameter 2) and generally power cycled at specified intervals (parameter 4).

An end-item level ESS can remove defects introduced at all levels of fabrication or overhaul. At this point in the process, the permissible stress level may not adequately exercise certain specific parts. However, these high level ESS's are considered important, even if it is thought that the lower level screens have eliminated all defective parts and assembly or unit level defects. The installation of the remaining components and the assemblies and units into the end-item or into the final system cannot be assumed to be free of failure-producing defects. Good parts may be damaged in final assembly, workmanship errors can occur, and system level design defects may be present. Typical reliability and quality defects found in the final system include overstressed parts, improper solder joints and cracked wires due to insufficient strain relief. Despite QC inspections, equipment has even been produced with parts missing. Special burn-in and temperature cycling facilities are required, but little expensive specialized test equipment is necessary because the completed end item or system can be exercised in a nearly self-testing mode.

As with intermediate level testing, a temperature-cycle ESS or a high temperature burn-in can be effective in revealing workmanship and process induced defects as well as those part defects which escape detection at the part and intermediate level screens.

The use of random vibration, which provides simultaneous excitation of many modes in contrast to the single frequency sine vibration test, has proven to be an effective screen for isolating workmanship defects found in electronic systems. A screening program developed by the US Navy[16] incorporates random vibration as a manufacturing screen. It requires that the equipment under test be hard-mounted to a shake table capable of reproducing random vibration having the power spectral density characteristics shown in Figure 4–25.

A key step in planning an effective ESS program is the identification of

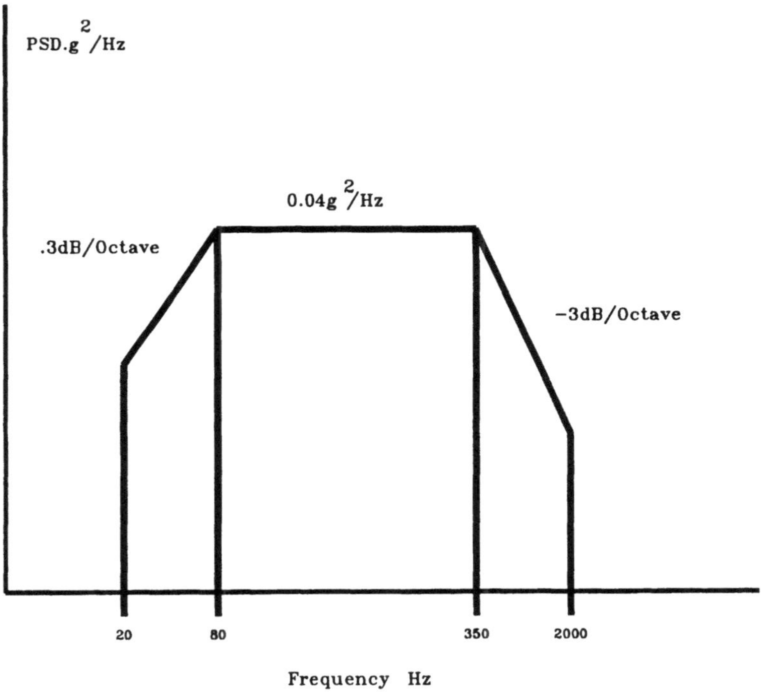

Figure 4–25 Random Vibration Spectrum (taken from NAVMAT P-9492)

the kinds of failure modes that can occur and the assembly level at which they may be induced. An appropriate ESS is that which is most effective in accelerating the identified modes, whether they are intrinsic to the part or induced by the manufacturing or overhaul process. Some of the more common screens are listed below, with an indication of their effectiveness.

Temperature Cycling — Extremely effective at all levels of assembly; reveals part/Printed Circuit Board (PCB) defects, solder problems, bond separations, tolerance drifts, mismatches and changes in electrical characteristics.

High Temperature Burn-in (Power Cycling) — Effective at all levels of assembly, will reveal time/stress dependent part and process defects.

Vibration, Random — Effective primarily at unit or end-item level; reveals solder problems, part/PCB defects, connector contact problems, intermittents, loose hardware and structural problems.

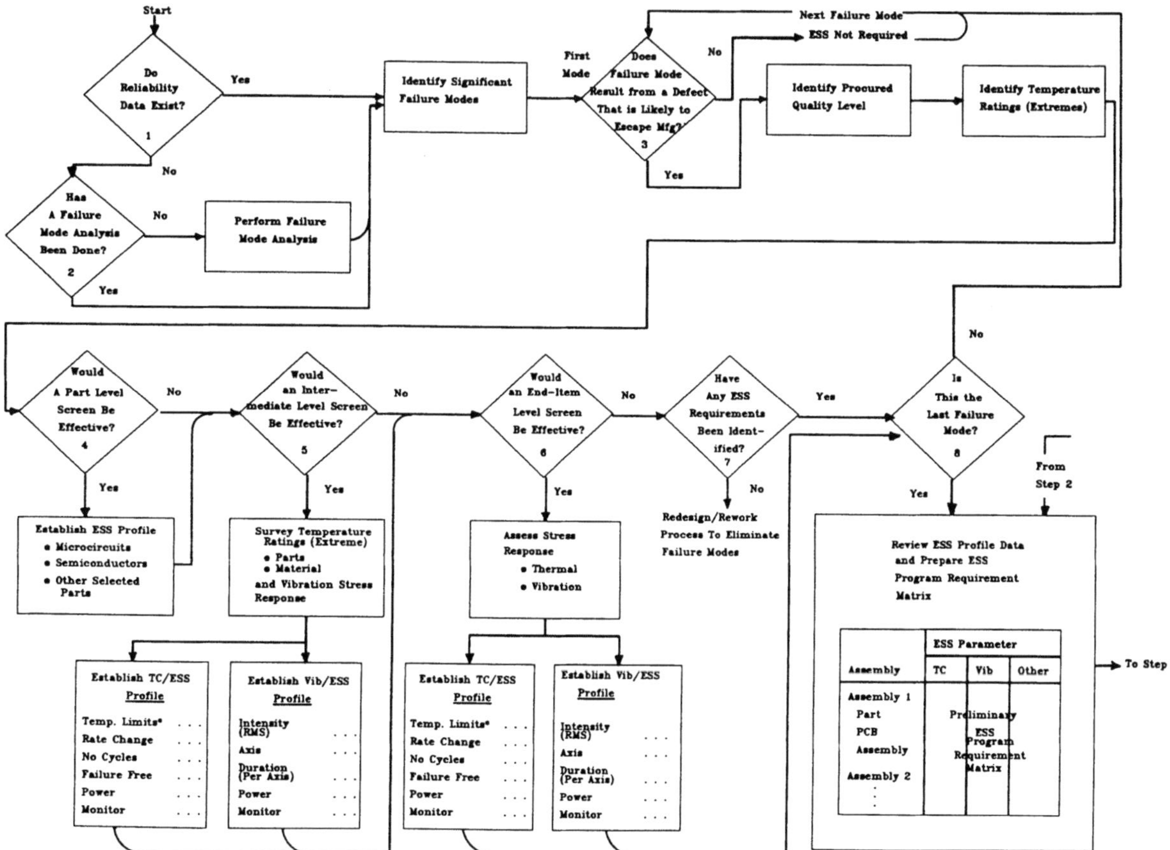

Figure 4–26 ESS Decision Logic: Step 1 ESS Profile

High Temperature Storage — Relatively inexpensive screen that can be applied at any level of assembly to reveal time/dormant stress (non-electrical) dependent defects.

Thermal Shock — Relatively simple screen that can be applied at the part or assembly level to reveal cracking, delamination and electrical changes due to moisture or mechanical displacement.

Vibration, Sine Fixed Frequency — Applied at end-item level to reveal loose hardware, connector contact problems and intermittents.

The planning process involves exercising various options and comparing expected failure or fall-out data with the number of possible defectives that may escape as estimated from an analysis of production inspection failure data and field data on similiar systems and equipment.

The failures that result from screening provide an indication that the process is effective and that potential field failures are being removed from the hardware. If screening shows very few failures, it is either insufficiently severe, or the item being screened has very few defects. The failure rate is a measure of the number of parts, subassemblies or units deviating from specification.

Experience has shown that the failure rates can vary widely depending on hardware technology, complexity, process variance (lot by lot) and the nature of the ESS applied. It should be noted, for example, that, if a large number of microcircuits are used in a subassembly, removal of even a small number of defects at the part level can have a significant economic impact. It also should be noted that failure rates can vary widely from lot to lot. It is not uncommon for a mature and normally reliable device (where most lots typically show no defects) to occasionally have a lot with an extremely high failure rate. The ESS can prevent the bad lot from causing severe problems later in production or overhaul.

Figure 4–26 and Figures 4–28 and 4–30 presented later in this section provide a detailed decision logic that can be used to facilitate the establishment of the most cost-effective ESS program for a hardware item. Step 1 of the ESS decision logic process, shown in Figure 4–26, is to develop preliminary ESS requirements based on complexity, part technologies, maximum ratings and thermal/vibration response considerations for the assemblies under evaluation. The answers to the Step 1 questions guide the development of the preliminary ESS requirements. The results are recorded on the worksheet illustrated in Figure 4–27. The worksheet is keyed to the questions and tasks identified in the decision logic diagram for Step 1. At the completion of this step a preliminary ESS program requirements matrix is developed.

<table>
<tr><td colspan="10">ESS PROCESS WORKSHEET (STEP 1) PAGE OF</td></tr>
<tr><td>ASSEMBLY:</td><td colspan="3">PREPARED BY:</td><td>PREPARING ORGANIZATION:</td><td>DATE:</td><td colspan="4">PART/PCB DESCRIPTION:</td></tr>
<tr><td colspan="10">(1&2) PART/PCB FAILURE MODES
A. B. C. D.</td></tr>
<tr><td>LOGIC QUESTION</td><td>FM</td><td>Y</td><td>N</td><td>COMMENTS</td><td colspan="5">TASK RESULTS</td></tr>
<tr><td rowspan="4">(3) Do Failure Modes Result From Latent Defects That Are Likely To Escape Manufacturing Test/Inspection?</td><td>A</td><td></td><td></td><td></td><td colspan="5" rowspan="2">Procured Quality Level</td></tr>
<tr><td>B</td><td></td><td></td><td></td></tr>
<tr><td>C</td><td></td><td></td><td></td><td colspan="5" rowspan="2">Part Temperature Ratings (Extreme)</td></tr>
<tr><td>D</td><td></td><td></td><td></td></tr>
<tr><td rowspan="4">(4) Would A Part Level Screen Or Rescreen Be Effective?</td><td>A</td><td></td><td></td><td></td><td colspan="5" rowspan="4">Part Level ESS Profile</td></tr>
<tr><td>B</td><td></td><td></td><td></td></tr>
<tr><td>C</td><td></td><td></td><td></td></tr>
<tr><td>D</td><td></td><td></td><td></td></tr>
<tr><td rowspan="7">(5) Would An Intermediate Level Screen Be Effective?</td><td rowspan="2">A</td><td rowspan="2"></td><td rowspan="2"></td><td rowspan="2"></td><td colspan="5">Temperature Ratings (Extremes)</td></tr>
<tr><td colspan="5">PCB Temperature Cycle ESS</td></tr>
<tr><td rowspan="2">B</td><td rowspan="2"></td><td rowspan="2"></td><td rowspan="2"></td><td>Temp. Limits (°C)</td><td>Rate of Change (° C/min.)</td><td>No Cycles/ Failure Free</td><td>Power</td><td>Monitor</td></tr>
<tr><td></td><td></td><td></td><td></td><td></td></tr>
<tr><td rowspan="2">C</td><td rowspan="2"></td><td rowspan="2"></td><td rowspan="2"></td><td colspan="5">PCB Vibration ESS</td></tr>
<tr><td>Intensity (grms/g ²/Hz)</td><td>Axes</td><td>Duration (min.per axis)</td><td>Power</td><td>Monitor</td></tr>
<tr><td>D</td><td></td><td></td><td></td><td></td><td></td><td></td><td></td><td></td></tr>
<tr><td rowspan="7">(6) Would An End-Item Level Screen Be Effective?</td><td rowspan="2">A</td><td rowspan="2"></td><td rowspan="2"></td><td rowspan="2"></td><td colspan="5">Assembly Stress Response</td></tr>
<tr><td colspan="5">Assembly Temperature Cycle ESS</td></tr>
<tr><td rowspan="2">B</td><td rowspan="2"></td><td rowspan="2"></td><td rowspan="2"></td><td>Temp. Limits (° C)</td><td>Rate of Change (° C/min.)</td><td>No Cycles/ Failure Free</td><td>Power</td><td>Monitor</td></tr>
<tr><td></td><td></td><td></td><td></td><td></td></tr>
<tr><td rowspan="2">C</td><td rowspan="2"></td><td rowspan="2"></td><td rowspan="2"></td><td colspan="5">Assembly Vibration ESS</td></tr>
<tr><td>Intensity (grms/g ²/Hz)</td><td>Axes</td><td>Duration (min.per axis)</td><td>Power</td><td>Monitor</td></tr>
<tr><td>D</td><td></td><td></td><td></td><td></td><td></td><td></td><td></td><td></td></tr>
<tr><td rowspan="4">(7) Have Any ESS Requirements Been Identified?</td><td>A</td><td></td><td></td><td rowspan="4">If No Redesign Hardware/ Rework Process To Eliminate Failure Mode Or To Minimize Underlying Defect</td><td colspan="5" rowspan="4"></td></tr>
<tr><td>B</td><td></td><td></td></tr>
<tr><td>C</td><td></td><td></td></tr>
<tr><td>D</td><td></td><td></td></tr>
<tr><td rowspan="4">(8) Is This The Last Part/PCB Failure Mode Sequence?</td><td>A</td><td></td><td></td><td rowspan="4">If No Repeat Process Until All Failure Modes Have Been Evaluated, If Yes Prepare Preliminary ESS Program Matrix</td><td colspan="5" rowspan="4"></td></tr>
<tr><td>B</td><td></td><td></td></tr>
<tr><td>C</td><td></td><td></td></tr>
<tr><td>D</td><td></td><td></td></tr>
</table>

Figure 4–27 Process Worksheet (Step 1)

A list of significant failure modes is generated from questions 1 and 2. As shown in the diagram, if experience data do not exist, a failure mode analysis is performed to identify those part failure modes considered most critical. The US Army aircraft safety analysis model described in Chapter 6 can be used to determine part failure mode criticality. The critical part failure modes are then grouped for processing in sequence through the remainder of Step 1. The remaining Step 1 questions are repeated below with some guidance to help in answering the questions.

Question 3: *Do failure modes result from latent defects that are likely to escape manufacturing*?

This is determined through a review of the failure modes, the failure rates and hardware complexity/technology factors in light of the planned manufacturing and inspection process. ESS is not required for failure modes which do not result from latent defects. If ESS is required, the procured quality level of the part/PCB and the maximum part temperature rating must be identified.

Question 4: *Would a part level screen or rescreen be effective*?

Screens for known failure modes should be performed at the lowest possible level where higher stress levels are permissible and more cost-effective screens can be applied. The following guidelines will aid in determining if part level ESS is effective in stimulating the failure mode:

1. Screens are very effective on all active components. They are especially important with new high technology parts: e.g., hybrids, microprocessors, large scale integrated circuits and special mechanical items. Unfortunately, it is both difficult and expensive to dynamically test many of these items at incoming inspection.
2. ESS is also effective on other part types for which subsequent test and field experience shows excessive failure, as well as on mature parts to assure consistent component quality.
3. Screens are applied to expose specific failure mechanisms unique to a given part type (and lot variance) that may not be exposed by standard screens. The screens are selected by evaluating their effectiveness in exposing anticipated failure mechanisms.

A 'YES' answer means that a screening profile must be established. Microcircuits, for example, can be procured to specified levels of quality as defined by US MIL-STD-883, 'Test Methods and Procedures for

Microelectronics'.[17] This standard provides screening and inspection test specifications for optional levels of microcircuits' quality. Included are specifications for internal visual precap, stabilization bake, temperature cycling, centrifuge, hermetic seal, electrical parameter verification and external visual inspection. The higher quality levels include screening based on thermal and mechanical shock. They also include a longer, 240 hour, burn-in followed by a special 72 hour reverse bias burn-in, and radiographic examination.

In specifying part level ESS requirements, the ESS sequence should be arranged, if possible, so that those individual screens that would most likely precipitate the highest number of defects are performed first — thus allowing early notification to minimize losses and delays. The supplier's normal commercial quality ESS program should be used if considered adequate.

In general, microcircuits, semiconductors and other critical parts should be rescreened at incoming inspection. This would include verifying the functional performance of all lots of microcircuits and semiconductors at both low and high temperature. Only those component parts (e.g., some commercial components) that are specified at room ambient only should also be rescreened at room ambient temperature. Normally, the high temperature for performance verification is either +125 or +150°C, and the low is - 55°C. The specified rescreening temperature limits are to be set such that they do not exceed the part's specification limits (maximum ratings). The characterized upper temperature limit of most components and materials is at least 125°C.

On a sample basis, the performance of each lot of diodes should be verified as described above. Additionally, all lots of microcircuits, transistors, diodes, electromechanical relays, switches and circuit breakers should be subjected to destructive physical analysis, solderability and hermeticity. Lot tolerance percent defectives (LTPD's) should be consistent with normal incoming practices.

For those electromechanical relays, switches and circuit breakers that exhibit a unsatisfactory failure history, either in factory test or field, particle impact noise detection (PIND) and perhaps constant acceleration testing on a selected case-by-case basis should be considered.

Parts rescreening may be reduced to sample testing on an item-by-item basis when the initial parts rescreening indicates that the parts are satisfactory. However, in the event that any lot fails, the parts manufacturing process is changed in any way or a vendor change is made, the initial set of screening levels should be reinstated immediately.

The need and extent of any further ESS, by the supplier or rescreening at receiving inspection, should be determined by considering:

- The percent of defects detected
- Achievable failure rate reduction and reliability improvement
- Test cost and cost avoidance

Note that cost avoidance can be substantial. Even a small percentage of part defects can result in a high number of board or assembly rejects and excessive rework. Part rescreening is applied to reduce defects that escape the supplier's ESS program, to a very low level, thus minimizing the influence of part defects on higher assembly rejects and reducing the rework cost.

It must be emphasized that part defects (and failures) occurring throughout the manufacturing process (as well as in the field) should be monitored and failures analyzed to determine their failure mechanisms and conditions to identify patterns and to determine if improved screens can be applied at the part level to expose and remove those failure mechanisms. Parts that fail during higher assembly screens are indicative of the efficiency of the part level screening process.

Question 5: *Would an intermediate level screen be effective*?

The following guidelines will aid in determining if an intermediate level ESS is effective stimulating the failure mode:

1. Screens are applied to expose and remove:
   — Parts damaged during assembly of the hardware
   — Latent defects that escaped part screening
   — Assembly induced defects such as poor solder joints or cracked plated through holes on printed circuit boards
2. Screens should focus on high density/high technology assemblies or boards
3. Screen stresses should not exceed the design capabilities of the individual parts and material that comprise the applicable assembly or board
4. Standard components, that use parts with proven reliability/life characteristics, should not be screened unless subsequent tests and field experience show excessive failure with these items

A 'YES' answer means that the parameters of the ESS must be defined. A general recommendation for electronic assemblies is to run 15 temperature cycles, with the last cycle failure free, from −50 to 95°C at a rate of change of 15°C/minute and with no power applied. It is also recommended that a vibration screen be applied at an intensity of 0.04 g/Hz from 20 Hz to 2 kHz for 10 minutes on all three axes with no power applied.

In developing intermediate level temperature cycling ESS requirements, automatic functional testing at ambient temperature should be carried out and the exact number of cycles should be dependent on the hardware density and the technology of the parts. Thermal cycling is more important for boards and assemblies with high part and wiring density since they are more susceptible to process, workmanship and temperature induced defects due to smaller error margins, increased rework difficulty, and thermal control problems resulting from the proximity of the parts and interconnections contained within a small area. For assemblies with large mass components, considerable care must be taken with the fixturing to prevent unnatural deflections and bending modes.

The temperature limits and rates of change to be applied are initially determined by surveying the parts and material specifications to identify part temperature ranges and allowable rates of temperature change. These specified maximum ratings must not be exceeded during the ESS. Also, rates of changes greater than 20°C/minute should be avoided because they can cause failure in good bonds and excessive solder cracking. A thermal survey is performed to ensure that the desired ESS profile is accomplished as well as to determine chamber cycle parameters. A vibration stress response survey is also performed to set the exact vibration intensity that ensures that the desired ESS profile is accomplished. For electronic assembly boards, ESS should, generally, be performed with power off because the additional defects precipitated by the application of power typically do not justify the added cost, particularly when considering the difficulties in implementing a power-on test, e.g., problems with connections, operating speeds, board to board interaction, etc.

The number of cycles (or ESS duration), failure free criteria and other requirements should be adjusted based on results of subsequent ESS and field performance. Failed items should be analyzed to identify cause of failure and to institute corrections to the design, vendor selected or the manufacturing process. Failure data should be analyzed to identify trends and failure modes that account for the greatest percentage of failures and to prioritize corrective action efforts. Working status charts should be prepared and maintained, showing number of failures by mode, cause and corrective action. Failure reports should be prepared in a format usable for design and manufacturing and should classify design, manufacturing and supplier (as well as software) discrepancies. Non-productive screens should be redefined or eliminated.

Question 6: Would an end-item, equipment or complete system level screen be effective?

Screens at the end-item level are designed to expose and remove:

- Parts, assemblies and units damaged during final fabrication
- Latent defects that escaped part and lower assembly screens
- Harnessing/connector intermittents, not properly mated connectors and other defects induced during subsequent assembly
- Defective chassis mounted component including poor solder joints

A 'YES' answer means that the parameters of the ESS must be defined. An initial baseline requirement may include running 12 temperature cycles, with the last cycle failure-free, from −50 to 70°C at a rate of change of 10°C/minute if possible (but no less than 5°C response) with the power on and the assembly monitored to detect momentary operational changes. The initial vibration requirement is generally the same as for lower assemblies except that power will be on and the item monitored. Power on during vibration is essential to identify intermittents.

Note that these are baseline requirements where the exact number of cycles and limits are determined by considering the thermal/vibration stress response of the assembly. In general, there are many paths along which a stress may be transmitted in the assembled hardware item. Each path is characterized by unique combinations of conduction, isolation, attenuation and amplification which will alter the characteristics of transmitted temperature and vibration stresses to an extent generally not predictable with precision. Therefore, the application of the ESS must be adjusted to the stress transmission characteristics of the hardware design as reflected by the placement and location of parts, the mounting of parts and the designed-in thermal protection techniques. If the adjustment is not made, critical parts may not be subjected to the proper stresses and seemingly rigorous screening process may be benign and ineffective.

This adjustment is accomplished during Step 3 by performing both a thermal and vibration survey prior to the step-stress tests sequence to set the exact levels that will ensure that the major parts are adequately stressed. However, during this step it is suggested that the baseline requirements be grossly adjusted to reflect a rough evaluation of the thermal/vibration response of the design (see Table 4–4). The adjusted requirements are then used as the starting point for the step-stress test. This will facilitate the conduct and the effectiveness of the tests and help ensure that the maximum ratings of the assembly's parts and materials are not exceeded.

In preparing end-item level ESS requirements, emphasis should be given to an operating ESS with power/temperature cycling (and perhaps the last few cycles or hours failure free) where the temperature extremes are set by optimizing test efficiency vs facility cost, space requirements and other cost/logistic constraints. Power should be on only during increasing temperature conditions. Functional testing both during and after screening should be performed.

Failure free criteria should be set based on assembly requirements to assure that screening has been effectively completed. Defects precipitated during screening provide an indication that the ESS is effective and that potential field failures are being removed from the hardware. During the failure free period defects precipitated indicate that the screen is not complete or the failure cause not removed and that additional screening should be performed. The extent of additional screening should be determined from an evaluation of the failure rates and assembly reliability objectives/risk factors. Failures should be analyzed, prior to rework and additional screening, to identify causes and to institute appropriate changes in design and manufacturing and to determine the possibility of earlier detection where ESS costs and rework are less.

Question 7: Have any ESS requirements been developed?

If a part, intermediate or end-item level screen cannot effectively remove the latent defect, then a design or manufacturing process change must be implemented to eliminate the defect or reduce its frequency of occurrence.

Question 8: Is this the last failure mode?

If not, repeat process until all failure modes have been evaluated. After the last failure mode has been sequenced through the process, review all ESS profile data and prepare an ESS program requirement matrix.

Step 2 of the ESS logic process (Figure 4–28) is to estimate the

TABLE 4-4
Thermal/Vibration Response Evaluation Guidelines

Thermal

Semiconductor devices

(a) Are thermal contact resistances between a device and its mounting minimized by using large areas and smooth contracting surfaces and by specifying thermal gaskets or compounds as required?
(b) Are the devices located remote from high temperature parts?
(c) Are heat sinks used with fins positioned vertically and in the direction of air or coolant flow? Are painted or coated surfaces used to improve radiation characteristics?

Capacitors

(a) Are capacitors located remote from heat sources?
(b) Are they insulated thermally from other heat sources?

Resistors

(a) Are resistors located for favorable convection?
(b) Are mechanical clamping or encapsulating material provided for improved heat transfer to heat sinks?
(c) Are short leads used whenever possible?

Transformers and inductors

(a) Are heat conduction paths provided for transfer of heat from these devices?
(b) Are they located favorably for convection cooling?
(c) Are cooling fins provided where appropriate?

Printed wiring boards

(a) Are larger area conductors specified where practicable?
(b) Are heat producing elements segregated from heat sensitive components?
(c) Are intermediate core layers used in multilayer systems and are good conduction paths provided from these layers to support members and intermediate heat sinks?
(d) Are protective coatings and encapsulants used for improving heat transfer to lower temperature supports and heat sinks?

Vibration

(1) The location of the components relative to the supporting structure (i.e., at the edge, corner, or center of the supporting structure)
(2) The orientation of the parts with respect to the anticipated direction of the shock or vibration forces
(3) The method used to mount the parts

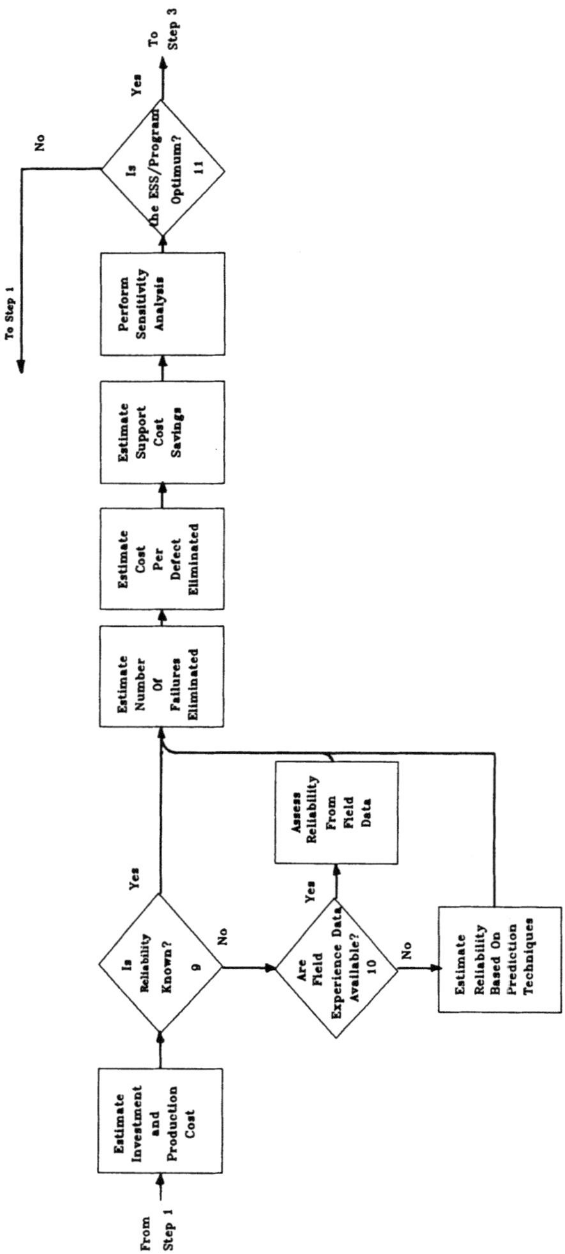

Figure 4–28 ESS Decision Logic: Step 2 Cost–Benefit Analysis

cost–benefit of the ESS program formulated in Step 1, including production costs, potential reliability improvement and ESS cost savings as a result of the improvement in reliability due to the implementation of ESS. The results of this step are then iterated with the results of Step 1 to optimize the program.

As shown in the diagram, prior to determining the cost–benefit of an ESS program the investment and production costs of the ESS must be estimated. The costs of ESS are combined with an estimate of the number of defects eliminated by ESS and used to determine the cost per defect eliminated. Estimating the ESS investment and production costs involves first determining the fixed screening costs for each level of assembly. These costs are one-time expenditures necessary to conduct screening at a particular assembly level and include the cost of facilities, the cost of test equipment and fixtures, the cost of program planning and the preparation of procedures, and the cost of training. The cost of facilities, test equipment and fixtures should be apportioned to the program for which the cost analysis is to be performed.

The variable costs for the total number of items to be screened at each assembly level are estimated next. These costs are recurring costs which are different for each level of assembly and depend upon the number of items to be screened. During early production, when the number of defects would be expected to be higher, repair and rework costs would be a large cost driver. During later production, when the number of defects would be expected to be lower, the primary driver would be the cost of labor to conduct the screens and their associated tests. The latter situation would be expected when failure-free screens and tests are employed at the assembly level. The costs to conduct failure-free acceptance tests and associated screens would thus be heavily dependent on the labor costs for screening and testing. Recurring costs include the cost of labor to conduct screens and tests, the cost of labor for screening program management, the cost of labor to conduct failure analysis and the cost to record and analyze screening program data.

After the cost data have been estimated the logic process proceeds by answering the decision logic evaluation questions.

Question 9: Is the reliability of the hardware item known?

A 'NO' answer means that reliability must be assessed through a review of field experience data or estimated based on its design characteristics.

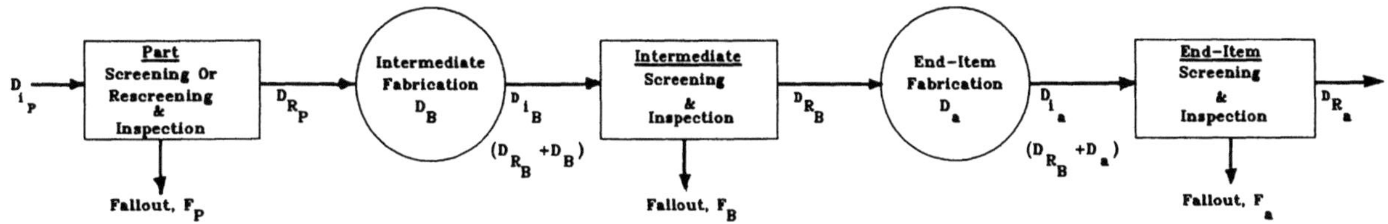

where: $F = S \times E \times D_i$

$D_R = D_i\,(1-SE)$ and is the outgoing number of latent defects

and: S, screen strength

E, inspection efficiency

D_i, incoming number of latent defects (from reliability analysis)

D_{i_P} – based on parts screened or rescreened at incoming inspection

$D_{i_B} = D_{R_P} + D_B$

D_B – reflects failure rates of parts and interconnections not included in D_{i_P}

$D_{i_a} = D_{R_B} + D_a$

D_a – reflects failure rates of parts and interconnections not included in D_{i_P} or D_B

Figure 4–29 Production ESS Defect Elimination Model

Question 10: *Are field experience data available*?

If 'YES', assess reliability based on an evaluation of the field experience data. If 'NO', estimate reliability using one the methods described in Section 4.1.

Once the reliability of the hardware item, including its constituent parts, has been established, a screening process and inspection analysis is performed to estimate the number of latent defects eliminated by the ESS program. The analysis model, illustrated in Figure 4–29, takes into account the strength of the applied screens to precipitate latent defects to failure and the efficiency of the quality inspections.

The incoming number of latent defects, D_i, can be related to the failure rate from the reliability assessment, λ_0, by the following equation from the chance defective exponential (CDE) model in US DoD-HDBK-344, 'Environmental Stress Screening of Electric Equipment': [18]

$$\lambda_s(t) = \lambda_0 + \frac{D_i}{N}\bar{\lambda}_D \exp(-\bar{\lambda}_D)$$

where $\lambda_s(t)$ is the failure rate of the assembly before ESS at time t, N is the number of assemblies to be produced and $\bar{\lambda}_D$ is the average failure rate of a latent defect in the field environment.

In order to distinguish a part with a latent defect from a 'good' part (with an average failure rate of approximately 1 failure per 10^6 to 10^7 operating hours), the latent defect will have to have a failure rate in the field environment of greater than 10^{-3} failures per hour, so assume $\bar{\lambda}_D = 10^{-3}$ per hour. If it is assumed that the failure rate from the reliability assessment is about 30% of the failure rate of the system before ESS at the start of product life, $\lambda_0 = 0.3\ \lambda_s$ (at $t=0$). Putting these assumptions into the CDE model equation and solving for D_i yields:

$$D_i \simeq 2300\, N\lambda_0$$

The analysis involves the following activities:

(*a*) Construct a process and inspection flow diagram that identifies the various processes, screens and inspections which take place during production and that describes how each process flows into the next process, screen or inspection.

(*b*) Establish values for screen strength, S, and inspection efficiency, E. Screen strength is defined as the probability that a stress screen will transform a latent defect into a hard failure (given that there is a latent defect present) and that the failure will be detected by the screen. Values for S should be developed from experience factors on the same (or a

TABLE 4-5
Screen Strength Factors (adapted from US DOD-HDBK-344)

Temperature

		Temperature range (°C)					
No. of cycles	*Temp. rate of change (°C/min)*	*80*	*100*	*120*	*140*	*160*	*180*
	$\dot{T}$						
	5	0·70	0·75	0·79	0·82	0·84	0·86
6	10	0·90	0·93	0·95	0·96	0·97	0·98
	15	0·97	0·98	0·99	0·99	0·99	0·99
	$\dot{T} \geqslant 20$: 0·99						
	$\dot{T}$						
8	5	0·80	0·84	0·87	0·90	0·91	0·93
	10	0·96	0·97	0·98	0·99	0·99	0·99
	$\dot{T} \geqslant 15$: 0·99						
	$\dot{T}$						
10	5	0·87	0·90	0·92	0·94	0·95	0·96
	10	0·98	0·99	0·99	0·99	0·99	0·99
	$\dot{T} \geqslant 15$: 0·99						
	$\dot{T}$						
12	5	0·91	0·94	0·95	0·97	0·97	0·98
	$\dot{T} \geqslant 10$: 0·99						
	$\dot{T}$						
14	5	0·94	0·96	0·97	0·98	0·99	0·99
	$\dot{T} \geqslant 10$: 0·99						
	$\dot{T}$						
16	5	0·96	0·97	0·98	0·99	0·99	0·99
	$\dot{T} \geqslant 10$: 0·99						

Random vibration

	Acceleration level (grms)													
Duration per axis (min)	*0·5*	*1·0*	*1·5*	*2·0*	*2·5*	*3·0*	*3·5*	*4·0*	*4·5*	*5·0*	*5·5*	*6·0*	*6·5*	*7·0*
5	0·007	0·023	0·045	0·07	0·10	0·14	0·18	0·22	0·26	0·30	0·35	0·39	0·43	0·47
10	0·014	0·045	0·088	0·14	0·20	0·26	0·32	0·39	0·45	0·51	0·57	0·63	0·68	0·72
15	0·021	0·067	0·13	0·20	0·28	0·36	0·44	0·52	0·60	0·66	0·72	0·77	0·82	0·85
20	0·028	0·088	0·17	0·26	0·36	0·45	0·54	0·63	0·70	0·76	0·81	0·86	0·90	0·92

similar) process. When such data are available and assuming perfect tests, then the screen effectiveness can be determined by use of the observed fall-out from the screen and the number of latent defects initially present, i.e.:

$$\text{screen strength} = \frac{\text{fall-out}}{\text{number of initial latent defects}}$$

If the screen effectiveness was known precisely, then the number of incoming latent defects could be calculated directly using the observed fall-out from the screen. The remaining number of latent defects would also be known.

Such idealized conditions are difficult, if not impossible, to realize in practice. Therefore, a modeling approach must be used in which screen effectiveness (strength) is based upon estimates derived from a combination of the actual screening program data, experiments and the published literature.

Values for S applicable to new hardware processes are given in US DoD-HDBK-344. Table 4–5, adapted from the handbook, provides values for S as a function of screen type, levels and duration that can be used with the ESS defect elimination model.

The efficiency of an inspection can be expressed in terms of a defect detection probability. A perfect or error-free inspection would have an associated numeric value of unity. DoD-HDBK-344 also provides values for inspection efficiency E which may be applied with stress screens. Table 4–6, taken from DoD-HDBK-344, provides values for E for use with the ESS defect elimination model.

Tables 4–5 and 4–6 are used with the ESS defect elimination model to estimate screen strength, S, and inspection efficiency, E. The values in the tables should be scaled upward or downward when prior knowledge or experience data are available. Adjustments can be made based on:

- Complexity of the part/assembly under test (e.g., simple part, easy access to measurement)
- Screen/inspection design adequacy (e.g., designed to detect specific failure modes? Does it reflect accept/reject criteria for significant parameters of unreliability as defined by FMECA?)
- Screen/inspection procedure adequacy
- Test equipment complexity
- Inspector experience (e.g., highly qualified, several years in quality control)
- Time allocated for inspection

TABLE 4-6
Inspection Efficiency vs Test Types (adapted from US DoD HDBK 344)

Test type	*Detection efficiency*
Production line GO–NO GO test	0·85
Production line in-circuit test	0·90
High performance automatic tester	0·95

Weighting factors can be applied to each of the above attributes and used with the tables to establish appropriate values for S and E.

(*c*) Compute the outgoing defect rate based on the reliability data and the screen strength/efficiency values [from (*b*)] using the process flow diagram and associated equations developed during (*a*).

The average in-house cost of labor and materials to repair a failed item is determined and then used with the fall-out estimates to calculate the various screening repair costs and assembly level screening costs and is combined to determine the cost per defect eliminated. The total expected fall-out, i.e., the total number of defects precipitated and detected by the screening program, is used in conjunction with a threshold cost of field repair to compute the support costs savings.

A sensitivity analysis is performed to optimize the ESS program. This involves varying the screens starting at the part level while holding the screens constant at the higher levels. The screen types and other parameters are iterated to find the optimum screen profile. The higher level screens are then optimized following the same procedure. Costs per defect eliminated data as well as support costs savings data are computed for each variation. The data are plotted and used as the basis to select the specific program requirements, including application levels and parameters, that show maximum cost savings.

Question 11: *Is the ESS program optimum*?

If 'NO', repeat the Step 1 and 2 process. If 'YES', revise the ESS program matrix prepared at the completion of Step 1 to reflect the optimized requirements.

Step 3 of the ESS process (Figure 4–30) is to perform ESS stress response surveys, to conduct step-stress tests in order to finalize the requirements and to run the finalized ESS in order to assure that the screens are effective and do not damage good hardware. Questions 12 through 15 address temperature cycling and questions 16 through 19

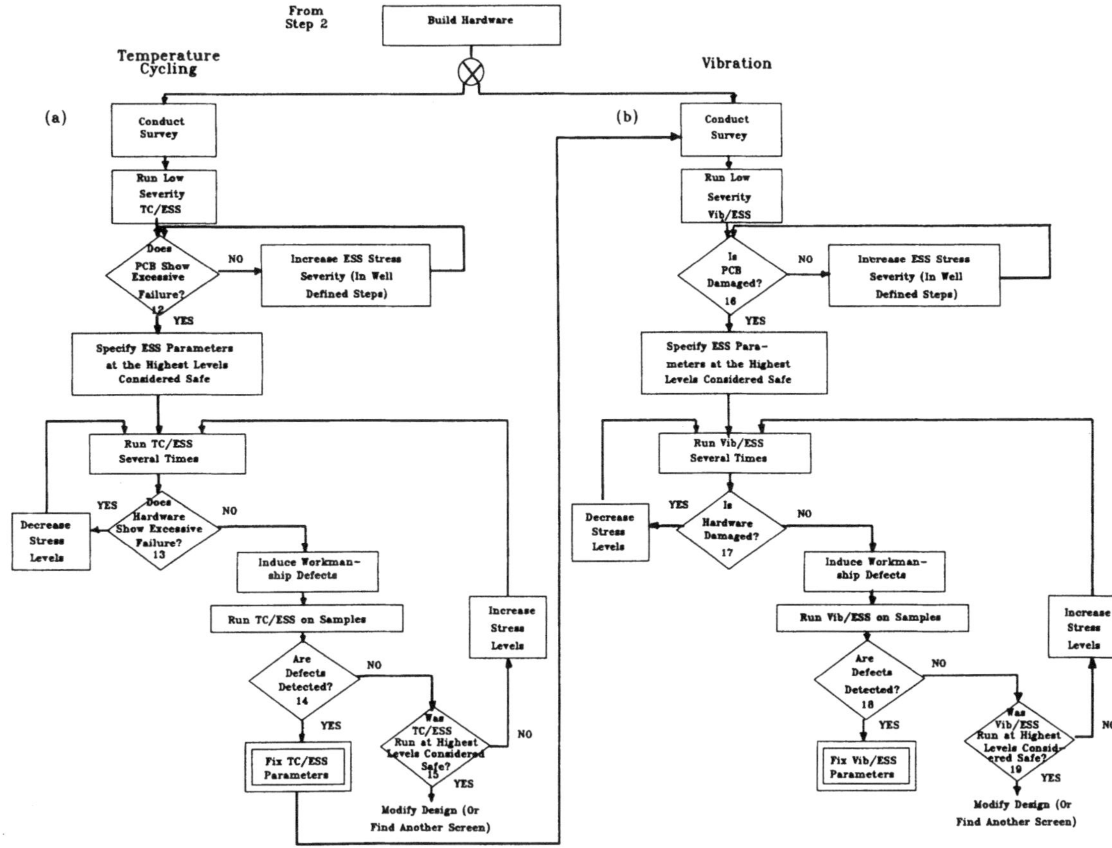

Figure 4–30 ESS Decision Logic: Step 3 ESS Development and Validation

address vibration. At the completion of this step the ESS program matrix developed during Step 1 will be finalized.

It is important to emphasize that the baseline values established during Step 1 and optimized during Step 2 should not be imposed as hard, inflexible requirements. The step-stress to failure approach integrated in the logic process must be applied to determine the failure mechanisms and to calculate the safe severity ESS levels.

As shown in the Step 3 sequence, hardware is first built using parts procured to the specified quality level and rescreened at the factory to the profile established during Step 1. Prior to the step-stress sequence, surveys are performed to determine the hardware thermal and vibration response; the chamber conditions and vibration profile are then adjusted to reflect the response characteristics. The adjustments must also take into account the number of items in the chamber, the transmission characteristics of the vibration fixture and whether the items to be tested are active or passive.

A step-stress test is then conducted as follows:

(1) Run low severity screen; i.e., for *temperature cycling*: narrow temperature excursion, lower temperature limits, lower rate of change and a reduced number of cycles from that given in the preliminary ESS matrix; for *vibration*: lower intensity and shorter duration than that given in the preliminary ESS matrix.
(2) If the screen does not show excessive failure, then increase the stress severity levels in several, small, well defined steps up to and beyond that given in the preliminary ESS matrix. Be careful so as not to exceed rate of change and temperature limits.
(3) Specify the ESS parameters at the highest level considered safe, e.g., where the test unit is not damaged (at the step just prior to the level that may have damaged or weakened the unit) and does not exceed part and material ratings.

After the step-stress test, the ESS is then run several times on the same sample to see if there is excessive failure or if the item is damaged. An abbreviated qualification test may be performed to determine the acceptability of the hardware item. If there is excessive failure or actual damage, then the stress levels are reduced and the process repeated on new samples.

The selected ESS must also be proven to be effective in detecting most of the latent defects that might be in the hardware which would cause a problem in the field while not damaging good hardware. The approach incorporated in the logic process is to detect latent defects using samples built with known latent defects. These samples are run through the ESS and, if most of the latent defects are detected, then the ESS is concluded to be effective. For those latent defects not found in ESS, the probability and consequence of the field environment precipitating them must be assessed. That is, the latent defects might not even show up in the field and so would be of no consequence. ESS clearance is accomplished by repeatedly running the same assembly (or parts of the assembly) through the ESS and then running a lifetime simulation on the same finished assembly. If no failures are found to be due to overstimulation, then the ESS is cleared or proven to be 'non-damaging'. The proof of ESS should be repeated from time to time to assure that subtle changes in the assembly are not allowing the ESS severity to move into the damaged area.

The ESS program, once implemented in the factory or at the depot, must be continually challenged relative to the effectiveness of the overall program as well as the individual tests. ESS results and field experience data must be evaluated to determine the need to modify individual ESS levels and parameters and to identify the possibility of earlier detection where the test costs are lower and the potential for cost avoidance is greater. It should be emphasized that the initial program represents a baseline for applying the ESS's. A screen-like any quality inspection, must be considered as a dynamic test, its duration (or number of cycles) adjusted depending on the results of subsequent higher level test or field performance. However, the extent and nature of any changes will be determined only through careful review and analysis of the subsequent failures.

A data system supported by failure analysis and corrective action should be implemented to maintain visibility over the effectiveness of the overall program and of each individual screen. The data system would compile, reduce and disseminate essential experience (operating/failure) data for monitoring the effectiveness of the overall screening program and would facilitate the modification, optimization and refinement of the ESS parameters through a sustained engineering activity. The ESS fall-out data would be continuously analyzed to ensure that failure causes are properly established and to provide essential data needed to monitor the effectiveness of the program, enable effective modification of the requirements and to support application of the ESS cost optimization model. Adjustments would be made as necessary to minimize cost and maximize screening effectiveness. The data system would provide the following information:

(1) Identification of the assemblies subjected to ESS;
(2) Total operating time for each item screened, the last operating time interval of failure-free operation, and the ESS completion data;
(3) Number of assemblies subjected to ESS;
(4) Type and number of latent defects identified including descriptions of the type of defect and the ESS type, level and conditions applied;
(5) ESS strength and inspection efficiency factors derived from the fall-out data;
(6) Identification of inspection and functional parameters measured before, after and during ESS;
(7) Failure analysis results, identifying the root causes of the defects, the responsible failure mechanisms and the times to failure relative to the start of the ESS program;
(8) Corrective actions taken to eliminate the cause of the defects from the assembly and/or process;
(9) Cumulative plots of ESS fall-out data vs ESS type, level, conditions and duration.

CHAPTER 5

The Application of RCM within Depot Maintenance

This chapter discusses the application of RCM within the depot maintenance process. It covers US Army aircraft depot maintenance and inspection methods, including the depot maintenance work requirement (DMWR) document and the procedures and standards for processing a component or a complete aircraft system through the depot. It describes a special on-condition maintenance technique to assess, in rank order, the actual condition of fielded aircraft, thus providing a means for identifying those aircraft most in need of repair or overhaul to prevent degradation of reliability and safety.

The maintenance of Army aircraft is implemented at three organizational levels: aviation unit maintenance (AVUM), aviation intermediate maintenance (AVIM) and depot maintenance. The *first level*, AVUM, performs preventive maintenance (such as inspection, lubrication and cleaning), makes minor repair or adjustments and replaces easily accessible parts. Frequent maintenance checks and servicing are performed based on systematic procedures to detect early indications of failure and to correct deficiencies before more costly and time-consuming repair is needed at higher maintenance levels. AVUM activities are staffed and equipped to perform the high frequency, 'on aircraft' maintenance tasks that are required to maintain the aircraft in a serviceable condition. The AVUM is limited by the amount and complexity of ground support equipment, facilities required, and the number and skills of available techniques.

The *second level*, AVIM, performs all of the maintenance tasks performed by AVUM, plus replacement and repair of modules/

components and end items which can be efficiently accomplished with available skills, tools and equipment. AVIM activities emphasize the repair of equipment for rapid return to use and to maintain operational readiness. AVIM establishes a direct exchange program for AVUM organizations by repairing selected items for return when such repairs cannot be accomplished by AVUM. AVIM organizations have capability to inspect, troubleshoot, test, adjust, calibrate and align aircraft system modules and components. They also have capability to determine the serviceability of specified modules/components removed prior to expiration of the time between overhaul (TBO) or finite life. Unserviceable modules/components and end items beyond the capability of AVIM are sent to depot maintenance.

The *third level*, depot maintenance, performs maintenance tasks exceeding AVUM and AVIM capabilities, including complete aircraft and component overhaul. Depot maintenance is controlled by funding limitations. This requires that the aircraft be evaluated for condition and prioritized annually in order to determine the number of aircraft and components to be overhauled within available funding.

Depots are equipped with the skills and equipment to completely disassemble an aircraft or component to its piece parts, repair or replace these parts, manufacture parts that are out of production, and rebuild the aircraft or component to a serviceable condition. Depot maintenance is carried out in accordance with prescribed requirements and instructions. A depot maintenance work requirement (DMWR) document and its change orders define the minimum procedures and standards required to process a component or end item through the depot.[19] It provides the necessary instructions for the complete overhaul of the item, including conversion/modification criteria and repair procedures covering the worst-case condition of applicable parts. It includes specific requirements for disassembly, cleaning, inspection, repair, reconditioning, rehabilitation, modification, reassembly, servicing, testing, and packaging/preservation of aircraft, engines, aircraft components, and related ground support equipment.

5.1 THE DEPOT MAINTENANCE PROCESS

Figure 5–1 illustrates the depot maintenance process for Army aircraft equipment. As shown in the figure, all hardware items are grouped into three main categories:

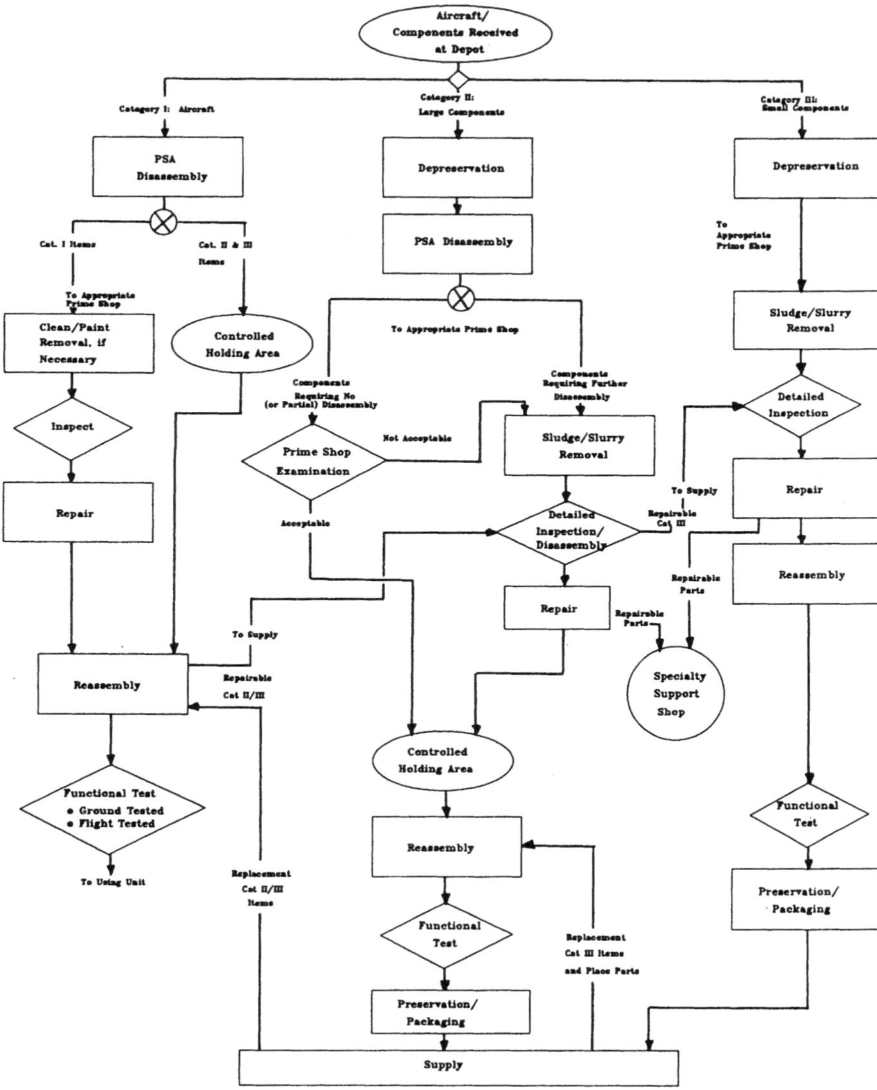

Figure 5–1 Depot Maintenance Process

(1) *Category I: Aircraft* — the total aircraft; e.g., the airframe, electrical wiring, seats and doors
(2) *Category II: Large Components* — large components and major assemblies; e.g., engines and transmissions
(3) *Category III: Small Components* — small components and accessories; e.g., generators, hydraulic pumps and oil coolers

In general, depot maintenance can involve a repair action, overhaul or a complete rebuild. A repair action restores serviceability by correcting specific damage, fault, malfunctions or failure in a part, subassembly, module (component or assembly), end item or system. An overhaul restores an item to a serviceable condition as prescribed in the DMWR and, normally, is the highest degree of maintenance performed at the depot. A complete rebuild restores an item to a like-new condition in accordance with its original manufacturing standards. Rebuild is the highest degree of maintenance and after an item is rebuilt its age is considered zero.

An aircraft sent to the depot for repair normally contains serviceable components (e.g., engine, transmission and gearbox) which are removed only to allow access for repair of the basic airframe and are subsequently reinstalled. If a discrepant component is discovered during assembly or functional testing and the discrepancy cannot be easily corrected, the component is turned into supply and a serviceable component issued. Engines returned to depot for overhaul normally contain serviceable accessories (for example, fuel control, overspeed governor and igniter unit) which are functionally tested (bench tested) and are subsequently reinstalled on the engines. Accessories which do not meet the functional test requirements and cannot be corrected by minor repairs are either turned into supply as 'repairables' or inducted into the component overhaul program. Some components require upgrading to a later configuration and are automatically inducted for repair.

When a complex unit becomes defective in the field, and is beyond field repair capability, it is often more cost-effective to repair the defect, rather than perform a complete overhaul, at the depot. A repair consists of the minimum maintenance necessary to correct the specific discrepancy that caused the item to be returned to the depot along with other applicable tasks associated with reassembly, testing and preservation. The repair changes the status of the unit from repairable to serviceable but does not increase its potential longevity. The decision to repair or overhaul is made at the depot during preshop analysis.

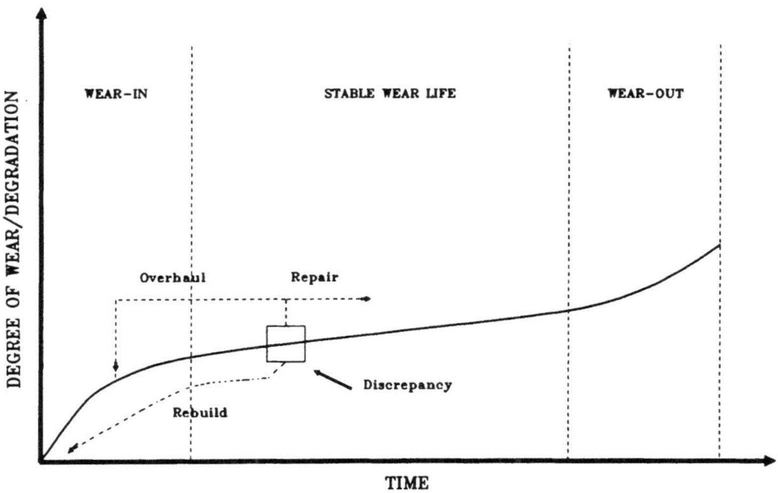

Figure 5–2 Impact of Repair on Potential Longevity

Figure 5–2 illustrates a longevity curve which depicts an item's wear/degradation behavior with respect to time. The slope of the curve during wear-in and wear-out varies from item to item as does the length and slope of the stable wear life part of the curve. A discrepancy changes an item's status from serviceable to repairable and acts as a roadblock in preventing the item from progressing down its longevity curve. When the discrepancy arises, usually either overhaul or repair is performed. Overhaul is performed to recover used-up wear life, whereas a repair removes the roadblock and allows the unit to progress; it does not recoup any of the used up wear life and therefore does not change the item's position on the longevity curve. If rebuild is performed, the item is new and is returned to the starting point of the longevity curve.

As shown in Figure 5–1, *depreservation* is generally the initial task to be performed in the depot process. Basically, the task is to remove the item from its incoming packaged state. Aircraft (Category I) do not require depreservation, except for the tail boom or the basic airframe (i.e., the aircraft without any Category II or III items) which may require deboxing. Large components (Category II) require decanning and/or other unpackaging. Any preservation fluids/compounds (e.g., oil and grease) need to be removed using vapor degreasing, emulsion degreaser and steam/detergent techniques. Small components (Category III) requires deboxing and/or other unpackaging. Preservation fluids/ compounds also need to be removed using degreasing, emulsion degreaser and steam/detergent techniques.

After depreservation, *preshop analysis* (PSA) is performed. A PSA involves inspecting an item for condition while it is being disassembled. The analysis focuses on the reason the item was sent to the depot and its remaining operating life. Defined weak spots within the component are inspected for specified historically common deficiencies. This includes identifying small structural cracks that can grow as a result of fatigue and corrosion, the main dangers as an aircraft ages. The analysis determines the extent of further disassembly and repair or overhaul which need to be performed at the appropriate prime shops. Components that do not need repair or overhaul and can be sent directly to a holding area for assembly are identified.

The aircraft PSA includes:

1. Determining which panels and other structure require arbitrary removal from the airframe and which are removed only for cause.
2. Identifying specific structural weak points.
3. Determining if an alignment check is required.
4. Reviewing the specified overhaul process to determine if any repair task can be eliminated, streamlined or accomplished only for cause.
5. Determining the extent of repair for the tailboom.
6. Determining the accessories/components which require functional test or special inspection.
7. Defining any maintenance actions that should be performed because of convenience.

After removal of all Category II and III components, tailboom, appropriate panels and doors, and the aircraft PSA is completed, the airframe and components are routed to their appropriate prime shop.

For Category II components, PSA is conducted while removing all accessory items and disassembling the basic components to the subassemblies and modules. Piece parts which should remain together for reassembly (where practical) are identified. The PSA is conducted in accordance with the applicable DMWR, which provides instructions for checking and testing the component. The overhaul process is also reviewed during PSA to determine if any repair actions can be eliminated, streamlined or accomplished only for cause.

The analysis identifies the subassemblies or modules that do not need further disassembly. There items are forwarded to their respective prime shops for inspection and minor repair, if needed. If major deficiencies are found, the items are turned into supply and scheduled for maintenance. The subassemblies and modules needing further disassembly and inspection are forwarded to their respective prime shops and repaired as needed.

In the case of the Category III, small components, a PSA and any further disassembly are generally not performed. These components are normally sent to the prime shops where they are completely repaired or overhauled. Those parts requiring further repair (e.g., machining, plating, welding or perhaps ion implantation and plasma coating) are routed from the prime shop to a specialty shop. Although disassembly is not generally performed on small components, a PSA, if performed, would identify parts that require further inspection and in the case of bearings if they should be disassembled and inspected in a clean room environment.

Instructions for PSA are included in the applicable DMWR's. These instructions take the general form and content shown in Table 5–1.

Cleaning and paint removal tasks are performed only if necessary to facilitate inspection and/or repair. When performing these tasks, caution must be taken to avoid skin contact and prolonged inhalation of vapors and dust. Also, care must be taken in the cleaning/paint removal method, particularly on ceramic-coated, aluminum-coated, carbonized, painted, nitrided, magnesium, steel and low-alloy steel type parts, in order to avoid process induced damage.

The paint on an aircraft is removed by using paint remover, vapor blasting (for example, glass bead, light abrasive or plastic media) or hot-alkali soak. The sludge or slurry on Category II and III components is removed by first masking all openings to prevent clogging by deposits and then using a solvent immersion, dry-cleaning solvent, vapor blasting (care must be taken to prevent wearing away of metal), hot-alkali soak or a periodic-reverse cleaning process.

After cleaning and paint removal detailed *inspection* is performed at the prime shops using NDI techniques, as appropriate. Various NDI techniques are described in Section 5.2. Before NDI, those components cleaned by a dry-cleaning solvent (with an oily film left by the process) are removed. Category I aircraft are inspected for cracks, corrosion, holes, bends, delamination, distortion, wiring defects, linkage wear, hydraulic leaks, missing fasteners, etc., as defined by the applicable DMWR. Table 5–2, taken from component DMWR's, provides a list of defects and their probable causes that can be identified during inspection. Category II and III components are inspected for cracks, bends, excessive wear, heat

TABLE 5-1
PSA DMWR Section Format and Content

Section I—Purpose, test and analysis standards

Purpose. This paragraph will state the purpose of the PSA, e.g., 'The purpose of preshop analysis is to determine, at the highest assembly level possible, the work required to return the item to a serviceable condition as specified herein. If inspection at the highest level of assembly is precluded by missing, damaged or diagnosed defective assemblies, consideration will be given to techniques that would allow continued inspection at that level. If this is not possible, inspection will proceed at the next lower level. A preshop analysis checklist will be used to record the results of the analysis and any required maintenance.'

General instructions. The paragraph will provide general instructions including:

Inspection of forms—instructions for checking all tags and forms attached to the end item, assembly, subassembly or component to determine the reason for removal from service, other discrepancies and the accumulated operating time.

Removal of end item, assembly, subassembly or component from shipping container, package or storage—instructions for removal from shipping container and package. Storage and electrostatic discharge control measures will be provided.

External inspection—instructions for external inspection of the end item, assembly, subassembly or component to determine completeness of end item, assembly, subassembly or component, evidence of damage and remaining life.

Cleaning—instructions for cleaning of the end item, assembly, subassembly or component for inspection.

Test—instructions for PSA including verifying tests to confirm damage/malfunction report/historical record/remaining life/data (unless damage/malfunction is obvious), nondestructive or operational check/performance functional tests and troubleshooting/fault isolation procedures to support testing.

Temporary preservation/protection—instructions (or references) for temporary preservation/protection of the end item, assembly, subassembly or component pending performance of maintenance required.

Special handling or condemnation procedures—instructions for special handling or condemnation procedures for such reasons as precious metal content, high dollar value, critical or hazardous material. Any pertinent available documents on handling or condemnation will be referenced.

Analysis maintenance actions—provide general standards for analysis to determine maintenance actions which are based on results of the PSA.

Section II—Preshop analysis checklist

A PSA checklist will be provided in this section. The checklist will contain, in working sequence, visual external inspections, tests and analysis maintenance actions required of the end item, assembly, subassembly or component at the highest level of assembly. The checklist will be used to evaluate the end item, assembly, subassembly or component to determine the extent of overhaul operations required to make the end item, assembly, subassembly or component completely serviceable as specified in the DMWR. Complex end items may require removal of assemblies (powerplant/engine and transmission, etc.) for preshop analysis in accordance with the applicable publication which will be listed under the recommended maintenance action column of the PSA checklist. Detailed test procedures or reference to the final test procedures, which include verifying test to confirm damage/malfunction, report/historical record/data (unless damage/malfunction is obvious), nondestructive or operational/flight test/performance test (with AOAP sampling and troubleshooting/fault isolation), will be listed on the PSA checklist.

TABLE 5-2
Component Defects and Probable Causes (taken from Army aircraft component depot maintenance work requirement documents)

Abrasion—Roughened surface, varying from light to severe.
Probable Causes: Abrasive material between moving surfaces.

Banding—Typified by parallel bands of discoloration. Occurs on bearing component part rolling contact surfaces. Original surface is not broken.
Probable Cause: Result of oil varnish or oxide film formation on bearing surfaces. Generally caused by high temperature bearing operation.

Bend—Distortion in a part (differs from local change in conformation).
Probable Cause: Exposure to heat or excessive force.

Blister—Raised portion of a surface separated from the base. Generally found on surface-treated parts, such as plated or painted surfaces.
Probable Cause: Poor original bond, excessive heat, or pressure.

Break—Separation of a part.
Probable Cause: Severe force, pressure, or overload.

Brinelling, false—Occurs only at rolling contact surfaces of bearing rings. It is a specialized form of fretting. Recognized by presence of a series of surface blemishes in loaded side of ring at each ball or roller position. Indentations are usually polished or satin finished in appearance. Due to very slight rotational movement, indentation will frequently be flatter than the roller or ball curvature.
Probable Cause: Result of continuous nonrotational shaft oscillation. Vibration caused by engine transportation may cause false brinelling.

Brinelling, true—Occurs at rolling contact surfaces of bearing rings. Recognized by presence of shallow, smooth indentations in ring at each ball or roller position on loaded side of bearing. Since original surface material has not been removed, indentations have the same surface appearance as surrounding surface area. Indentation contour is the same as the roller or ball curvature.
Probable Cause: Result of high shock loads, leaving a permanent impression of roller or ball in the ring contact surface.

Buckling—Large-scale deformation of part contour.
Probable Cause: Pressure or impact with a foreign object, unusual structural pressures, excessive localized heating, or any combination of these causes.

Burning—Melting or loss of material.
Probable Cause: Excessive heat.

Burnishing—The smoothing of a metal surface by mechanical action, but without loss of material. Surface discoloration is sometimes present around the outer edges of the burnished area. NOTE: Normal burnishing from operational service is not detrimental if coverage approximates the carrying load and if there is no evidence of burns.
Probable Cause: Rubbing.

Burr—A rough edge or sharp projection.
Probable Cause: Excessive wear or poor machining.

Chipping—Breaking away of small metallic particles.
Probable Cause: Heavy impact of foreign object.

TABLE 5-2—*contd.*

Corrosion—Surface chemical action that results in surface discoloration, a layer of oxide, or, in the advanced stages, removal of surface metal.
Probable Cause: Improper corrosion-preventive procedures and excessive moisture.

Corrosion discoloration—Chemical discoloration of bearing surfaces without removal of surface metal; recognized by red or black colored clusters (not to be confused with corrosion pitting which is actual metal removal). If not arrested, corrosion discoloration will advance to corrosion pitting.
Probable Cause: Result of any adverse chemical action due to water, acid, lubricant or a corrosive atmosphere, and generally caused by improper preservation procedures or lack of precaution during installation, removal, inspection or storage.

Corrosion fretting—Discoloration where surfaces are pressed or bolted together under pressure. Color of residue on steel parts is usually reddish brown while that on aluminum or magnesium parts is black.
Probable Cause: Incomplete adhesion of metal or excessive loads.

Corrosion pitting—Irregular surface depressions having ragged edges due to metal removal.
Probable Cause: Corrosion substance adhering to exposed surfaces.

Crack—A break in material.
Probable Cause: Severe stress from overloading or shock.

Crazing–Minute cracking which tends to run in all directions. It is often noticed on coated surfaces.
Probable Cause: Uneven cooling or thermal shock.

Dent—A small, smooth depression.
Probable Cause: A sharp blow or excessive pressure.

Distortion—A change from original shape.
Probable Cause: Exposure to severe heat.

End loading—Defect on face of gear tooth near end of tooth.
Probable Cause: Axial misalignment with mating gear.

Erosion—Wearing away of metal and/or surface coating.
Probable Cause: Hot gases, corrosive liquids or grit.

Fatigue pitting—Relatively deep irregular surface cavities resulting from the breaking away of portions of the surface.
Probable Cause: Advanced corrosion condition or fatigue generated by high-stress conditions.

Flaking—Loose particles of surface metal or surface covering.
Probable Cause: Imperfect bond or severe load.

Fracture—Separation of a part.
Probable Cause: Severe force, pressure or overload.

Fretting—Discoloration of contacting parts resulting from the removal of original surface material.
Probable Cause: Movement between two contacting surfaces.

Frosting—Minute indentations within a localized area.
Probable Cause: Generally a wear-in process.

Galling—Recognized by presence of metal from one part remaining attached to another. Occurs at poorly lubricated surfaces that are in sliding contact.

TABLE 5-2—*contd.*

Probable Cause: Results of localized breakdown of lubrication, causing friction, intense heat and part fusion.

Gouging—Removal of surface metal typified by rough and deep depressions.
Probable Cause: Protruding object, misalignment.

Grooving—Found on rolling contact surface of ball or roller bearings. Recognized by presence of depressions in elements of rolling contact surfaces.
Probable Cause: Results from overload lubrication breakdown and skidding.

Heat discoloring—Characterized by a discoloring film. Color varies from straw, tan and light brown to red-purple, purple and blue.
Probable Cause: High-temperature operation.

Inclusion—Foreign matter enclosed in metal.
Probable Cause: Occurs during manufacture of the metal.

Indenting—Smooth surface depressions. Evidenced by metal displacement, not metal removal.
Probable Cause: Loose material flattened by rolling action will create smooth, shallow indents.

Lack of braze—Interruption (air pocket) in joint or filled cross-section braze material.
Probable Cause: Improper braze-repair.

Metalization—Molten metal coating of a part.
Probable Cause: Molten particles sprayed through the engine.

Nick—A sharp-bottomed depression that may have rough outer edges.
Probable Cause: Impingement of foreign object on surface.

Peening—Flattening or displacement of metal.
Probable Cause: Repeated blows. A surface may be peened by continuous impact of foreign objects or loose parts.

Pickup—Transfer of one material into another.
Probable Cause: Insufficient lubrication, unbroken edges of press-fitted parts and seizure of rotating parts during operation.

Pitting—Small indentations in a surface; usually smooth-bottomed.
Probable Cause: (1) Chemical pitting. Oxidation of surface or electrolytic action. (2) Mechanical pitting. Chipping of surfaces caused by improper clearances and overloading, and by pressure of foreign material.

Scoring—Deep scratches following path of travel.
Probable Cause: Breakdown of localized lubrication between sliding surfaces of foreign material.

Scratch—A very shallow furrow or irregularity; usually longer than wide.
Probable Cause: Movement of a sharp object across the surface.

Scuffing—Surface damage of pieces of a plated or finished surface.
Probable Cause: Rubbing off of fine particles of metal by slight movement.

Seizing—Advanced stages of galling. Recognized by welding of one bearing component to another, preventing rotation.
Probable Cause: Result of localized breakdown of lubrication, causing friction, intense heat and part fusion.

TABLE 5-2—*contd.*

Spalling end loading wear—Large particles or chips that break out of tooth surfaces, usually along the flank area and near the ends.
Probable Cause: Excessive internal stresses due to heat treatment or overloads.

Stress failure—Metal failure.
Probable Cause: (1) Movement of a sharp object across the surface. (2) Compression. Action of two opposed forces that tend to squeeze a part. (3) Tension. Action of two directly opposed forces that tend to stretch a part. (4) Shear. Actions of two parallel forces acting in opposite directions. (5) Torsion. Action of two opposed forces around a common axis. (6) Shock. Instantaneous application of stress.

Stripped thread—Nut, stud, bolt or screw damaged by tearing away part of thread form.
Probable Cause: Improper installation or mismatching thread size.

Tear—Parting of parent material.
Probable Cause: Excess tension, created by an external force.

Unbalance—A condition that usually results in vibration.
Probable Cause: Unequal distribution of mass about a rotating axis.

Void—A continuous lack of braze material through a braze joint cross-section.
Probable Cause: Improper repair.

Wear—A loss of material from contacting surfaces. The degree of wear is dependent on such factors as gear rpm, load, lubrication and alignment.
Probable Cause: Contacting surfaces abrading one another.

damage, fatigue, etc. The turbine blades in an engine, for example, are inspected for deterioration and need for balancing. The linkage in a fuel control is inspected for wear and damage.

The *repair or overhaul tasks* are performed at the appropriate prime shops in accordance with the instructions given in the DMWR's. A project work directive (PWD) identifies the repair action needed and the applicable DMWR. Also, the specific work is described on a 'shop traveler' attached to the item. A piece part may be sent to a support shop for a specialized repair action if needed.

Reassembly of components is performed in accordance with the DMWR instructions. The airframe is riveted together and painted. All Category II and III components, including transparencies, seats, doors and wiring, are installed. The final assembly of Category II components is normally accomplished in stages, with each subassembly or module assembled in its respective prime shop and then installed in the component. The subassemblies and modules often require bench testing, load testing, balancing, gear patterning, gear backlash check, measurement of fit/clearance/alignment, etc., upon assembly and/or during installation into the component. Category III components are assembled in the prime shops.

A *functional test* is performed on the aircraft's systems and on components and their accessories. This consists of subjecting the items to a series of tests to verify compliance with specifications. The aircraft, after final assembly, is subjected to ground testing and any needed adjustments are made. The aircraft is then flight tested.

The engines, transmissions, gearboxes, etc., are functionally tested in a test cell. Aside from the various parameters that are measured during testing, the operator looks for oil leaks, air leaks, abnormal sounds, etc. Some transmissions and gearboxes require partial disassembly to check pinion gear tooth patterns before completing the functional test. Some accessory items, such as engine fuel controls, are adjusted to achieve compatibility with the engine during engine testing. The small components are bench tested and needed adjustments are made to achieve performance requirements.

Preservation and packaging tasks complete the depot repair process. Preservation prepares the item to withstand effects of decomposition caused primarily by moisture and is especially applicable to items to be stored at the depot. Packaging prepares the item to withstand damage from travel as well as to inhibit storage decomposition effects. The selection of packaging techniques depends on such factors as: susceptibility of the item to damage, normal hazards to which the item will be exposed and the length of time the item will remain in storage and in the package.

Most aircraft are flown from the depot to the user and consequently require no preservation. Aircraft that are airlifted outside the continental United States are palletized. For example, for certain aircraft this involves removal of the main rotor blades, the rotor head and mast, and the tailboom. The main rotor blades, along with the rotor head and mast, are installed in a holding fixture and attached to a pallet beneath the aircraft. The tailboom is installed in a fixture and attached to the side of the aircraft.

Preservation is accomplished on Category II components and accessories immediately after they are functionally tested. For example, after checking pinion gear tooth pattern on a gearbox, corrosion resistant oil is used in the gearbox for the remainder of the testing. Basic components, such as an engine or gearbox, are installed in reusable metal transport/storage containers. Most of these containers are sealed airtight and contain bags of desiccant which absorb moisture from enclosed air. For example, an engine may be enclosed in an airtight ziplock bag, with desiccant located inside the bag, and then placed in vented metal transport/storage containers. The Category III components are preserved with an oil film and/or other compounds, possibly bagged, and boxed for transport.

The DMWR, as mentioned earlier, provides complete instructions and acceptance criteria for depot maintenance based on the process shown in Figure 5–1. The DMWR identifies minimum acceptable standards and (where applicable) provides PSA guidelines for determining the extent of repair. It is normally provided as the 'Statement of Work' for each item contracted or programmed for depot level maintenance. It is a 'how to do' type of document which provides the necessary instructions for the complete overhaul of an item, including conversion/modification criteria (parts, subassemblies and assemblies required to convert to latest item configuration as specified in depot program notices and the modification of parts, subassemblies and assemblies required by engineering change proposals) and piece part reclamation procedures for the worst-case condition of applicable parts.

The DMWR is the result of an intensive effort to determine the processes required to achieve maintenance standards and incorporate those processes into a usable document. The DMWR is prepared in a manner which will enable it to be used to produce a quality product that meets serviceability requirements. An effective DMWR will result in minimizing the resource and materiel expenditures required to restore and retain the reliability and safety of the hardware.

The condition of an item sent to the depot, as determined by PSA, along with the reason it was sent to the depot and its operating time, will dictate the minimum amount of depot maintenance that is needed. As previously indicated, PSA identifies the extent of disassembly and repair or overhaul required at the appropriate prime shop(s) using visual inspection, diagnostic testing, nondestructive testing, dimensional inspection, and other methods determined as appropriate in conjunction with available experience and historical data. It is performed at the highest possible level of assembly to prevent unnecessary disassembly.

The DMWR's are reviewed and updated continually to reflect the latest on-condition PSA requirements. This review and updating activity for a particular item involves determining if the item can be restored by minor repair or if overhaul is necessary. A decision logic that can be used to help determine if minor repair is feasible is given in Figure 5–3. If a minor repair is performed, the time since new or the last overhaul is not changed.

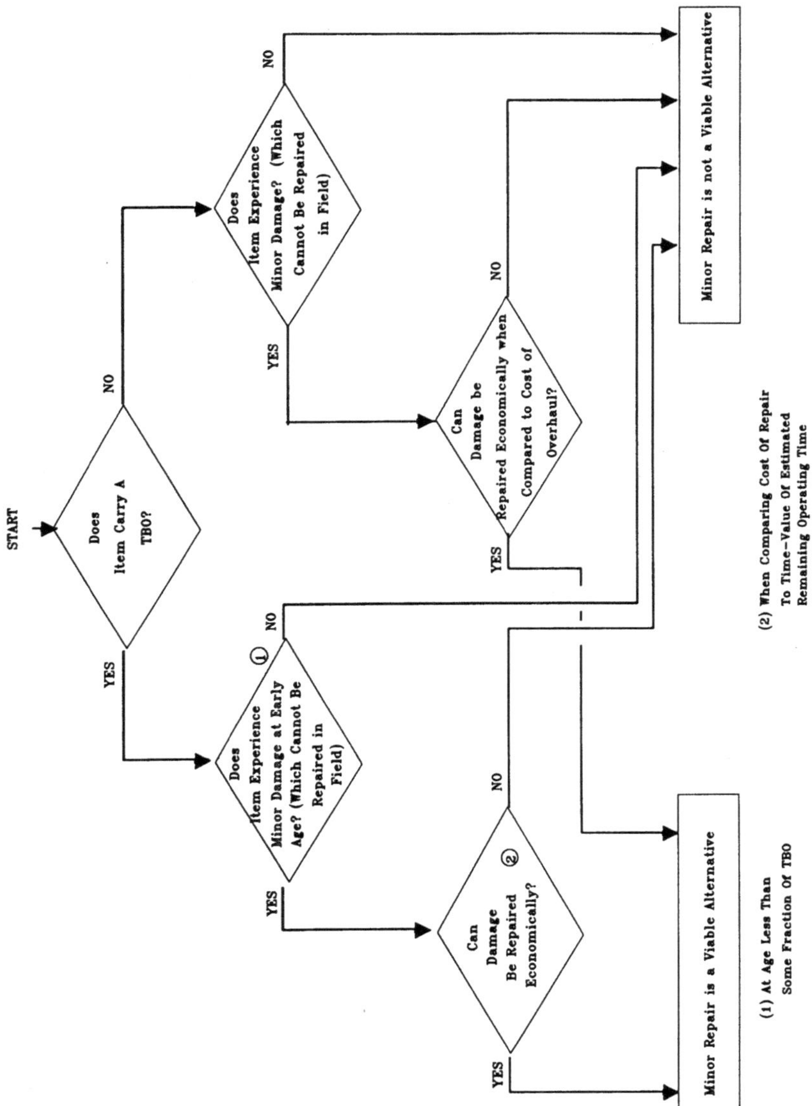

Figure 5–3 Minor Repair Feasibility Decision Logic

If an overhaul is performed, the item's records are adjusted to reflect zero time since overhaul.

The replacement parts needed to support the depot maintenance process are, in general, produced by readily available manufacturing facilities. However, there are many critical low volume, high technology parts which are difficult to acquire and could cause work stoppages due to shortages which would have a significant impact on operational readiness and cost.

These replacement parts are continually reviewed to determine if readiness can be achieved and cost reduced by 'breakout' for purchase from other than prime contractors while maintaining the integrity of the systems and equipment in which the parts are to be used. An often overlooked aspect of spare part procurement is the assurance that the spares are qualified and have equivalent 'as-delivered' reliability to the original hardware. In many cases it has been found that spares have not been rigorously qualified and did not receive a conformance inspection and screening equivalent to that accorded the originally manufactured part. Consequently, spares with poor quality and reliability have been delivered and used for replacement. A rigorous qualification program must be applied to spares, for those parts identified as safety critical, in a similiar manner to that for the initial parts.

The replacement part problem becomes much more crucial if a critically short part also is a flight safety part (FSP). A flight safety part is any part, assembly or installation whose failure, malfunction or absence would cause loss of or serious damage to an aircraft and/or serious injury or death to the occupants. In addition to involving small lot size and a required rapid response to demand, a critically short flight safety part requires high quality and reliability and, generally, has unique design requirements. These critically short flight safety parts need to be produced quickly, with high quality and low cost, and on time.

One approach to meeting this urgent need involves the use of a *flexible manufacturing system* (FMS) in the production of these parts. An FMS, illustrated conceptually in Figure 5–4, is an integrated computer-controlled complex of numerically controlled machine tools, automated material and tool-handling devices, and automated measuring and testing equipment.

The FMS can process, with a minimum of manual intervention and short change-over time, any hardware item belonging to certain specified families of items within its stated capability and to a predetermined schedule. Such systems permit the continuous manufacture of different parts within a family of parts in small batches within a dedicated facility.

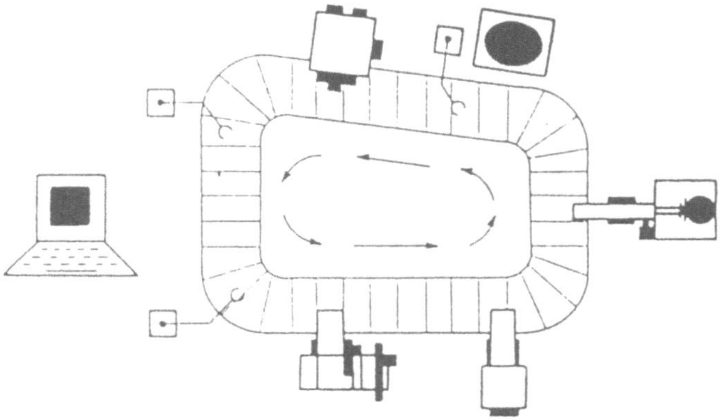

Figure 5–4 Flexible Manufacturing System (FMS) Concept

They use the concept of integrated raw material storage, robot part picking, part transportation by conveyors and direct numerical control machining. Everything is linked together in such a way that the parts being worked on are completely unattended, i.e. travel from raw material storage to finished goods storage in different sequences is under the full control of computers.

A central computer schedules and tracks all production and material movement in the FMS. Based on a family of similar parts, an FMS can be reprogrammed quickly through downloaded instructions from the central computer to individual machines, conveyors and robots to perform a new set of tasks.

An FMS, then, is an automated production system for the manufacture of mid-volume and mid-variety products (or components) with minimal setup times. It consists of several numerically controlled machines integrated with automated workpiece and tool-transfer and handling systems, which are connected to some form of automated warehouse and tool-storage system. All the subsystems of the FMS are controlled by the central computer, which downloads numerical control programs to individual machine tools, controls workpiece flow, and generates performance reports. The functions of scheduling, part-program selection, cutting-abnormality detection, tool-breakage detection, tool-wear compensation, pallet retraction, measuring and self-diagnosis are all carried out automatically.

An FMS is particularly applicable to the manufacturing of critical flight safety parts used in Army aircraft systems and components. It allows more effective control over the entire manufacturing process, the management of changes, and the process parameters and their technical characteristics including:

- Throughput
- FSP Flexibility (Response to Change)
- Variability (Number of Variations within FSP Family)
- Quality (AQL — Reject Rate/Rework, MTBF — Outgoing from Production, Product Life)
- Batch Setup Time
- Turn around Time
- Downtime
- Efficiency (Machine, Human)
- True Cost (Capital Investment, Operating, Inventory)

Of particular importance is that an FMS allows better and more consistent reliability and quality control and reproducibility, including improved inspection, statistical process control, environmental stress screening, material control, failure control and data recording/feedback. An FMS allows small lots to be processed quickly, economically and to various requirements (within a defined capability) and provides diversification of parts in batches.

In addition to maintaining operational readiness by providing a practical means to produce critical flight safety parts, there are several other advantages in using an FMS. In an FMS, production can be continually adjusted to changing needs and to new products, largely by software reprogramming. This allows continuous incremental adaptation to changing requirements for products and systems that otherwise would require major retooling and downtimes. The high entry costs for new product manufacture are greatly reduced because a dedicated facility at partial capacity is no longer required. Therefore, machine utilization is increased and there is quick reaction to market and design changes. There is a reduced time to market for a product.

An FMS allows just-in-time manufacturing and delivery. The right material will be available at the right place and at the right time. Work in progress, lead-times, inventories and setup times will be reduced to an absolute minimum in order to obtain low cost, high quality, on-time production. An FMS will substantially reduce the cost of inventory. Also, cost savings will be realized through a reduction of direct labor.

However, the cost savings will, most likely, not be reflected in the organization's financial statements unless the accounting methods, particularly the method for allocating indirect cost, are adjusted to truly

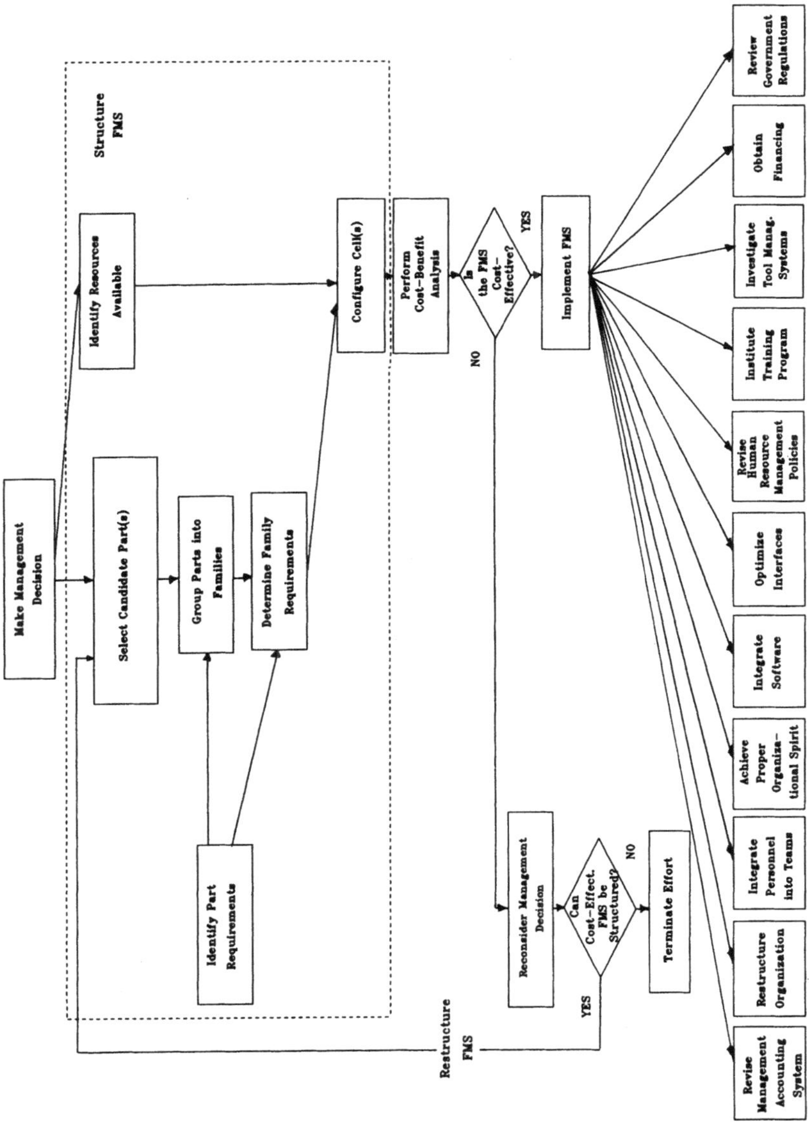

Figure 5–5 A Standardized Methodology for the Structuring and Implementation of a Flexible Manufacturing System

reflect the benefits of an FMS. Since implementation of an FMS reduces direct labor hours per product unit, but increases indirect costs, the common practice of allocating indirect costs by direct labor hours would not be realistic. In an FMS, the direct labor is more for setup and supervision than for actual processing of output. An FMS not only offers significant direct labor savings, but, as previously indicated, also offers considerable improvements in quality, inventory and floor space reduction, great reductions in throughput and lead-times, and flexibility to accommodate product redesigns and new generations of products. Also, the flexibility of the FMS technology ensures that the useful economic life will be much longer than that of traditional dedicated equipment. The accounting methods must be changed to reflect these benefits.

There are many other challenges, both technical and management, that must be overcome to successfully implement an FMS. A key challenge involves the changes that the production personnel must make in order to implement the FMS. The human considerations that must be addressed include the amount of integration required between the operating production departments and the management layers — both horizontally and vertically — the perspectives and skills required to perform tasks in a new way, and the level of understanding that is needed to successfully maintain and operate an FMS.

Figure 5–5 provides a standard methodology that can be applied to facilitate the establishment of the most cost-effective FMS for the fabrication of selected depot replacement parts. Through the methodology, appropriate parts for FMS fabrication can be selected, the corresponding manufacturing computer-controlled machine tools required to produce the selected parts can be determined, the structure of an FMS to produce the parts can be defined, the cost–benefit of the FMS can be determined, and the challenges to implementing it can be addressed. The standardized methodology, but not necessarily each component of it, is applicable to the critically short flight safety replacement parts needed at the depot as well as to other products with similiar characteristics.

The methodology is divided into two major parts, where the first is to structure the FMS and the second is to perform a cost–benefit analysis and to implement the FMS using the analysis as guidance. The first part of the methodology can be iterated, to optimize the cost-effectiveness of the FMS, prior to implementation.

The entire process is initiated by a management decision to investigate the structuring and implementation of an FMS. Such a decision has probably been preceded by a feasibility study, but at this point no serious consideration has gone into determining what an FMS would actually consist of or look like. First, candidate parts are selected. The best parts to be produced by an FMS are those mid-volume and mid-variety parts which are manufactured in small batches themselves, but which fall within a family of parts, with the total family constituting an economical level of production. Other candidates are those parts for which changes are foreseen or which will probably be replaced with parts that will be able to fit within the same family. Depending on the individual circumstances, either all of the parts produced in a given facility or, perhaps, only some of them can be produced using an FMS.

Once the parts have been selected, they must be grouped into families with similar geometric and manufacturing characteristics. The criteria used in segregating parts into families involve: (1) the type of tooling or holding fixture required, (2) the physical size of the part and (3) the geometric shape of the part.

When the parts have been grouped into families, it is then necessary to analyze the function and capacity of each of the required manufacturing and inspection operations to determine the specific requirements for fabricating each family of parts. Since the manufacturing and inspection requirements of each individual part were identified as part of the grouping process, this is primarily a matter of drawing together these individual part requirements into family requirements. From these requirements the manufacturing operation sequence for each family of parts is determined.

In parallel with the determination of the part family requirements, the resources available for supporting the implementation, setup, and on-going operation of the FMS must be identified. This can include existing machines which can be modified for use in the FMS, resources available for purchasing new machines, material handling capability, a facility or portion of a facility for location of the FMS, numerical control programming and other software support including CAD/CAM, personal computers to serve as central computers for integrating the FMS, and fixturing and tooling support.

The part family requirements and the resources available for the FMS are then put together to configure the FMS cell(s) for fabricating the families of parts. At this stage this will probably be just a graphical representation of each cell showing the machines and equipment in the proper operational sequence. Specifications for the machines to be used in the cell(s) configured are then prepared with enough detail in order to perform a cost–benefit analysis. Each cell may be configured to fabricate just one family of parts, or each cell may be able to handle more than one family of parts.

A cost–benefit analysis is then performed on the FMS structure. If the cost–benefit analysis shows that it is not cost-effective to design, install and use the FMS structured, the management decision to implement an FMS must be reconsidered. It is possible that the FMS can be restructured, perhaps by producing a different mix of parts, so that it is cost-effective. In this case the 'Structure FMS' portion of the methodology can be redone with a new set of management guidelines and objectives. Iterations can continue until a cost-effective FMS is structured. If it is not possible to structure a cost-effective FMS, based on the information gained during the 'Structure FMS' portion of the methodology, then the FMS effort is terminated.

If the FMS structured can be cost-effectively designed, installed and used, then the FMS is implemented. The technical changes and management challenges are addressed and specific implementation plans are prepared describing the actions that must be taken and the necessary coordination and scheduling activities. Resources are then obtained, and the appropriate specifications are developed.

5.2 FAILURE MODE INSPECTION TECHNIQUES

The major failure modes which are encountered during the depot maintenance process are:

(1) Corrosion
(2) Wear
(3) Skin Damage
(4) Fastener Damage
(5) Erosion
(6) Cracking
(7) Wiring Damage

Table 5–3 provides an overview of some of the inspection and repair methods considered applicable to these failure modes. Each of these failure modes is described in the following paragraphs in terms of probable causes, applicable NDI methods, indicators of failure or oncoming failure, and applicable repair actions.

(1) *Corrosion*

Corrosion is caused by the presence of salt in moist air, certain chemicals and/or microbes in water, elements in metal, treatment of parts and contact of dissimiliar metals. High temperature and moisture are drivers of fungus and bacterial growth which produce acids and other products which expedite corrosion etching of surfaces and oxidation. Corrosion is normally not as prevalent on painted, clad or plated surfaces. There are various forms of corrosion, including:

- Superficial Corrosion — This type is the least serious on aluminum clad parts. After deposits are removed, an etching is noticeable which results in the clad surface having a series of hills and valleys. Provided the etching has not reached the core, the effect on the strength of the metal is negligible. Corrosion of this type on non-clad alloy parts is serious.
- Intergranular Corrosion — This type of corrosion is not easily detected. It is caused by imperfect heat treatment and occurs mostly in unclad structural aluminum alloy parts. It is the most dangerous form of corrosion for sheet stock because the strength of the metal is lowered without visible structured indicators.
- Stress Corrosion — This type occurs in a part along the line of grain flow if the part is stressed too highly without proper heat treatment.
- Galvanic Corrosion — This type of corrosion occurs when dissimilar metals are in contact and an electrolyte is present at the joint between the metals. For example, aluminum and magnesium skins riveted together form a galvanic couple if moisture and contamination are present. When aluminum pieces are attached with steel bolts or screws, galvanic corrosion occurs between the aluminum and the steel. Figure 5–6 presents a galvanic series chart taken from the US Army aircraft ACE maintenance pamphlet. Metals close together, as illustrated in the table, have no strong tendency to produce galvanic corrosion and are relatively safe to use in contact with each other. The coupling of metals and the distance from each other in the table dictate the galvanic or accelerated corrosion of the metal higher on the table. The farther apart the metals are in the table, the greater is the galvanic tendency, as determined by measurement of the electrical potential difference between them.

TABLE 5-3
Applicable NDI Methods and Repair Tasks For Depot Failure Modes

	Failure mode						
Contributing factors	*Corrosion* ● *Moisture* ● *Chemical action* ● *Dissimilar metals*	*Wear* ● *Gear RPM* ● *Load* ● *Lack of lubrication* ● *Alignment* ● *Vib./flexing* ● *Environmental conditions*	*Skin damage* ● *Environmental conditions* ● *Flexing* ● *Cyclic changes from tension to compression*	*Fastener* ● *Over-torquing* ● *Overloading* ● *Vibration/flexing* ● *Environmental conditions*	*Erosion* ● *Foreign object damage* ● *Environmental conditions* ● *Contaminants*	*Cracking* ● *Flexing* ● *Overloading* ● *Vibration* ● *Thermal/cycling* ● *Over-torquing* ● *Over pressure*	*Wiring* ● *Excessive heat* ● *Aging environmental conditions*
NDI/Repair task							
Liquid penetrant				●		▲	
Magnetic-particle				●		▲	
Electromagnetic						▲	
Ultrasonic			▲			●	
Penetrating radiation	●		●			●	
Visual	▲	▲	▲	▲	▲	●	▲
Dimensional	▲	▲			▲		
Hardness		●					
Vapor degrease					■		
Emulsion degrease					■		
Solvent immersion					■		
Dry-cleaning solvent					■		
Vapor-blast					■		
Hot-alkali soak	■						
Periodic-reverse clean	■				■		
Paint-remover					■		
Chrome plate		■					
Nickle plate		■					
Cadmium plate		■					
Helical coil insert				■			
Chase/tap				■			

Electro-plating tape						■	
Doubler and filter						■	
Repair gusset						■	
Oversized fastener						■	
Glass cloth			■			■	
Weld			■			■	
Metal patch			■			■	
Abrasive-blast	■						
Wire brush	■						
Grind	■						
Polish	■				■		
Lubricate		■					■
Sanding	■				■		
Flame descale	■						
Phosphoric acid treatment	■						
Replacement	■	■	■	■	■	■	■

▲, Primary NDI. ■, Applicable task. ●, Secondary NDI.

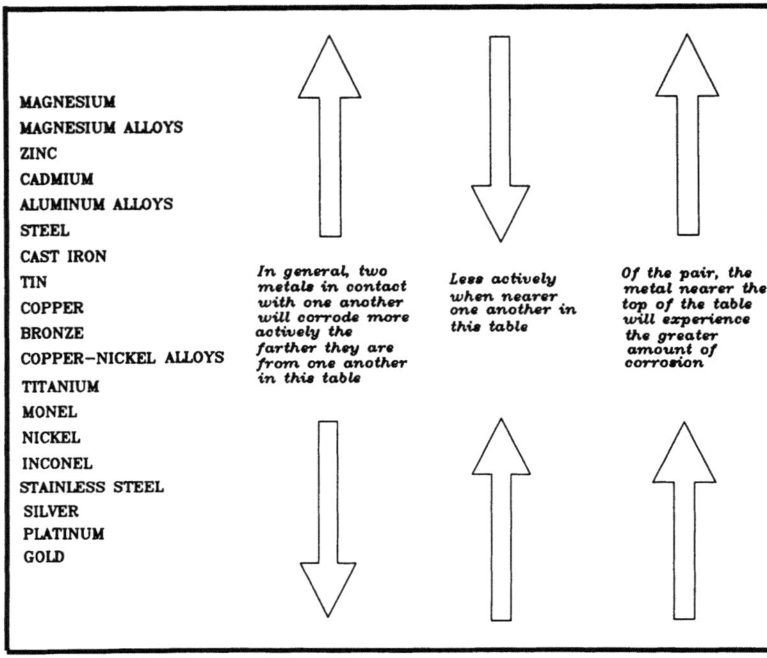

Figure 5–6 Galvanic Series

- Hydroscopic Material Corrosion — This type of corrosion is caused by such materials as sponge rubber, felt and cork absorbing water and holding it in contact with the part.

The extent and form of corrosion are determined primarily through visual and dimensional NDI. A secondary NDI method is penetrating radiation (radiography), but this method is not commonly used. When examining corrosion a fine pointed instrument is used to test the area, but caution must be taken not to further damage the area. It may be necessary to remove scales and powdery deposits before examination can occur.

Indicators of corrosion include brinelling, fretting, scuffing, slatting, galling, etching and abrading. For example, electrolytic action causes the formation of slats and deep etching of surfaces. This occurs at riveted and bolted joints, bearings, slides and screw threads.

The initial action to be taken in corrosion removal is determining if its state is beyond repair. To remove corrosion from steel alloys, the following methods are applicable: hot-alkali soak, abrasive-blast, wire brush, polish, sand, grind, phosphoric acid treatment and flame descale. Caution is needed when removing corrosion to prevent possible dust explosion. Goggles or a face shield should be worn when utilizing wire brush, grinder or abrasive-blast removal methods.

To remove corrosion from aluminum-base alloy materials, chromic acid treatment is applicable. Anodic treatment by the chromic acid process increases corrosion resistance and provides a surface that ensures proper adherence of finishes (for example, paint).

The following are general steps in the corrosion removal process:

(1) Mask all fittings and decals.
(2) Perform cleaning method.

After the part is clean of corrosion, the following actions are applicable, if appropriate:

(1) Coat clean parts, except oil system components that are to be stored, with Rust Nox or Rust Foil.
(2) Coat oil system components with lubricating oil.
(3) Cover all unprotected openings with caps, plugs or other suitable covers.
(4) Package all oil system components in plastic bags.

(2) *Wear*

Wear is indicative of two contacting surfaces abrading one another. The result is a loss of material from contacting surfaces. This loss of material may occur as microscopic particles or as large particles. Wear is an expected occurrence as no surface is free from friction or foreign matter. The following factors contribute to wear:

- Gear rpm — Rapid, continual contact between moving parts accelerates the wearing process. Gears are made of hardened material, for example, hardened steel, to resist this action.
- Load — Weight of one item upon another creates a frictional force. The end result of friction is a continual wearing process when any movement is present.
- Lack of Lubrication — The lack of a lubricant allows for excessive frictional heat build-up. The end result of friction is a continual wearing process.
- Misalignment — Improper positioning or adjustment of parts in relation to each other causes unexpected contact and eventual wear.

- Vibration/Flexing — Unnecessary and unexpected random movement between two surfaces results in frictional forces between the surfaces and initiates wear.
- Environmental Condition — Conditions promoting the contamination of surfaces with particles increases the abrading process. The particles act as cutting stones, etching the contaminated surfaces.

The presence and extent of wear is determined through dimensional, hardness and visual NDI. When examining parts for wear, the part surfaces under question should be free of dirt and other substances.

Two major actions associated with wear are: (1) fretting — discoloration of contacting parts resulting from the removal of original surface materials and (2) frosting — minute indentations between two contacting surfaces. Other results of wear are grooving, gouging, and burning. As an example of wear, rivets and bolts may wear the skin, spar and frame holes so that there is not a correct fit in the holes for adequate strength in joints or attachments of a wing section. This can occur due to continued flexing of components from use or because of severe stress due to unusual operating conditions in turbulent weather or an adverse landing. This condition may also result in radial cracks from bolt holes.

To correct the effects of wear, action is taken to build up the area(s) lacking material. Several processes are applicable: metal spraying, used on bushings, sleeves and skins; welding; and metal plating. Of these corrective methods, metal plating is the most common. The major metal plating processes utilized at the depot are chrome plating, nickel plating, cadmium plating and silver plating. Each process includes three steps:

(1) Plate to desired thickness
(2) Bake in accordance with specifications
(3) Machine to dimension

Unless otherwise specified, parts harder than Rockwell C-40 which have been ground after heat treatment are suitably stress-relieved before plating. Plating of carbonized areas is not to be attempted due to potential hydrogen embrittlement. A part suspected of being carbonized is tested for hardness. Noncarbonized areas have a Rockwell 15–N range of 75–82; carbonized parts are usually above 90.

(3) *Skin Damage*

Skin damage affects the overall aircraft structural strength and aerodynamic stability. Excessive skin damage is capable of leading to a safety of flight incident. The following factors contribute to skin damage:

- Environmental Conditions — Conditions leading to skin damage include extreme values of cold and heat. Varying temperature ranges cause the skin to expand and contract, eventually breaking down the original skin strength.
- Flexing — Continual flexing breaks down skin structural strength and permits an acceleration of defects.
- Cyclic Changes from Tension to Compression — Continual action of tension to compression breaks down skin structural strength and permits an acceleration of defects.

Skin damage is easily detected through both visual and ultrasonic NDI. Utilizing ultrasonic NDI requires trained interpretation of results. Common indicators of skin damage include the following defects: nonuniformity, gaps, cracking, holes, blistering and delamination. Delaminated areas have a whitish appearance through translucent piles; if the piles are not translucent, delamination may be detected by lightly tapping the radome surface area with a small metal object (coin tap test), such as a 25 cent piece, taking care that the tapping is not vigorous enough to damage the part. The sound of such tapping is a clear metallic ring over well bonded areas and a dull thudding sound over delaminated areas. A relatively large delaminated area results in a soft, flexible or ballooned characteristic of the debonded laminate and is detected visually or by exerting thumb or hand pressure to the area. To perform a repair, the damaged area is removed and replaced with approved materials in such a manner that normal stress can be carried over the area.

(4) *Fastener Damage*

Fasteners including screws, threaded inserts, internal threads and studs are damaged because of:

- Over-Torquing — Excessive torque puts unexpected strain on fastener threads and can lead to the stripping of threads.
- Overloading — Excessive load is capable of exceeding component safety factor. The safety factor is the fastener's minimum strength vs maximum stress.
- Environmental Conditions — Conditions to which fasteners are exposed promote decomposition and eventual loss of strength, leading to breakage.
- Vibration/Flexing — Erratic action and stress imposed upon fasteners promote wearing characteristics and cracking.

Damage to fasteners is usually evident through the performance of visual inspection. If damage is suspected, but not detected visually, the further inspection using liquid penetrant or magnetic-particle NDI should be performed.

Fastener damage is indicated through the presence of crossed threads, looseness, breakage and stretching. For example, a fastener hole may become oversized due to vibration and flexing; this dictates that the fit of the fastener be corrected to provide a tight immobile bonding of parts.

Often, it is more cost-effective to replace a fastener rather than repair it. This determination is based on fastener acquisition cost, man hours needed to perform repair, type of repair needed, etc. To repair a damaged thread, a suitable chasing tool, tap or die is used. If damage is too great, a screw thread insert (helical coil) is used. Broken or loose studs in a tapped hole are repaired by using an oversized stud or screw thread insert (helical coil).

(5) *Erosion*

Erosion is indicative of an abrasive substance wearing away a part. This action is most evident on the airframe and within the engine. The following factors contribute to erosion:

- Foreign Object Damage (FOD) — Foreign objects hitting the aircraft create both denting and chipping of airframe surface.
- Environmental Conditions — Conditions promoting erosion include rain, sand storms, etc. These conditions abrade away surface coatings.
- Contaminants — The injection of contaminants into the engine air intake, for example, sand and dust, promotes the wearing away of components such as the main rotor, compressor or stator blades.

Erosion can be detected by visual or dimensional NDI. Discrepancies due to erosion consist mainly of scars, scratches, surface abrasions, and excessive wear. For example, continued effects of rain erode away airframe protective coating, i.e., paint, exposing the airframe to corrosion effects and removing aircraft camouflage characteristics.

(6) *Cracking*

The formation of cracks is due to excessive stress vs material strength. Cracking results in loss of component stability and is capable of leading to a safety of flight incident if not detected and corrected. Both dynamic and static components (for example, gears, stringer cutouts and tubing) are susceptible to cracking. The following factors contributing to cracking are encountered at the depot:

- Overloading — Excessive load is capable of exceeding component safety factor. The safety factor corresponds to component minimum strength vs maximum stress.
- Vibration/Flexing — Erratic action and stress cause material strength breakdown.
- Thermal Cycling — Varying temperature extremes cause expansion/contracting actions, leading to material strength breakdown.
- Over-Torquing — Excessive torque puts unexpected strain on components, leading to material separation.
- Over Pressure — Excessive/unexpected pressure causes material deformation, inducing material separation.

The detection of cracking is achieved through visual, magnetic-particle, liquid penetrant, ultrasonic, electromagnetic or radiographic NDI. When material is stressed at a maximum limit at a high temperature or is repeatedly flexed, fracture of material occurs in the form of thin hairline cracks. This is the most difficult service type failure to detect by radiographic NDI. Normally cracks which are easily detectable by radiographic NDI are visible to the naked eye.

Materials treated at high temperatures often develop intergranular or transgranular cracks. Other indications of cracking are fracturing, crazing and breaking.

Welding techniques are commonly used to repair cracked parts. Methods include fusion welding, spot and seam resistance welding, and electron-beam welding. For cracking in cowling or airframe skin, the cracked area is cut out and replaced with a welded-in patch made from the same type of material. Upon completion of any crack repair that involves welding, a hardness NDI is performed on the weld to ensure resistance to expected stresses.

(7) *Wiring Damage*

Aircraft wiring is vital to total electrical system operation. It is imperative that damaged wiring be corrected to help ensure mission capabilities and safety. The following factors contribute to wiring damage:

- Excessive Heat — Heat build-up in wiring causes insulation materials to melt.

- Aging — Long periods of time eventually cause wiring insulation to dry out, become brittle and develop cracks.
- Environmental Conditions — Conditions of changing environments, e.g., from hot to cold, create a breakdown in insulation characteristics, allowing the insulation to become brittle and fall apart.

In the event that wiring damage is not readily visual, other action is taken to isolate the malfunctioned component in the system. Troubleshooting actions should be supported by printed procedures, diagnostic circuits, test points and diagnostic routines.

Typical problems that arise in electrical systems are positive lead shorted to ground, open circuit, shorted relay contacts and low power. The following are general criteria for wiring replacement:

(1) Wiring damaged to the extent that the primary insulation has been broken.
(2) Wiring having weather-cracked outer insulation.
(3) Wiring exposed to battery acid or on which the insulation appears to be, or is suspected of being, in an initial stage of deterioration due to the effects of battery acid.
(4) Wiring showing evidence of overheating.
(5) Wiring insulation saturated with engine oil, landing gear lubricant or hydraulic fluid.
(6) Wiring bearing evidence of having been crushed or severely kinked.
(7) Shielded wiring where the metallic shield is frayed or corroded. Cleaning agents or preservatives are not to be used to minimize the effects of corrosion or deterioration of wire shields.
(8) Wiring bearing evidence of breaks, cracks, dirt or moisture in the plastic sleeves placed over wire splices or terminal lugs.
(9) Sections of wire where splices occur at less than 10–foot intervals.

Any wiring that is suspected of not being of high quality is to be replaced. Replacement wiring is to be of similiar quality as original wiring. A wiring diagram should be addressed to ensure that all proper connections are achieved and system operations have not been altered.

The methods described above are essential in the detection and repair at the depot of damaged or defective parts. For example, gear and splined parts are inspected, in accordance with appropriate NDI methods, for conditions that signify failure or the onset of failure. Discrepancies that occur in gear and spline units are cracks, damaged threads, excessive wear, nicks, dents, scores, burrs and scratches. Table 5–4, taken from an Army

TABLE 5-4

Gear Discrepancies – Visual 4 × Magnification (Sample Inspection Criteria – taken from an Army Aircraft Engine DMWR)

Condition	*Applicable gear*
Corrosion	
Any rust corrosion noted after cleaning is unacceptable	All gears
Cracking (unacceptable)	All gears
End loading	
Normal and moderate end wear is unacceptable	Primary and secondary gears
Pitting end spalling due to end loading is unacceptable	Output gear and sun gearshaft
Light end wear is acceptable; all pitting or spalling due to end loading is unacceptable	Accessory gears
Frosting	
Light uniform frosting is acceptable; any other condition of frosting is unacceptable	Primary planet gear and sun gearshaft
Light frosting is acceptable; any other condition of frosting is unacceptable	Accessory gears
Light to medium frosting is acceptable; any other condition of frosting is unacceptable	Secondary planet gear and output gear
Pitting	
Light and initial pitting is acceptable at pitchline; any other condition of pitting is not acceptable	Primary and secondary planet gears and output gear
Initial and light pitting is acceptable; any other condition of pitting is unacceptable	Accessory gears
Pitting is unacceptable	Sun gearshaft
Scoring	
Light scoring is acceptable; any other condition of scoring is unacceptable	All gears
Spalling (unacceptable)	All gears
Wear	
All light wear of a uniform pattern around the circumference is acceptable; any other wear condition is unacceptable	Primary planet gear and sun gearshaft
Light to moderate wear of a uniform pattern around the circumference is acceptable; any other wear condition is unacceptable	Output gear
All normal and moderate wear of a uniform pattern around the circumference is acceptable; any other condition is unacceptable	Secondary planet gear
All normal and moderate wear is acceptable; heavy and irregular wear is unacceptable	Accessory gears

aircraft engine DMWR, provides examples of gear discrepancies and sample acceptance criteria. Gear and spline deficiencies are repaired by:

- Using blending techniques to repair minor nicks, burrs and scratches on other than working surfaces of gears or splined components.
- Using chrome plating techniques to restore worn bearing or seal journals on gearshafts (i.e., noncarbonized or nonnitrided gearshafts).
- Machining splined ends of shafts off and replacing by electron beam welding techniques.

Table 5–5, also taken from the Army aircraft engine DMWR, provides examples of bearing discrepancies and acceptance criteria. Common discrepancies that occur in bearings are pitting, scoring, wear, corrosion, and brinelling. Magnetic-particle and/or fluorescent-penetrant nondestructive inspection methods are most effective in detecting these discrepancies. When performing inspections, blade oxide finishes should not be confused with overheating. Also, bearings which are subject to heavy load or high speeds are not to be used when condition is questionable. Bearing inspection should be performed in a well lighted room with the room and workbench being vibration-free and maintained at a maximum practicable level of cleanliness. The following are general guidelines for detecting bearing discrepancies:

(1) Clean bearings thoroughly.
(2) Allow all bearings to stand for a minimum of 3 hours before inspection. This permits dimensional stabilization.
(3) Use 0 × magnification to visually inspect all bearings and retainers. If necessary, bearing rings may be inspected using aid retainers.
(4) Utilize a radius probe to determine the severity of various surface defects.
(5) Inspect the thickness of bearing plating using a nondestructive thickness tester.
(6) Perform hardness inspection, particularly if heat discoloration is evident.

The application of NDI techniques is a key part of the depot maintenance process. An NDI is performed to determine an item's integrity; composition; physical, electrical or thermal properties; or dimensions, without causing a change in any of these characteristics.

TABLE 5-5
Bearing Discrepancies – Visual 0 × Magnification (Sample Inspection Criteria – taken from an Army Aircraft Engine DMWR)

Condition or damage	*Limits*
Banding	Acceptable
Brinelling, false	Acceptable
Brinelling, true	Noticeable depressions unacceptable; minor ring brinelling, so slight that it only can be detected in reflected light, will not be cause for rejection
Corrosion discoloration	Corrosion effectively removed by standard cleaning methods is acceptable. Remaining isolated pitting on active surfaces is acceptable provided: a. It cannot be felt with a 0·020 inch radius probe b. No more than three visually evident individual pits exist in an 0·25 inch diameter area
Fretting	Acceptable, provided rings do not have fretting on more than 20% of their inactive areas. Fretting on ring faces may be removed by minor lapping rework, provided rings meet dimensions' requirements
Frosting	Acceptable, provided it cannot be felt with a 0·040 inch radius bearing probe
Galling	Not acceptable
Greasing	Not acceptable
Heat discoloration	Varnishing is acceptable, provided any heavy varnish films can be removed by standard cleaning. Staining is acceptable, provided stain is not caused by acid etch as observed after standard cleaning. Heat discoloration: Bearings discolored straw or brown color are acceptable. Bearings discolored red-purple, purple, or blue are acceptable, provided hardness inspection at three locations on both inner and outer ring faces is within limits. (When there is any doubt as to a tempered discoloration condition, hardness inspections should be performed)
Indenting	Isolated indenting acceptable, provided it cannot be felt with a 0·040 inch radius bearing probe
Nicking	Small isolated nicks with no projections are acceptable on active surfaces if they cannot be felt with a 0·040 inch radius bearing probe. Minor nicks on inactive surfaces are acceptable
Pitting, fatigue	Not acceptable
Retainer damage	a. Any cracking is unacceptable b. Overheating, as evidenced by melting or flowing of silver plate, is unacceptable c. Wear of ball pocket and/or land-riding surfaces exposing any base metal up to 0·063 inch measured in any direction (not depth) is acceptable. Wear beyond this limit is unacceptable
Roller end wear	Not applicable
Scoring	Isolated axial scoring acceptable, provided it cannot be felt with a 0·040 inch radius bearing probe
Scuffs and scratches	Acceptable, provided they cannot be felt with a 0·040 inch radius bearing probe. A maximum of two scratches per square inch of active surface is allowed
Seizing	Not acceptable
Wear	Determined by inspection of radial internal clearance, axial end play contact angle, and other dimensional inspection requirements

Note: Minor scoring, fretting or wear of the outer ring outer diameter may result from spinning or movement of the bearing in its housing. This ring outer diameter wear is acceptable, provided outer diameter measurements taken at various locations (in the wear area) are within limits.

NDI methods in the hands of trained and experienced technicians are capable of detecting flaws or defects with a high degree of accuracy and reliability. It is important for maintenance engineers to be fully knowledgeable of the capabilities of the NDI methods and it is equally important to recognize the limitations of the methods. No NDI method is ever considered conclusive. To be reliable, a defect indication detected by one method must be confirmed by another method.

NDI equipment is highly sensitive and is capable of detecting discontinuities and anomalies which are of no consequence in the service for which a component is used. Limits for acceptance and rejection are therefore as much a part of an inspection as the method itself. For example, ultrasonic inspection equipment is fully capable of detecting normal grain boundaries in some cast alloys. Therefore, inspection criteria are designed to overlook these 'normal' returns and to discriminate in favor of those discontinuities which will affect the component in service.

The NDI methods commonly used during depot maintenance are described in the various Army DMWR's, technical manuals and in information compiled from the Nondestructive Testing Information Analysis Center (NTIAC) managed by the Southwest Research Institute, San Antonio, Texas, USA. These methods are summarized in the following paragraphs.

Liquid Penetrant NDI

Dye-Penetrant Inspection

This inspection method detects surface discontinuities, such as cracks, in ferrous and nonferrous materials. It involves the following steps:

(1) Clean part utilizing the vapor degreasing method.
(2) Wipe part clean and display.
(3) Apply penetrant by immersion, spraying or brushing to completely cover all surfaces to be inspected.
(4) Allow penetrant to remain on surface for approximately 8 minutes, then wipe with clean, lint-free cloth. [Repeat step (3) if needed.]
(5) Apply developer by brushing or spraying thin, even coating over surfaces involved.
(6) Allow approximately 4 minutes for developer to dry to a thin layer before inspecting.
(7) Inspect parts under ordinary white light: interpret results as follows:

- No distinct pattern
  — no faults apparent
- A continuous line
  — indicates a crack
- An intermittent line
  — indicates a partially closed defect at the surface
- Rounded area
  — indicates deep crater cracks in welds or porosity caused by gas or pin holes
- Small dot
  — indicates a porous condition due to pinholes or coarse grains in casting.

(8) Clean part utilizing vapor degreasing method.

Fluorescent-Penetrant Inspection

Wear rubber gloves when performing this inspection; oil remaining on the skin may cause skin inflamation. Presence of penetrating oil on the skin is detected under ultraviolet (black) light. Developing powder is harmless if inhaled, but heavy concentrations can be annoying. This inspection involves the following steps:

(1) Clean all surfaces thoroughly.
(2) Immerse, spray or brush all surfaces of the part being inspected with fluorescent-penetrant.
(3) Allow 15–30 minutes for fluorescent-penetrant to enter defect.
(4) Immerse, spray or brush all surfaces of part with fluorescent-penetrant emulsifier.
(5) Allow sufficient time (3 minutes minimum) for emulsification of excess fluorescent-penetrant.
(6) Remove all penetrant from surface of part with warm water.
(7) Dry part in hot air dryer at 140–180°F.
(8) Apply fluorescent-penetrant dry developer powder to part by dusting or powder box immersion.
(9) Allow approximately 15 minutes for indications to develop.
(10) Examine part in a darkened enclosure under ultraviolet (black) light.
(11) Interpret results as follows:
  - No distinct pattern
    — no faults apparent
  - A continuous line

— indicate a crack
- An intermittent line
  — indicates partially closed defect at the surface
- Rounded area
  — indicates deep crater cracks in welds or porosity caused by gas or pin holes
- Small dot
  — indicates a porous condition due to pinholes or coarse grains in casting.

Magnetic-Particle NDI

Continuous-wet method

This inspection method is applicable *only* to ferromagnetic steel and is *not* effective on nonferrous materials. It involves the following steps:

(1) Clean all parts free of grease and foreign material.
(2) Plug all passages that are too difficult to clean.
(3) The conductor outer diameter should approximate the part bore inner diameter.
(4) Immerse part into the wet bath solution.
(5) Remove part from the wet bath solution and apply magnetizing current shot.
(6) Inspect at low-current application first and then at high-current application.
(7) Accomplish circular and longitudinal magnetization by using three current applications each ranging from 0.4 to 0.5 second duration. (Apply the wet bath during the first two current applications only.)
(8) Magnetize parts longitudinally into the magnetizing oil so that the major part axis, centers of long shafts and the axis of rotating parts are parallel with the coil axis. (Inspect nuts by circular magnetization.)
(9) Interpret results by visual inspection as follows:
  - No distinct pattern or gathering of particles
    — indicates no surface or subsurface cracks
  - A distinct sharp and well defined build-up of particles
    — indicates a surface crack
  - A broad fuzzy looking accumulation of particles
    — indicates a subsurface crack.
(10) Demagnetize part with a demagnetizing coil.

(11) Clean parts by rinsing in pure bath solution without magnetic particles.

Electromagnetic NDI

Eddy Current Inspection

Crack detection standards are necessary to assure that an eddy current test set-up is sufficiently sensitive to detect small cracks. Proper adjustment of an instrument for material type, lift-off adjustment, etc., does not necessarily indicate that the instrument is sufficiently sensitive to detect small cracks. A suitable crack standard to check instrument calibration is mandatory. It involves the following steps:

(1) Consult test equipment manual for operating and adjustment instructions.
(2) Check probe sensitivity using a crack standard.
(3) Inspect the surface in question by rotating the probe so that the pick-up coil passes over all the surfaces.
(4) If a crack exists, the instrument will show a sharp deflection as the probe passes over it. Do not confuse this sharp deflection with the slow minor changes in the conductivity of the base material.
(5) Adjust the test equipment balance control as required to bring the meter in scale as the coil approaches the damaged surface or edge of a hole.
(6) Repeat process, as needed, to verify meter readings.

Ultrasonic NDI

Pulse Echo Inspection

This technique is used for inspecting materials and parts to detect discontinuities detrimental to the serviceability of a part. It involves the following steps:

(1) Examine material or part for any surface irregularities (for example, burns and gouges) and remove with a sander capable of producing a surface finish equal to that required for a good response. Clean part of any loose scale, dirt or foreign material which will interfere with the transmission of the ultrasonic vibrations.
(2) Select the proper test frequency and search unit. The transducer must be capable of efficient operation at the same rated frequency as that to be used in the test. In general, a high frequency is used in the detection of internal defects of small magnitude; however,

this gives minimum penetrating power. A low frequency will give the greatest penetrating power but is less sensitive to small defects.

(3) Calibrate the device according to manufacturer's specifications.

(4) Use the required ultrasonic quality level stated in the applicable repair procedure as the material acceptance criteria.

For material areas identified as class AA:

— Discontinuity indications in excess of the response from an 0.047 inch diameter flat-bottomed hole at the estimated discontinuity depth are not acceptable.

— Discontinuity indications greater than 10% of the response from an 0.047 inch diameter flat-bottomed hole at the discontinuity depth are not to be closer than 1 inch or exhibit a length greater than 0.125 inch.

— Harsh or sonic noise is not to exceed 10% of the response height received from an 0.047 inch diameter flat-bottomed hole at the estimated discontinuity depth.

— With the instrument set so that the first back reflection from the correct test block is at 80% of the screen saturation adjusted for nonlinearity, the material is inspected for loss of back reflection. Any loss in back reflection in excess of 50% of full saturation of the screen is considered not acceptable.

For material areas identified as class A:

— Discontinuity indication in excess of the response from an 0.078 inch diameter flat-bottomed hole at the estimated discontinuity depth is acceptable.

— Multiple indications in excess of the response from an 0.047 inch diameter flat-bottomed hole are not closer than 1 inch.

— Elongated (stringer) type defects in excess of 1 inch in length are not acceptable, if, at any point along the length, the discontinuity indication is equal to or greater than 50% of the response from an 0.047 inch diameter flat-bottomed hole.

— Multiple discontinuities giving an indication less than the response from an 0.078 inch diameter flat-bottomed hole are acceptable only if the back reflection pattern is 50% or more of the back reflection pattern of sound material of the same geometry. The sound beam must be normal to the front and back surfaces to ensure that loss of back reflection is not caused by surface roughness, surface waviness or part geometry variation.

For material areas identified as class B:

— Discontinuity indications in excess of the response from an 0.125 inch diameter flat-bottomed hole at the estimated discontinuity depth are not acceptable.
— Discontinuity indications in excess of the response from an 0.078 inch diameter flat-bottomed hole at the estimated discontinuity depth are not to be closer than 1 inch.
— Elongated (stringer) type defects in excess of 1 inch in length are not acceptable, if, at any point along the length, the discontinuity indication is equal to or greater than the response from an 0.078 inch diameter flat-bottomed hole.
— Multiple discontinuities giving an indication less than the response from an 0.078 inch diameter flat-bottomed hole are acceptable only if the back reflection pattern is 50% or more of the back reflection pattern of sound material of the same geometry. The sound beam must be normal to the front and back surfaces to ensure that the loss of back reflection is not caused by surface roughness, surface waviness or part geometry variation.

For material areas identified as class C:

— Discontinuity indications in excess of the response from a 1 inch diameter flat-bottomed hole at the estimated discontinuity depth are not acceptable.

Material or parts failing to meet the above requirements shall be subject to rejection. In the inspection of machined parts discontinuity indications in excess of the specified ultrasonic quality level are permitted if it is established that such discontinuities are removed by subsequent machining operations. In such cases, a record of the ultrasonic test results is provided showing the location and size of indications by discontinuity class with respect to a bench-mark on one corner of the surface from which the material is scanned.

Penetrating Radiation NDI

Radiographic (X-Ray) Inspection

To guard operating personnel from possible danger of X-ray absorption, cover rear side of film holder with a sheet of lead thick enough to absorb fully any secondary reflected radiographic rays. As a further precaution, all personnel should wear a radiation detector-type badge or cylinder. This technique involves the following steps:

(1) Operate radiographic device according to manufacturer's instructions.
(2) Determine the degree of sensitivity of inspection desired and the required film density using a color densitometer.
(3) Use fine grain, high contrast, safety-type industrial radiographic film to ensure that all radiographs are clean and sharply define the existing discrepancy.
(4) Make measurements of the object and note differences in shape and size. This will allow determination of correct placement of the X-ray generator for an exposure or series of exposures. When a series of exposures is to be made, make up a position chart to map out the areas to be radiographed.
(5) Determine access to the object with regard to the X-ray generator.
(6) In areas of high stress, take a sufficient number of views to establish the nature and extent of the discrepancy.
(7) Visually inspect radiographs for evidence of material change, cracking, etc.

Visual NDI

This inspection should be used in conjunction with all inspection methods. It involves the following steps:

(1) Perform inspection in a well lighted room which is free of dust and dirt, if possible.
(2) Cover work benches with clean dry paper.
(3) Use naked-eye observance to detect faults.
(4) Look for:
 — Loose or missing parts
 — Cracks
 — Distortion
 — Wear
 — Erosion
 — Corrosion
 — Damage to surface coating
 — Nicks
 — Dents
 — Burned areas
 — etc.
(5) Use micrometers and special gauges where applicable.
(6) If subsurface flaws are suspected, perform a magnetic-particle, fluorescent-penetrant or radiographic NDI.

Dimensional NDI

This inspection is specifically designed for analyzing areas of material wear, warpage, erosion and corrosion. It involves the following steps:

(1) Clean part of any dirt, scale, etc.
(2) Identify indentations, gouges, and deformations visually.
(3) Use measuring devices as applicable.

Performing this inspection involves the utilization of measuring devices such as vernier or micrometer calipers. These devices must be calibrated at a fixed temperature because of thermal expansion and contraction. There is always some degree of error due to the difference in the coefficients of thermal expansion (CTE) between the measuring device and the part or material. 'Steel on steel' usually produces a negligible error unless the size of the dimension and the temperature are at extremes and the dimensional tolerance is very small. Magnesium and aluminum alloys have a high strength to weight ratio, making them popular materials in the design of aircraft systems. These alloys have relatively high CTE and when large dimensional measurements are made with a steel measuring device the parts and measuring device must be stabilized as close as possible to the calibration temperature or corrected based on temperature, for each individual dimension.

Hardness NDI

This inspection method is a test used to determine material hardness and provides information to evaluate accept/reject criteria. It is primarily used to evaluate part hardness characteristics for suspected metallurgical wear.

Rockwell Hardness Test

There are two types of Rockwell hardness testers, standard and superficial. The standard tester has a load range from 60 to 150 kilograms and is used for general aircraft parts. The superficial tester has a load range from 15 to 45 kilograms and is used mostly for surface-hardened and thin materials. The test involves the following steps:

(1) Prepare sample by filing, grinding and polishing to remove all scratches and variations that may affect reading.
(2) Select proper penetrator and place corresponding weight on weight plan.
(3) Place sample on anvil and, by turning hand wheel, raise it slowly until contact is made with penetrator. Continue turning until

pointer of indicator has made three revolutions and is within five divisions (plus or minus) of upright position. This applies the 10–kilogram or minor load on sample.
(4) Apply major load by means of handle.
(5) Release major load by returning handle to its original position and read hardness number directly on indicator scale.

Brinell Hardness Test

The Brinell hardness test consists of pressing a hardened steel ball into a flat surface of the metal being tested by application of a known pressure. The impression made by the ball is measured by means of a microscope with a micrometer eyepiece, and the Brinell number is obtained by dividing the load in kilograms by the area of the spherical impression made by the ball (load/area). The test involves the following steps:

(1) Prepare sample by filing, grinding and polishing to remove all scatches and variations that may affect reading.
(2) Place sample on anvil and elevate until hardened ball contacts surface to be tested.
(3) Apply load by pumping handle. A load of 3000 kilograms is required for steel, while 500 kilograms is used when testing softer metals. Apply load for 30 seconds. This time may be increased to 1 minute for hardened steels.
(4) Release pressure and measure area of impression with calibrated microscope.
(5) Calculate Brinell number (load/area).

5.3 AIRFRAME CONDITION EVALUATION (ACE)— AN ON-CONDITION MAINTENANCE TECHNIQUE

An on-condition maintenance (OCM) technique, called airframe condition evaluation (ACE), is used to select aircraft for depot maintenance. ACE is a meaningful and inexpensive OCM method for ranking aircraft, within a fleet, as candidates for depot level maintenance. It is a particular approach to OCM in which the condition of an aircraft is established from a carefully designed profiling technique.

A team of specially trained technicians is sent out each year to inspect airframes on every helicopter. During the inspection the aircraft panels and structures are evaluated in terms of certain specifically selected condition indicators and assigned a score. Those aircraft assigned a higher score will be inducted for depot overhaul.

Typical indicators include the condition of the main lift beam, the nose fuselage skin and the upper bulkhead, and the state of the corrosion protection. Weights are then assigned to each of the indicators using ranking and distribution techniques. The evaluation includes examining the basic aircraft structure for corrosion defects together with an assessment of the external areas of components, both structural and dynamic, for deterioration caused by corrosion.

ACE uses for its evaluation a representative list of indicators of structural condition selected for each aircraft type. Each indicator is further defined by condition codes which depict the condition of the indicator, e.g., no defect, cracked, buckled, etc., and the results are recorded on special worksheets.

The process of identification, selection and review of indicators is a key element of the ACE process. Aeronautical importance, the extent of deterioration and how fast a defect will cause further deterioration, and the repair capability at the depot are all considered in the establishment of cost-effective indicators. The indicators are also selected for ease of accessibility and inspectability. This is essential in order to avoid extensive airframe disassembly and the use of cumbersome and overly complex equipment and to be able to perform the evaluation quickly. Condition codes are then assigned to each indicator to denote the pertinent range of severity encountered. The indicators and codes are continually reviewed and updated to reflect current field experience and changing depot capability.

Indicators are ranked by their degree of criticality and safety and economic benefits to be derived if the symptom and, more importantly, its cause are eliminated by depot maintenance. A subjective technique is used, based on the Pareto distribution, to establish a logical balance between the various ranked indicators in terms of their relative criticality (Figure 5–7). The weight distribution for the indicators is determined by using the ratios of areas under the truncated curve. By proper choice of the constant A, weighting of the indicators can be adjusted to achieve the curve balance desired. The choice of A is a management decision and is usually related to the desired weight percentage of the first designated number of indicators. Once the indicators, condition codes, and weights are established, the process is ready for implementation.

A trained ACE team conducts an annual evaluation of each aircraft's condition, using the established indicator and condition codes. Any faulty

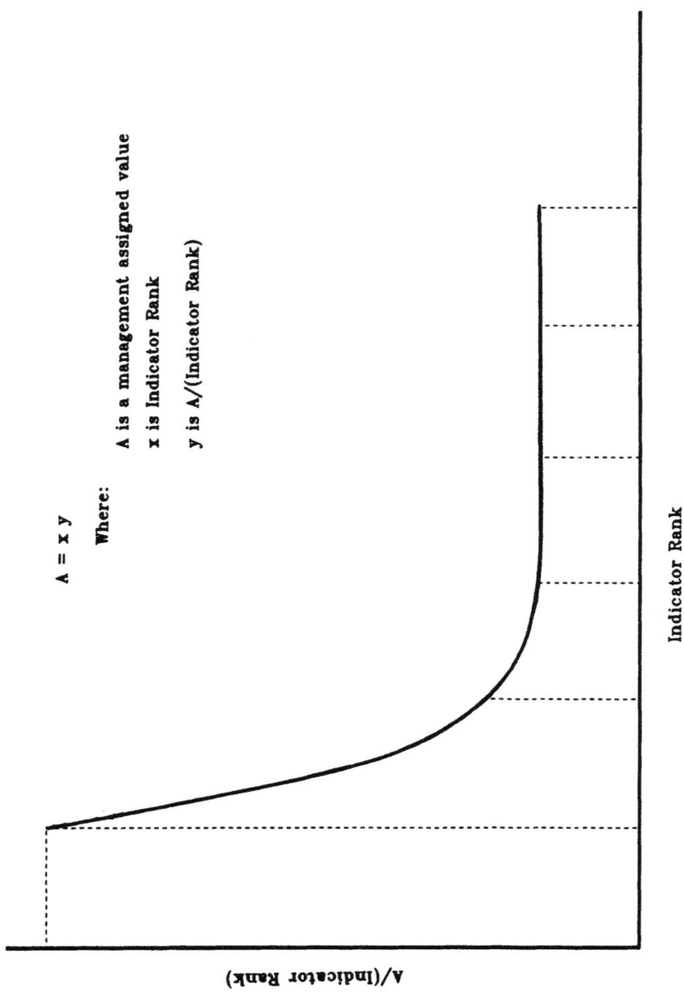

Figure 5–7 Pareto Distribution Curve

indicators in terms of their worst condition code (degree of severity) are noted. The team does not attempt to assign weights or make any other computation in the field. The profiling does not require a complete technical inspection of the aircraft and, therefore, does not duplicate any other scheduled inspections required to be performed by the owning unit's regularly assigned personnel, nor can it be construed as such. However, any safety-of-flight discrepancies noted by the team are verbally brought to the attention of the field unit, and responsibility for action rests there. The activities of the inspection team are limited to its specific defined function and do not constitute an evaluation of the field unit's maintenance capability or performance. The data collected by the ACE team create an information base whereby better management decisions and actions can be derived through engineering analysis.

After the field evaluation, the condition of each aircraft in the fleet is computed in terms of a single numerical value known as the profile index (PI). The higher the PI, the worse the condition of the aircraft.

With the aircraft ranked by their need for repair, a PI threshold is then established and used as the basis for determining the aircraft for induction into the depot. The establishment of a PI threshold is a key area in the ACE program since it determines the operational acceptance level for the airframes of the active fleet. If an aircraft's PI is at or exceeds this threshold, it becomes a candidate for depot repair. Various criteria are used to establish the threshold, such as safety, mission capability, readiness, reliability, depot facility availability, cost and funding limitations. The threshold is a powerful discriminator. The condition of the entire fleet as well as the money spent on depot repair is affected by the threshold value. If management decisions change, then the threshold must be re-evaluated.

Once a threshold based on engineering considerations has been set, the consequences are apparent. If, because of funding limitations, a different threshold is applied, then the range between the two thresholds defines a readiness gap in terms of the condition of the fleet and the cost of depot repair. To the extent that meaningful cost correlation data are available, the readiness gap and the associated PI data provide a basis for establishing or adjusting depot budgets in rational, operationally oriented cost–benefit terms.

The ACE Planning and Evaluation Cycle

ACE is performed through the course of a yearly cycle illustrated in Figure 5–8. During the course of the year, the ACE program is reviewed

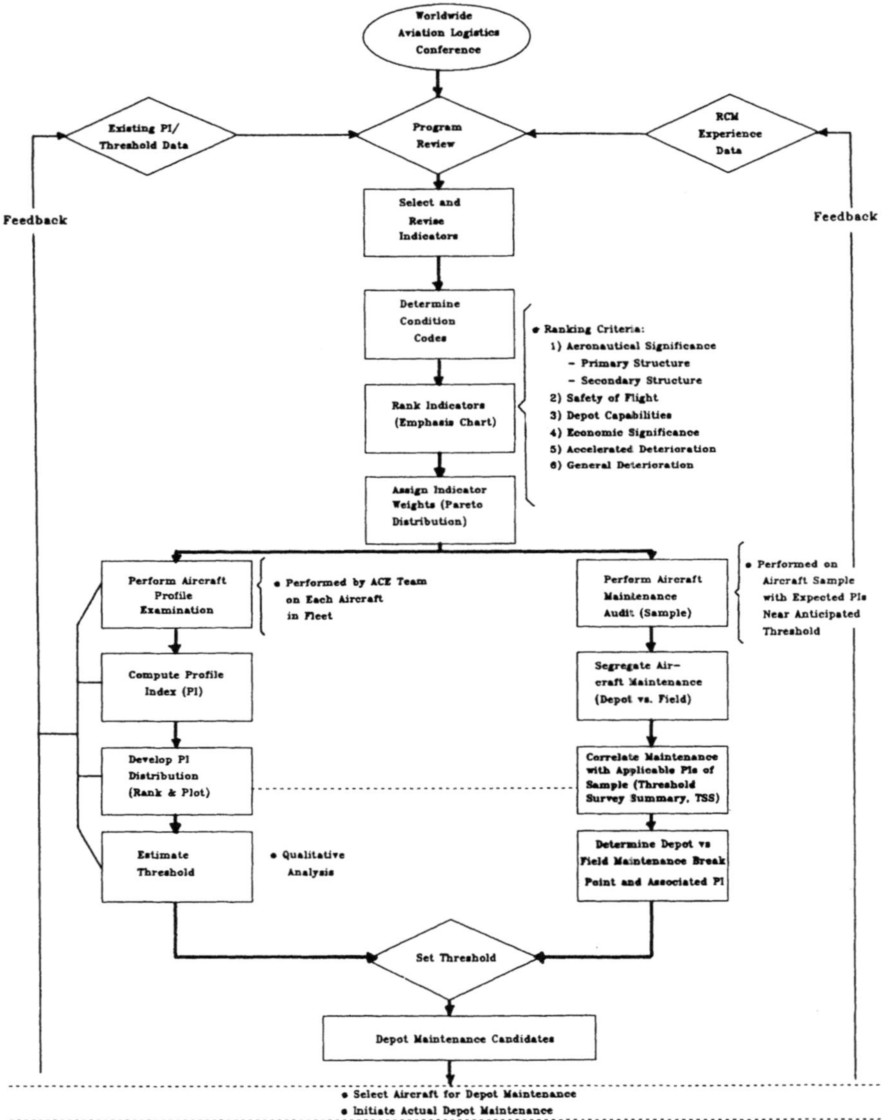

Figure 5–8 ACE Program Cycle

for improvements using readily available data sources to maintain an optimum level of program effectiveness. The application of the ACE methodology involves the following steps.

Step One: Select Indicators

Indicators are selected by conducting a thorough analysis of the aircraft to determine appropriate deterioration symptoms. This analysis also considers the impact on structural integrity if an identified section of airframe deterioration is not repaired. The indicators are annually reviewed and revised, as needed, to reflect current experience and changing depot capability.

The following criteria are used to guide the selection of the indicators:

Aeronautical Importance — The criticality of the indicated defect to aircraft availability. This includes determining the impact of the defect on:

a. Safety of flight
b. Mission essentiality
c. Interchangeability

Depot Capability — The need and economic impact of performing maintenance at the depot. This includes:

a. Man hours and material
b. Tools
c. Facilities
d. Procedures and processes
e. Expertise
f. Maintenance allocation chart (MAC)
g. Experience data

Accelerated Deterioration — The increase in deterioration if a needed repair is not performed.

General Deterioration — The expected deterioration if aircraft remains in the field until next profile.

The indicator selection process, outlined in Figure 5–9, consists of five tasks as follows:

(1) Review the significant aircraft maintenance and structural design characteristics including:

- The basic aircraft attach points for quick and easy component removal/replacement

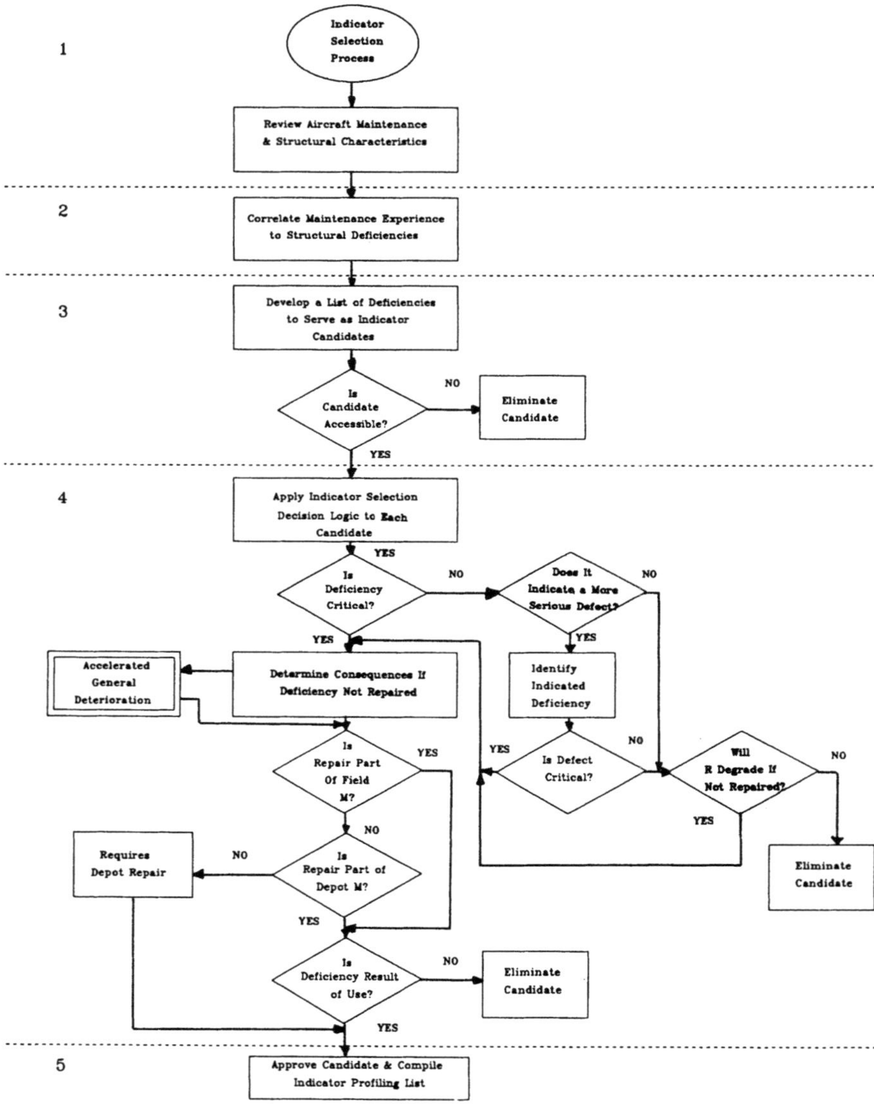

Figure 5–9 The Indicator Selection Process

- The structural components requiring specialized equipment for inspection, alignment, etc. Repair for these components may require depot attention as the only applicable maintenance level
- The specialized tasks needed to perform assembly and disassembly of the basic aircraft for example, jigs and fixtures used to position parts during their formation of a sheet metal assembly
- Any applicable portable equipment which is used to control parts locations while performing repair by depot teams sent to field units
- The general aircraft assembly procedures including any specialized equipment needed to perform such action
- The estimation of man hours and costs needed to perform basic repair tasks
- The examination of the various modular design features and the prescribed action taken by each maintenance level; e.g., a component removed from an aircraft at AVUM may need to be sent to AVIM; if AVIM cannot perform needed repair on the component, then the component is sent to the depot for repair or rebuild
- The examination of the various maintenance levels, capability to perform a specific repair procedure, taking into account skills available, man hours needed, costs, etc.
- The aircraft maintenance allocation chart (MAC) to identify maintenance assigned functions for AVUM, AVIM and the depot based on skills available, time required, and tools and test equipment required and/or available
- Maintenance tasks for each level of maintenance as described in the appropriate technical manual (TM)
- Depot level tools
- Depot level facilities
- Depot level procedures and processes
- Depot level expertise

(2) Correlate maintenance experience to aircraft locations which may display a deficiency indicative of a potentially more serious hidden structural deficiency.

(3) Develop a list of indicator candidates based on the aircraft maintenance characteristic/correlation data from tasks (1) and (2). Review each candidate to determine if it is easily accessible and discernible by a trained ACE profiler. Those candidates that are not easily accessible and discernible are removed from the list.

(4) Apply the selection decision logic to each indicator candidate. The information compiled during tasks (1) and (2) are used to help answer the decision logic questions.
(5) Prepare indicator list, with a description of each indicator and what structural defects may exist elsewhere in the aircraft, based on the observations. For example:
- The aft fuselage skin exhibiting buckling may mean misalignment of the longerons that support the tailboom and high local stresses
- The cargo door tracks exhibiting cracking and excessive wear may mean excessive vibration, high landing loads and extreme helicopter usage
- A pylon assembly exhibiting cracking, buckling and looseness may mean repetitive landing at or near the design limits
- A transmission support exhibiting cracking and looseness may mean excessive hard landing or excessive rotor vibration
- A center post assembly exhibiting buckling, cracking and looseness may mean hard landing or repeated landings that exceeded design loads over a prolonged period of time
- A door post exhibiting looseness and cracking may mean possible twisting of cabin structure
- The fuel cell exterior honeycomb exhibiting delamination, deterioration, puncture, corrosion and dents may mean that the passenger seat belts were banging and cutting into the honeycomb during flight with the doors removed
- The engine firewall exhibiting corrosion, buckling, looseness and cracking may mean excessive engine vibration
- The battery compartment exhibiting corrosion and looseness may mean battery spillage or poor servicing techniques
- The nose section exhibiting looseness, cracking, improper hardware and buckling may mean severe structural vibration

Also, the general characteristic of the indicators should be described to facilitate the subsequent assignment of condition codes.

Step Two: Determine Condition Codes

A list of condition codes is developed for each indicator to denote its varying degree of degradation, such as dented, delaminated and corroded, or good, fair and poor. Condition codes are also reviewed and revised, along with the indicators, to reflect current experience factors. The selection and specification of condition codes are done in conjunction with the indicator selection process.

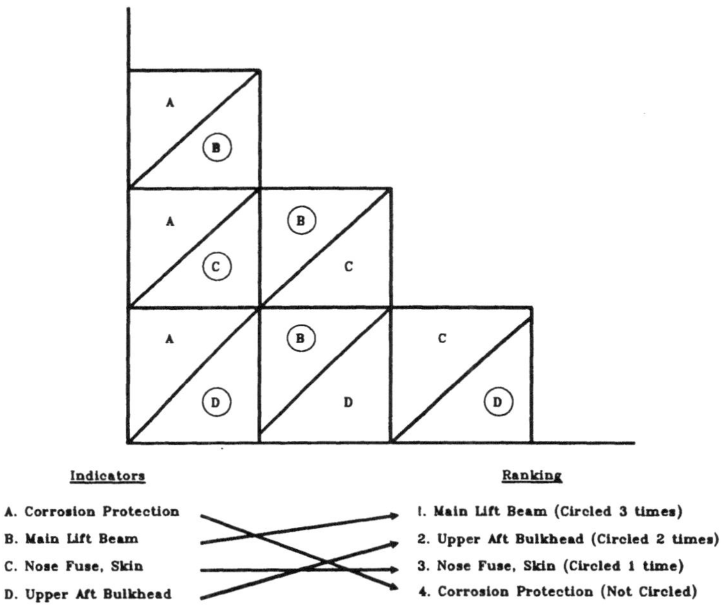

Figure 5–10 Emphasis Chart

Step Three: Rank Indicators and Condition Codes

The ranking of indicators is performed by listing each indicator and then comparing it against each of the other indicators involved. This ranking procedure is carried out using an 'emphasis chart' (Figure 5–10). The comparison of each indicator against each other indicator is based upon the following criteria:

- Could one of the conditions indicated itself be hazardous or could it progress to become hazardous?
- Could one of the indicators be the cause of customer rejection?
- Which of the indicators better shows accelerated airframe deterioration or consumption of components?
- Which of the indicators better shows fair wear and tear?
- The relative cost of item replacement.

In Figure 5–10, using these criteria, indicator A is compared against indicator B; then, indicator A is compared against indicator C; etc. In each case, the more critical indicator is circled and, when all indicators

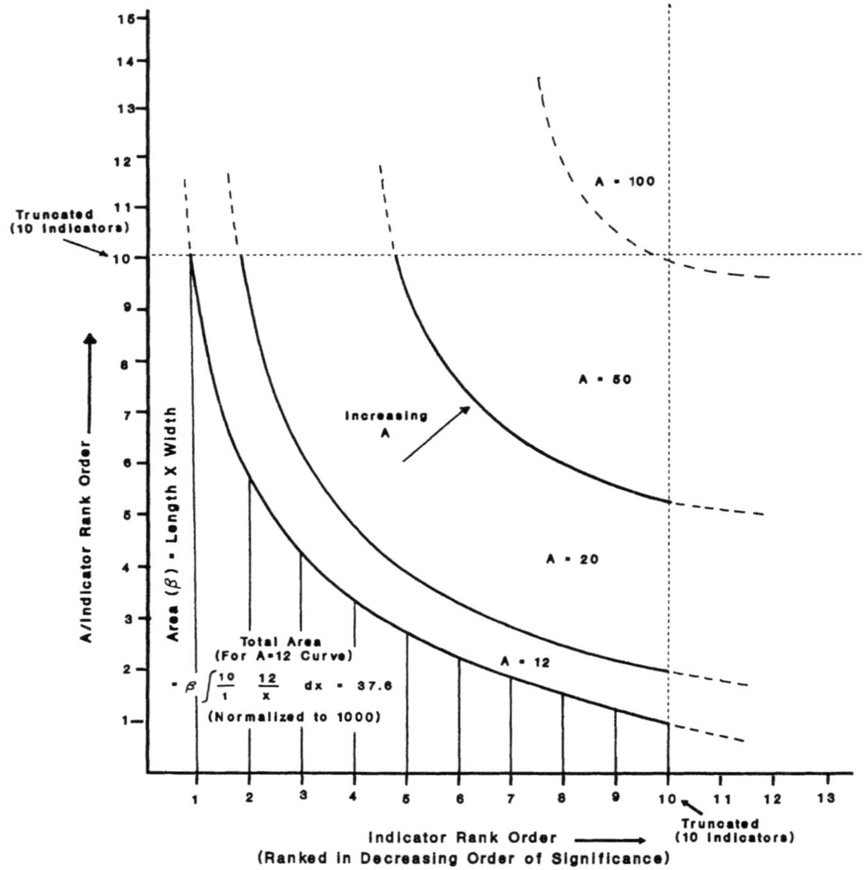

Figure 5–11 Pareto Distribution Curve

have been compared, the number of times an indicator has been circled is counted and noted. These numbers reflect the rank or importance of each item in relation to the other indicators with respect to the criteria.

Step Four: Assign Weights to Indicators

Indicator weights are assigned utilizing Pareto's principle of maldistribution, i.e., a small portion of the indicators will lead to a large portion of aircraft problems. Figure 5–11 illustrates the Pareto curve and the weight assignment process.

Pareto's distribution is expressed mathematically as a hyperbolic curve of the form $xy = A$ where x is indicator rank, y is A/(indicator rank), and A determines the shape of the curve and how significant the lower ranking indicators are. By proper choice of the constant A, weighting of the indicators can be adjusted to achieve the balance desired. This choice of A is a management decision and it is usually related to the desired weight percentage of the first designated number of indicators. The weight distribution for the indicators is determined by the ratio of the area under the curve in the respective indicator interval to the total area under the curve with truncation at x = number of indicators and y = number of indicators. The sum of all indicator weights is normalized to 1000. The following equation is used to determine an indicator weight:

$$\text{indicator weight} = \frac{\text{indicator interval area}}{\text{total truncated area}} \times 1000$$

The indicator weights for the example illustrated in Figure 5–11 with 10 total indicators and a shape factor, A, of 12 (resulting in a total truncated area of 37.6) are:

	Indicator interval area	*Indicator weight*
Indicator with rank order 1	area $(\beta) = 10 \times 1 = 10$	$W_1 = 266$
Indicator with rank order 2	area $= \int_1^2 \frac{12}{x} \mathrm{d}x = 8{\cdot}317$	$W_2 = 221$
Indicator with rank order 3 and so on	area $= \int_2^3 \frac{12}{x} \mathrm{d}x = 4{\cdot}866$	$W_3 = 129$

The total weight distribution for this example is:

Indicator (rank order)	*Weight,* W_i
1	266
2	221
3	129
4	92
5	71
6	58
7	49
8	43
9	38
10	34

TABLE 5-6
Condition Code Weight Distribution

No. of faulty codes	*% Total indicator weight for codes (listed worst to best)*					
	First code	*Second code*	*Third code*	*Fourth code*	*Fifth code*	*Sixth code*
6	100	50	20	15	10	5
5	100	50	25	15	10	0
4	100	50	30	20	0	0
3	100	60	40	0	0	0
2	100	60	0	0	0	0
1	100	0	0	0	0	0

Example: An indicator with a weight of 79 and three faulty codes.

Condition code:	Numerical value used in formulating PI:
1—Deteriorated (100%)	79 (100% of 79)
2—Poor (60%)	47 (60% of 79)
3—Fair (40%)	32 (40% of 79)
Good (0%)	0 (0% of 79)

Figure 5–11 also shows that, as the shape factor A increases, the weight distribution becomes more even. When A = (number of indicators)^2, all indicators have the same weight.

The weights established by the Pareto process represent the maximum possible values. If, during examination by the profiling team, an indicator shows the worse condition, then 100% of the indicator weight is applied to PI determination. If the aircraft indicator shows a nonfaulty condition, then 0% of the weight is applied to PI determination.

The proportion of the indicator weight used in formulating the PI depends on the total number of faulty condition codes for that indicator and the order of severity of the condition. Table 5–6 presents the weight distribution (by percent) associated with condition codes.

Step Five: Examine Aircraft

Once the indicators have been selected and weights assigned, then each aircraft within the fleet is profiled. This profiling involves assigning the appropriate condition codes for each indicator and recording them on worksheets for subsequent PI computation. A data base is created whereby optimum management decisions and actions can be derived through engineering analysis.

Profiling an aircraft involves carefully examining the airframe structure for symptoms of possible hidden defects. The pre-selected indicators and condition codes form the basis for the evaluation. Since the indicators were selected, in part, for ease of detection and accessibility the profiling should be accomplished within a few hours.

The indicators and applicable condition codes selected for each aircraft are used to evaluate the basic airframe in an effort to detect general progressive deterioration of the airframe, regardless of cause (i.e., normal wear, overstressing, climatic conditions, etc.), and to inspect the basic fuselage structural members and dynamic components, and component structures, in an effort to detect general progressive deterioration of the aircraft due to the effects of corrosion. The ACE team examines the aircraft in an orderly manner and records the results using condition codes pre-established for each indicator.

The condition codes are used to identify the condition of the indicator being evaluated. They identify 'what can go wrong' and/or 'how bad it is'. The selection of a condition code dictates the indicator weight (numerical value) used in formulating the aircraft profile index. The condition codes and criteria for selection are described in the ACE program regulations and are summarized in Tables 5–7 and 5–8.

An example of an indicator is paint condition. Paint is normally the only protective cover available for an aircraft; it must therefore fully cover all outer surfaces of the aircraft except for the transparencies. The exterior paint performs two primary functions: (1) protect the outer surfaces from the corrosive effects of the weather and (2) camouflage the helicopter. Paint in good condition also performs the secondary function of adding eye appeal and giving the appearance of a well maintained aircraft. Based on the criteria shown for paint in Table 5–8, the ACE team determines the applicable condition code to be circled.

The evaluation proceeds with respect to the order of indicators on the applicable worksheet(s), following a counterclockwise motion about the aircraft. The ACE team proceeds through the profile until an indicator deficiency is found. At this point, the applicable condition code is determined and circled on the worksheet. This process continues until the profile is completed.

Step Six: Compute Profile Index for Each Aircraft

The weights of the indicator condition codes selected for each aircraft are cumulated to give the profile index (PI) for that aircraft. This PI provides a quantification or numerical ranking of the condition of each aircraft as compared with other aircraft and thereby provides a means to rank the fleet in terms of need of programmed depot maintenance. For

TABLE 5-7
ACE Condition Codes

A—Worn excessively	M—Good
B—Buckled	N—Loose
C—Deteriorated	P—Bent
D—Corroded	Q—Minor
E—Cracked	R—No defect
F—Misaligned	S—Delaminated
G—Loose rivets	T—Improper hardware
H—Major	U—Dent
I—Oxidized	X—Scratch
J—Punctured	Y—Temporary repair
K—Poor	Z—Bolts in lieu of rivets
L—Fair	

example, an aircraft with a PI of 100 is in greater need of depot repair than one with a PI of 50. It should be recognized that an aircraft with only one major faulty indicator may outrank an aircraft with several faulty indicators because of the Pareto principle.

Step Seven: Set Threshold

To be selected for depot maintenance, an aircraft must surpass a specified threshold PI. This threshold is initially specified based on past depot maintenance and field experience data. Sample aircraft with PIs expected to be near the anticipated threshold PI (based on previous year's data) are first audited. A thorough examination of each aircraft is made to estimate the overhaul/repair cost and to determine if the maintenance must be performed at the depot or if it can be performed at the unit level with or without assistance from a depot team. Once the sample aircraft are segregated by maintenance need, i.e., depot or field, a threshold survey summary is developed which correlates the applicable PIs resulting from the ACE profile with the maintenance determination. The aircraft are ranked by PI and the point at which maintenance changes from field to depot identifies the audit threshold PI. Upon comparison of the audit and the general trend cut-off point of the PI distribution, a threshold is defined which accurately depicts the aircraft most in need of maintenance and which are candidates for depot repair.

The threshold is then evaluated by appropriate management personnel from a cost/budget standpoint. This evaluation takes into account the major costs impacting the ACE program including:

TABLE 5-8
Condition Codes Criteria (taken from the ACE Program Regulations)

Indicator nomenclature	*Condition*	*Code*	*Negligible defect*	*Recordable defect*
Bonded panels	Delaminated	S	Delaminated area no larger than 2 inches in size. No more than three delaminated areas in each pocket	Delaminated areas that exceed negligible defects
	Deteriorated	C	None. *Note*: Core corrosion will sound like delamination when tapped and is spongy to the touch	Internal cores corroded or contaminated with water, oil, fuel, hydraulic fluid, etc. If required, verify contamination by drilling a 0·040 inch hole at bottom of suspected deteriorated area. If contaminated, liquid will come out of hole
	Temporary repair	Y	Adhesive bonded or sealed patches which examine like the original panel are considered permanent repairs and shall not be marked	Injections or fills, larger than 3/4 in. in diameter. Unbonded lap patches. Delaminated patches. Improper patch material. Repairs of poor quality of workmanship. More than one repair per pocket
	Punctured	J	None	Opening in facing of panel (normally round hole)
	Corroded	D	Corrosion residue on surface that can be removed by rubbing with a cloth and no pitting of the metal can be seen	Cannot remove surface corrosion residue with cloth. Metal is pitted under corrosion
	Dents	U	Damage not exceeding 10% of surface area of one pocket. No more than five dents in any 3 square inch area with no voids. Size shall not exceed $1\frac{1}{2}$ inches in diameter or 2 inches in length. Depth shall not be greater than 0·060 inch	Damage exceeding negligible defects
	Loose	G	Single loose rivet in each row is acceptable	Indications of loose rivets, visible wear residue (dark ring) around head, deteriorated paint and primer around head, group of several consecutive rivets tipped in

TABLE 5-8—*contd.*
Condition Codes Criteria (taken from the ACE Program Regulations)

Indicator nomenclature	*Condition*	*Code*	*Negligible defect*	*Recordable defect*
				same direction, visible space under head, and/or move rivet with finger pressure. If rivet is missing, mark as loose rivet and make note of this in remarks section
Sheet metal, forgings, castings, mounts, & spars	Cracked	E	None	Any cracks in specified area. *Note*: For forgings, castings, mounts and spars, verify cracks by dye penetrant inspection if possible
	Loose rivet	G	Single loose rivet will be acceptable	Indications of loose rivets, visible wear residue (dark ring) around head, deteriorated paint and primer around head. Group of several consecutive rivets tipped in same direction, visible space under head, and/or rivet movable with finger pressure
	Buckled	B	None	Sheet metal that has been warped, kinked, deformed by an external force
	Temporary repair	Y	Repairs authorized by existing technical data and of good quality which correct a damaged area. These repairs are considered permanent repairs and shall not be marked	Repairs which are not equal to or better than the original structure
	Corroded	D	Surface corrosion which can be wiped away with a cloth and with pitting depth of 10% less of the material thickness. Area of corrosion shall be 10% or less of the surface area of the part	Corrosion which cannot be wiped away and/or pitting is over 10% of material thickness. Corroded area is more than 10% of the total surface of the part
	Bolts in lieu of rivets	Z	None. *Note*: If unacceptable bolts have been installed and	Close tolerance bolts are acceptable substitutes for hi-shear rivets and hi-lock fasteners. If any of these bolts

			have the code for improper hardware, use this code to show defect	areas or interchangeable (hard) points, mark this code
	Improper hardware	T	None	Any hardware used as a substitute for original hardware and not approved by existing technical data, e.g., unapproved bolts replacing hi-shear rivets or hi-lock fasteners
	Scratched	X	Superficial scratches which are not deeper than 10% of material thickness or less than 0·010 inch deep	Scratches which will exceed 10% of the material thickness or over 0·010 inch deep after clean-up
Wiring	Deteriorated	C	None	Coaxial cable insulation which is oil and hydraulic fluid soaked. Check cable by rolling insulation between thumb and index finger. If insulation rolls free of wiring, the cable is deteriorated Discolored wires (brown or grey) indicate overheating. If discoloration is more than 3 inches on any wire, mark wire deteriorated
	Cracked	E	Cracked insulation on one wire in any one bundle	Cracked insulation on 2 or more wires in any bundle
	Temporary repair	Y	Repairs which are of good quality and prevent contamination and shorts	Repairs not meeting negligible defect criteria. Wire bundles in which all wires have been spliced at one location
Paint	Good	M		Paint coverage complete. Few cracks on paint or rivet heads
	Fair	L		General cracking of paint on rivet heads. Most rivet heads still covered. Flat surfaces nearly completely covered
	Poor	K		Most rivet heads partly bare. Some chipping of paint on approximately 1/3 of exterior surface. Paint checked on flat surfaces on 1/3 of exterior
	Deteriorated	C		More than 1/3 of exterior exhibits chipping of paint. Most rivet heads bare. Paint oxidized with whitish cast. Extensive checkering

- Transportation — The costs involved in moving an aircraft to and from the depot.
- Overhaul/Repair — The costs involved in performing the needed maintenance (i.e., man hours, material, facilities, level of expertise and processes).
- Acquisition — The current cost of acquiring a new aircraft.

The threshold is then set to reflect the best balance between the engineering analysis (candidates) and management cost evaluation. This final threshold defines the aircraft to be actually inducted into the depot. The difference between the engineering and final management threshold defines a readiness gap which identifies the aircraft that are potentially not mission available. An aircraft PI distribution is then plotted, as shown in Figure 5–12. This type of presentation provides a concise ranking profile of the entire fleet and permits the necessary management decisions to be approached in a straightforward manner.

Step Eight: Select Aircraft for Depot Repair

The actual selection of aircraft for depot repair is made by management. The user decides which of the aircraft identified for depot repair are sent first, keeping in mind that the aircraft could remain in the field for some time before being returned to depot. Those aircraft identified for depot repair must continue to receive the same care and maintenance as all aircraft in the unit. Even when an aircraft is identified for depot repair, it may be displaced and not be called in as programmed.

It should be noted that the ACE program elements are interrelated. Changes in one must be carefully considered to determine if a corresponding change is necessary in another. For example, if new indicators are developed, then the current indicator ranking must be re-evaluated to correctly reflect the additions. This dictates that the Pareto curve be replotted and the weights recalculated. This further dictates that the applicable PI threshold be re-evaluated. Likewise, the addition of condition codes to indicators may warrant threshold re-evaluation.

The identification of corrosion in aircraft is an essential aspect of the ACE program and the subsequent depot repair and overhaul process. The ACE indicators must be continually reviewed and updated to enable the detection of corrosion at its earliest possible stage prior to its acceleration to a major aircraft defect. Fault tree analysis (FTA) can be used to further define the corrosion failure mode, evaluate the adequacy of the corrosion-related ACE indicators and determine if all essential corrosion mechanisms

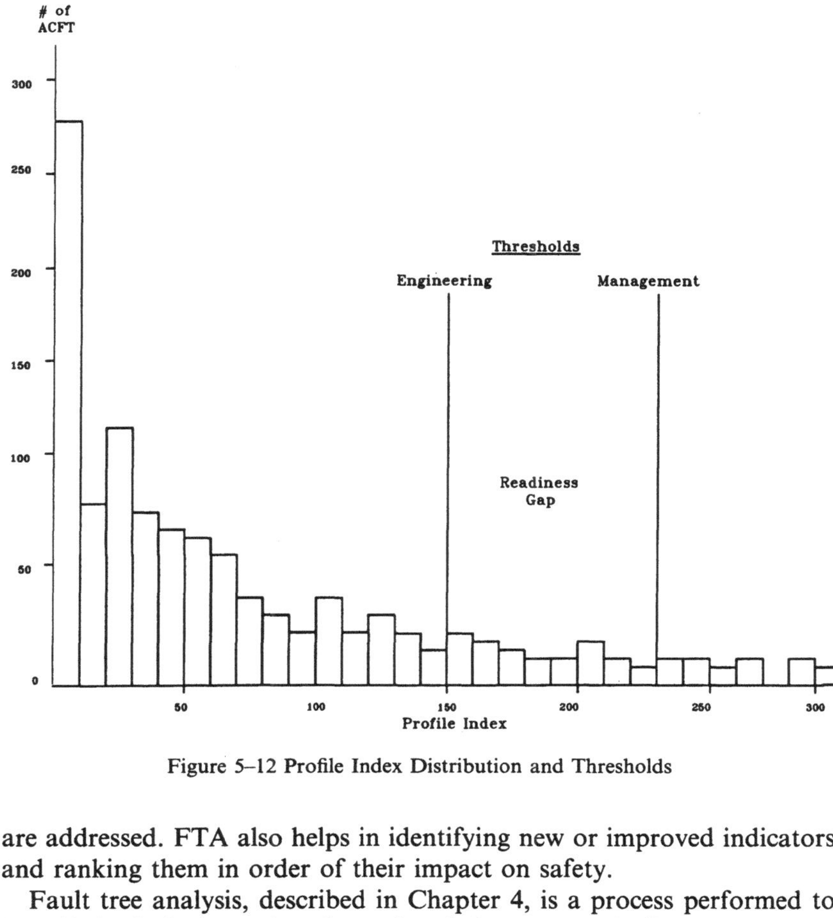

Figure 5–12 Profile Index Distribution and Thresholds

are addressed. FTA also helps in identifying new or improved indicators and ranking them in order of their impact on safety.

Fault tree analysis, described in Chapter 4, is a process performed to identify basic faults and to determine their causes and effects. It involves several steps, including the development of a highly detailed logic diagram which depicts basic faults and events that can lead to system failure and/or safety hazards.

Figures 5–13 through 5–15 present fault tree diagrams for a broad ensemble of possible corrosion-related faults that can lead to aircraft failure during flight. At any point in the diagrams the lower level events (i.e., component faults, maintenance actions, operating procedures and conditions, etc.) which must occur to precipitate a specific corrosion-related consequence are connected by basic logic elements ('and' gates,

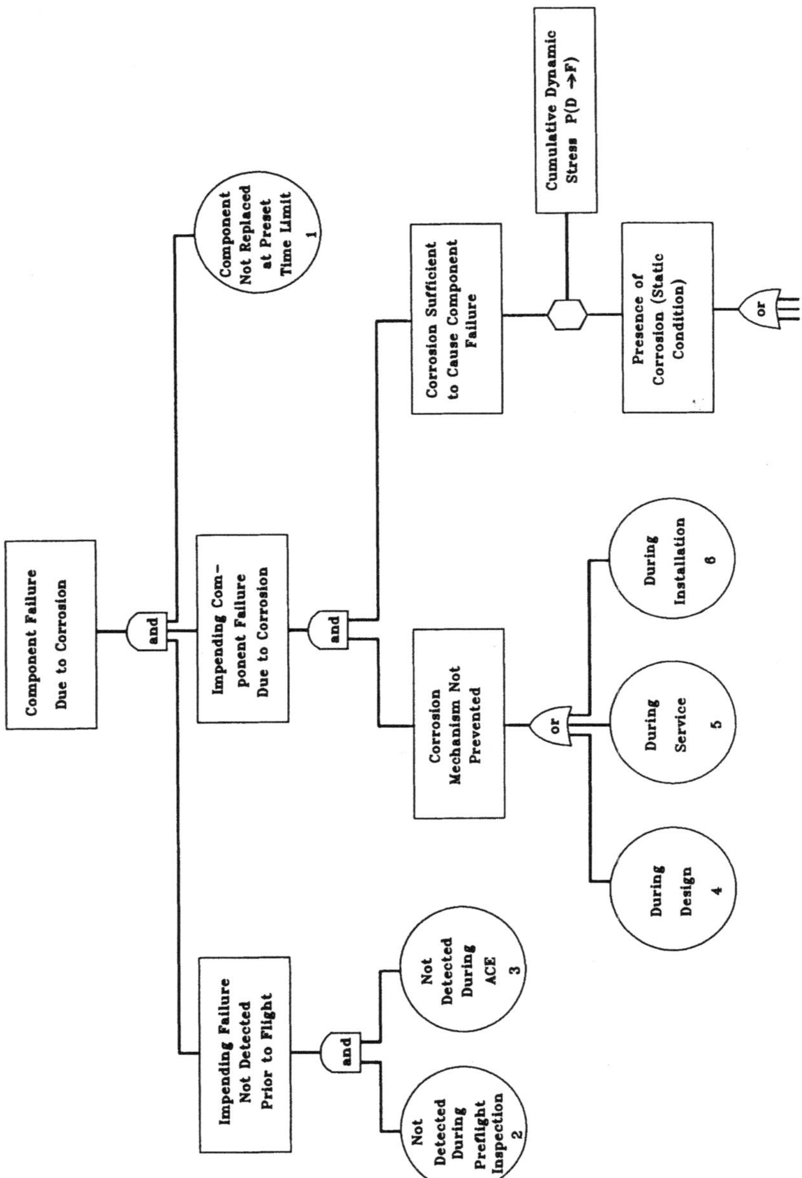

Figure 5–13 Corrosion-Related Fault Tree: Top Events

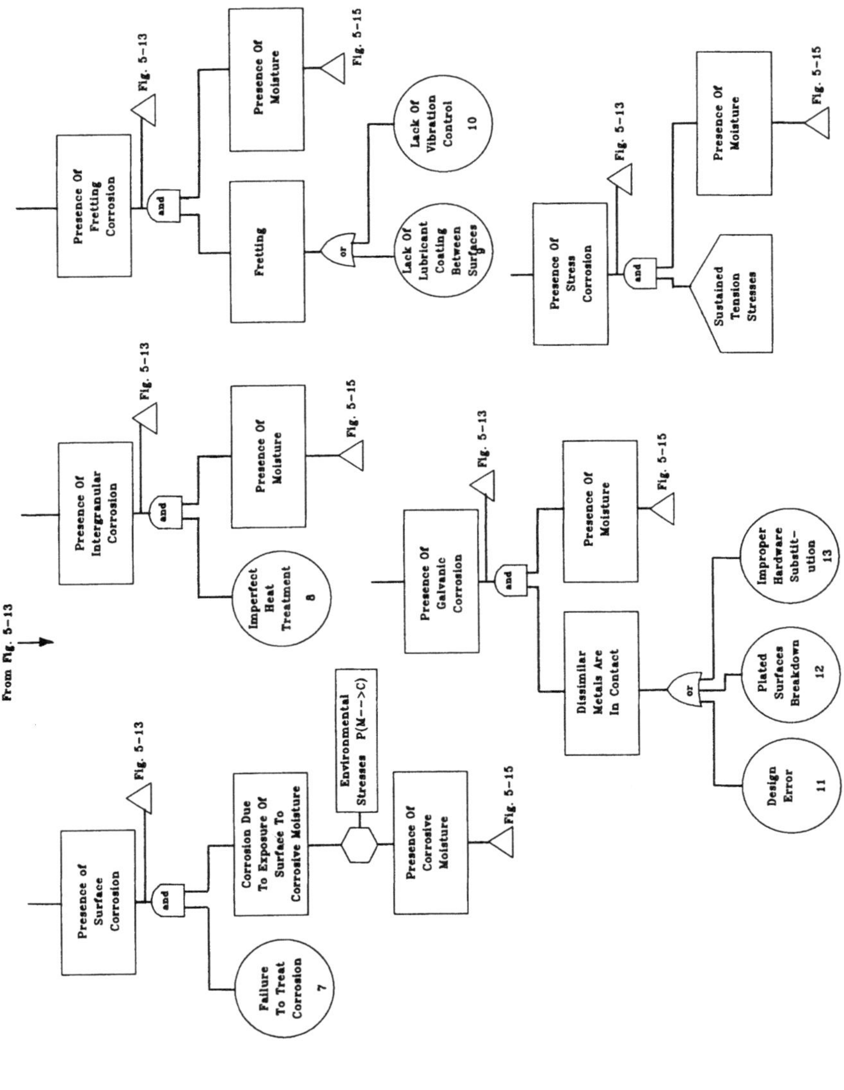

Figure 5–14 Corrosion-Related Fault Tree: Corrosion Types

'or' gates, etc.) which portray the essential causal relationships. A first-cut analysis of the safety level of a corrosion control system can be based on the relative occurrence of 'and' and 'or' gates at various levels within the tree.

Figure 5–13 shows that component failure due to corrosion can only be caused by the combination of (1) an impending failure due to corrosion, (2) a corrosion-related failure not detected prior to the flight during ACE and preflight inspection and (3) a component which has not been replaced at a preset time limit. An impending component failure due to corrosion can result from a mechanism which was not prevented by design or eliminated by inspections and repairs performed during installation or service. Corrosion sufficient to cause component failure can result from the effect of cumulative dynamic stresses on corrosion present in a static condition; this corrosion can be surface corrosion, galvanic corrosion, intergranular corrosion, stress corrosion or fretting corrosion.

Figure 5–14 shows the portions of the fault tree pertaining to each type of corrosion. Surface corrosion can result from the failure to treat corrosion due to the exposure of the surface to corrosive moisture and environmental stresses. Galvanic corrosion can result from the contact of dissimilar metals, through design error, a breakdown of plated surfaces or improper hardware substitution, in the presence of moisture. Intergranular corrosion can result from the presence of moisture at an imperfectly heat treated component. Stress corrosion can result from the effect of sustained tension stresses in the presence of moisture. Fretting corrosion can result from fretting, through a lack of either lubricant coating between surfaces or vibration control, in the presence of moisture.

Figure 5–15 shows the portion of the fault tree pertaining to the presence of moisture or corrosive moisture. In each case, moisture may be present due to either entrapment or entrance in the field environment. Entrapment may be induced during either manufacturing or depot maintenance. If moisture is present in the manufacturing environment, it may become entrapped and not be detected by the quality control inspection. If moisture is present in the maintenance environment, faulty repair or maintenance may lead to moisture entrapment which may not be detected through inspection. Moisture entrance in a field environment where moisture is present may occur as a result of defects induced during manufacturing with inadequate quality control inspection or during maintenance due to faulty repair or maintenance, which is not detected by inspection.

The basic faults identified on the fault tree diagrams can then be used

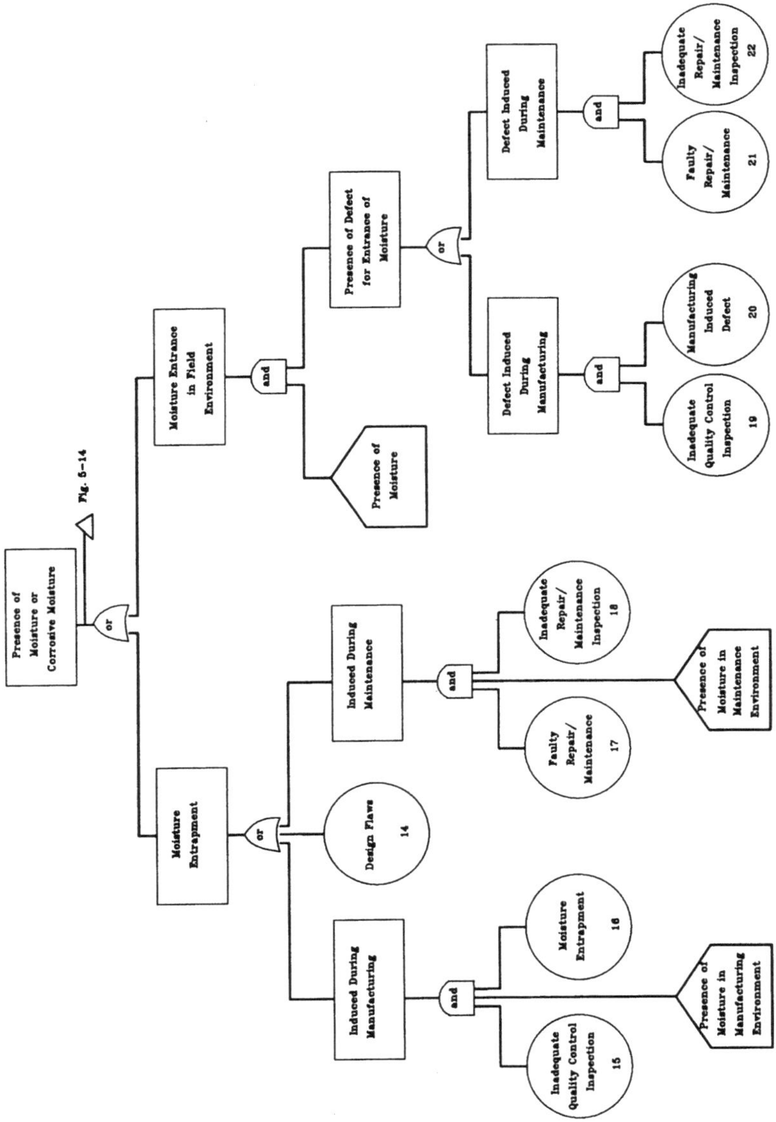

Figure 5–15 Corrosion-Related Fault Tree: Presence of Moisture

to help determine the adequacy and completeness of the ACE indicators with respect to corrosion and to identify the need for new or improved indicators. The quantification of the basic faults, following the FTA procedure described in Chapter 4, would enable the basic faults to be listed in order of flight safety criticality and, thus, provide criteria for assigning condition codes and ranking the indicators.

CHAPTER 6

The Army Aircraft Flight Safety Prediction Model

RCM is a technique involving the use of a decision logic to systematically analyze failure mode, rate and criticality data in order to establish the most effective maintenance support program for a hardware system. The RCM process is first applied during development once the system R&M parameters have been defined and an FMECA has been completed. The FMECA, by identifying part failure modes, their effects and criticalities, provides the essential data needed for the RCM logic analysis and for formulating the maintenance support requirements to be applied during operation. The logic analysis is reapplied after deployment during field operation using actual reliability–age experience data to optimize the process and update the maintenance support requirements.

This chapter describes an aircraft flight safety prediction model that can be used with field experience data, such as that compiled in the Army's RCM data bank, to reapply the RCM decision logic, as part of a sustained engineering activity. The model, when tailored to a specific aircraft system or component, provides a quantitative measure of the effect of material-caused failure or human error on flight safety. This quantification is proportional to the probability that the part failure or human error will cause an aircraft to operate in a hazardous mode. It is an extension of the fault tree analysis (FTA) procedure described in Chapter 4.

A hazardous mode is characterized by the fact that there has been a failure or error and the aircraft and crew are exposed to danger. The occurrence of an actual accident after exposure to danger is dependent to a large extent upon the skill and reaction of the air crew as well as other factors, such as weather conditions, runway length, and if the failure or error occurs during day or night operation. Direct control of the factors which determine whether an exposure to danger results in an accident is beyond the scope of this model. For this reason, the model does not predict accidents, but does provide probability estimates for a flight safety incident, i.e., the exposure of an aircraft and its crew to danger because of part failure or human error. It assumes that the aircraft is operated properly and within its mission envelope.

Included in this chapter are a description of the overall procedure, the basic assumptions for its use and the result of tailoring the general model to a US Army helicopter. Also, this chapter illustrates application of the model and the development of criticality data for a specific component and shows how the data can be used with the RCM logic to develop cost-effective maintenance requirements.

6.1 GENERAL DESCRIPTION AND ASSUMPTIONS

The Army aircraft safety analysis model is an extension and adaptation of the FTA procedure to rotary aircraft. The model establishes the logic relationships between the causes of failures and their consequences in terms of the probability of a flight safety incident. It defines the relationships of life-cycle intervention and control procedures that can be applied to mitigate the failures in a meaningful and mathematically tractable form and allows determination of their impact on safety. When fully implemented the model greatly facilitates the RCM logic analysis and increases its utility. In addition to its use with the RCM logic to select optimum maintenance tasks, the safety prediction model can be used directly to:

(1) Identify critical parts so that action can be taken prior to a safety incident.
(2) Determine the quantitative impact of part failure, human error and other potential causes on flight safety, thus providing a basis for identifying and prioritizing improvement modifications.
(3) Evaluate the effect of changes before and after implementation.

The model provides a uniform and generally accepted basis for making safety decisions.

Figure 6–1 pictorially identifies the steps to be taken in implementing the model.

Step 1 of the technique is to identify the aircraft functions and how these functions are interrelated.

Step 2 is to construct a fault tree diagram (FTD) for each aircraft functional group. A fault tree, as described in Chapter 4, is a detailed logic diagram that portrays the events that may lead to the condition under examination. In this model, the focus is on a flight safety incident. All events (i.e., component faults, human errors, operating conditions, etc.) which must occur to produce a flight safety incident are interconnected systematically through basic logic elements ('and' gates,'or' gates, etc.) to form the fault tree.

Based on a knowledge of the system design, its function and operating environment, the way it is operated, and its maintenance requirements, the diagrams are constructed, beginning with the defined hazard condition and proceeding downward, using a series of engineering judgments, to define, ultimately, basic input events. This process continues until each event chain, or branch, has been terminated in a basic fault. When the diagrams are complete all basic faults (hardware and human) whose occurrence, either alone or in combination, results in a safety incident are identified regardless of their apparent frequency of occurrence. The basic faults are defined such that they are independent events and failure rate (or human error rate) data are available or can be estimated.

One of the most important procedures in fault tree diagramming is to define the intermediate events such that the entire connection from each basic fault to the top event is logically correct. This requires that a careful distinction be made between a defect (a latent condition having potential for failure) and the actual occurrence of failure. This distinction is reflected in the aircraft fault tree diagrams by the use of inhibit gates where the gates condition define, quantitatively, the translation of a defect to an actual failure.

Step 3 is to assign fault tree codes to functions, subfunctions and other blocks (excluding logic symbols) of the fault tree diagram. This is generally carried out to the part level. It is also extended, wherever possible, to include pilot and inspection errors.

Step 4 is to compile data for each basic fault that comprises the fault tree. Part failure rate, pilot error probability and inspection efficiency data are necessary inputs for determining occurrence probabilities and assessing criticality.

In general, the part failure rates are determined through a review of historical or actual failure rate data. Where such data are unavailable,

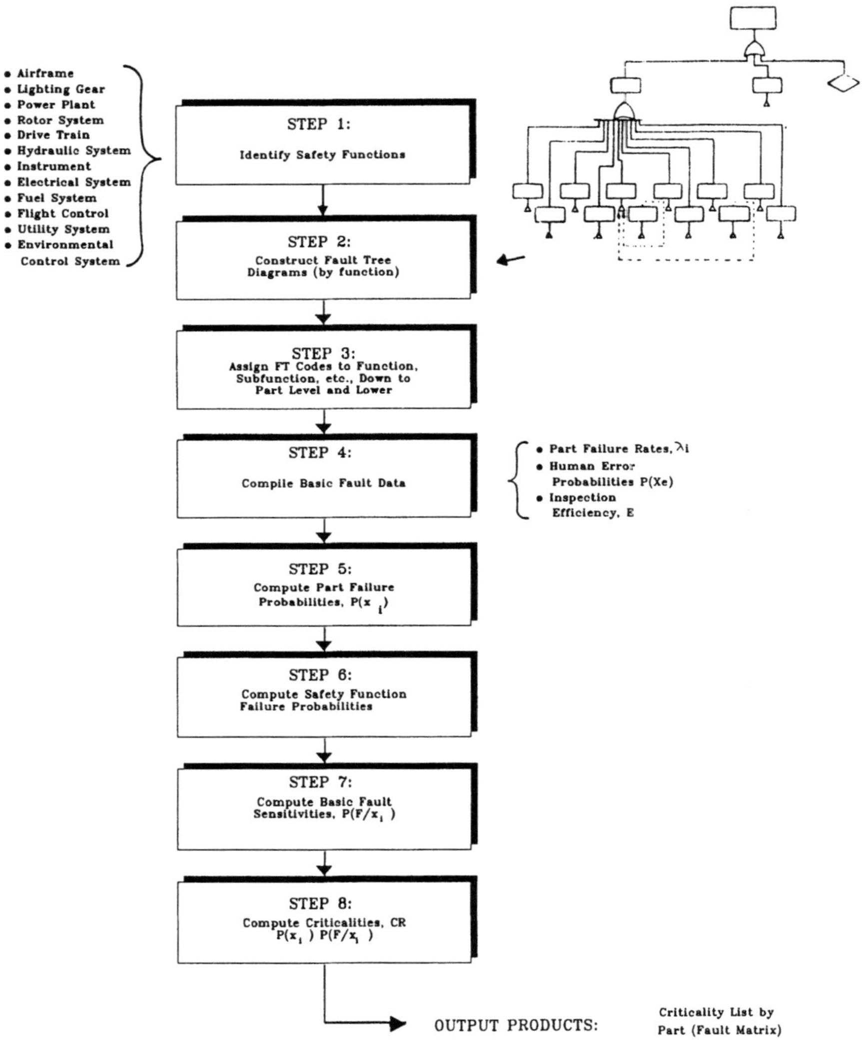

Figure 6–1 Army Aircraft Safety Analysis Model

failure rates are estimated using part failure rate models in conjunction with application and quality adjustment factors as discussed in Chapter 4.

Pilot error probability means the expected probability that a failure caused by a pilot takes place, whether intentionally or unintentionally. It is very difficult to obtain pilot error probability since very few data exist in this area. Since a large scale data base is lacking, pilot error probability is usually estimated by subjective techniques that take into account human performance attributes and are based on discussions with personnel familiar with the system operation environment.

Inspection efficiency is the probability that a given defect will be detected prior to a failure resulting in a flight safety incident. Inspection efficiencies are estimated by subjective techniques that take into account the inspection attributes.

Steps 5 and 6 are to compute the failure probability numerics. This involves, as described in Chapter 4, computing the occurrence probabilities for all basic faults, events and hazardous conditions (top faults) based on the combinatorial properties of the logic elements in the fault tree diagrams. Given a fault tree diagram (consisting of basic faults and output events that are properly interconnected) the output event probabilities are computed, starting with the lowest levels and continuing to the highest levels in the tree.

Step 7 is to compute the sensitivity of each basic fault. Sensitivity is the probability that an occurrence of a basic fault will cause a flight safety incident. A fault sensitivity is computed by assigning a probability of 1.0 to the basic fault and then recalculating all higher events to determine the resultant incident probability.

Step 8 is to compute the criticality of each basic fault. Criticality is a measure of the relative seriousness or impact of each fault on the top event. It is defined quantitatively by the following expression:

$$\mathrm{CR} = P(X_i)\,P(F/X_i)$$

where $P(F/X_i)$ is the conditional probability (or 'sensitivity') of a flight safety incident given that the basic fault, X_i, has occurred; and $P(X_i)$ is the probability of the basic fault occurring. $P(X_i)$ can be derived from RCM failure rate and mode data.

CR combines the effect of sensitivity and probability of failure. It provides a comparison of a part whose failure has a high impact on flight safety but which rarely fails, with a part whose failure has a lesser effect but which fails frequently. It thus provides a basis for ranking the faults in their order of criticality.

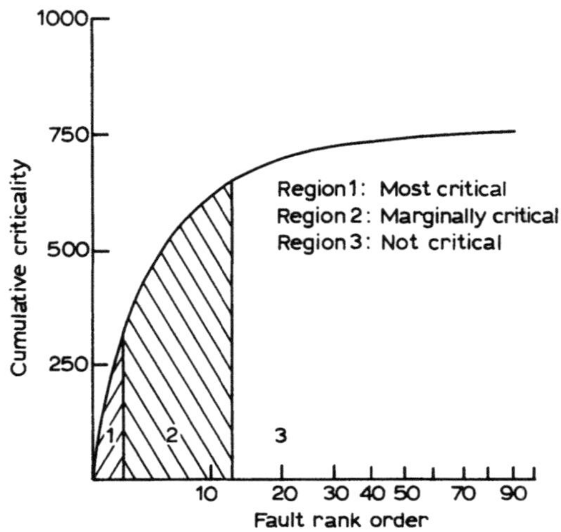

Figure 6–2 Criticality Curve

The criticalities for all failure modes can be ranked in descending order (the most critical failure mode is assigned to position 1, while the least critical failure mode is assigned to the last position). Associated with each ranked criticality value is a cumulative sum of all previously ranked criticalities. The cumulative criticality for the third ranking failure mode, for instance, is the sum of the criticalities for ranked failures 1, 2 and 3. A plot of the criticality ranking (independent variable, logarithmic axis) and cumulative criticality (dependent variable, linear axis) produces the relative criticality curve shown in Figure 6–2.

The curve can be divided into three distinct regions of criticality: the most critical region, the marginally critical region and the non-critical region. The position of a specific failure mode on the graph provides criteria for determining if the RCM decision logic is to be applied.

For those applications where a more accurate quantification of the effect of part failure on flight safety is important, the following two additional steps may be taken in the development of the FTD's and the subsequent computation of criticality data:

1. Divide the mission into the flight phases.
   The importance of a part failure or human error to safety is not constant throughout a flight. To deal with the problem of changing roles, a mission can be divided into flight phases. Part

failure rates and human error probabilities can be assigned to each flight phase, and fault sensitivities computed for each phase.

2. Evaluate external flight conditions and make appropriate adjustments.

   Some of the possible external influences which affect flight safety are icing conditions, adverse altitude and speed combinations, night operation, IFR conditions, rain and cold weather. These and other possible influences can be taken into account either by using inhibit gates with specific conditions applicable to the particular aircraft under study or by modifying the part failure rates/human error probabilities used as input to the analysis with suitable adjustment factors that account for their conditions.

6.2 US ARMY HELICOPTER APPLICATION

This section presents a helicopter flight safety prediction model based on the general procedure and assumptions described in the preceding section. Standard fault tree diagrams are presented for each helicopter function and, as such, serve as 'templates' that can be used in subsequent evaluations of specific aircraft or major components and the computation of criticality data. These FTD templates are provided to introduce discipline into the analysis procedure, standardize the process and facilitate its application (or tailoring) to specific aircraft systems or major dynamic components.

The standard fault tree diagrams reflect consideration of all elements that comprise the complete operational system including the interaction of man, machine and the operational and maintenance environment. This operational system consists of the helicopter, its flight and maintenance environment and the personnel involved in the flying of the aircraft (see Figure 6–3).

The diagrams focus on identifying potential component part failures and human errors that singly or in combination can cause a flight safety incident. These component failures have their root causes induced, or influenced, during the various system/component life-cycle stages, i.e., design, manufacturing, storage, and operation and maintenance (overhaul/repair).

In developing the standard helicopter fault tree diagrams (FTD's), the top event is a flight safety incident. In order to subdivide the aircraft

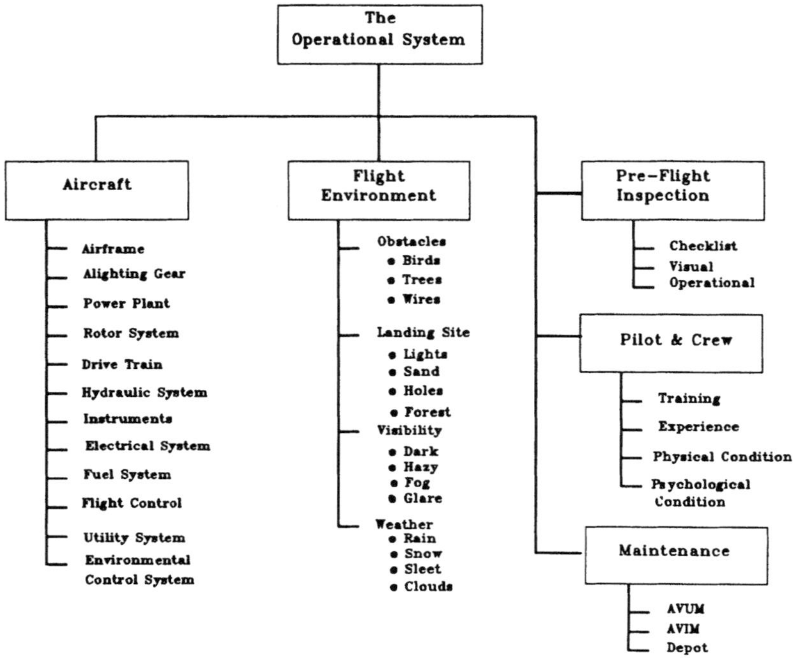

Figure 6–3 The Elements of the Operating System

systems and equipment into manageable segments against which fault trees could be constructed, and information compiled, functional groups were used. Within each functional group, the fault tree incorporates a component orientation where practical.

The FTD for each functional group was developed to the point where all basic faults at the part level are identified, including inspector errors, controls and other areas where corrective action could be effected. This included integrating into the diagrams basic component part failure modes and pilot errors as well as their significant life-cycle failure influencing factors. In general, the FTD's addressed stress–strength related failure as well as wearout failure caused by operational aging, manufacturing and maintenance degradation, and non-operational storage and dormancy factors.

The predominant reliability–age characteristics of the aircraft components, as depicted in Figure 6–4, guided the development of the FTD's

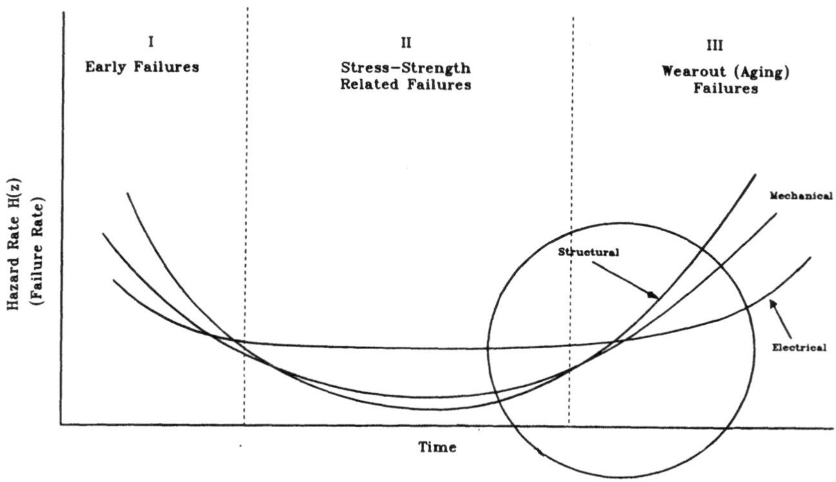

Figure 6–4 Aircraft Component Reliability–Age Characteristics

for each functional group. Area I in the figure represents the early life failure (infant mortality) period, where defects in manufacturing contribute to failure. Once the components have passed through this infant mortality period, they enter area II, where the incidence of failure is relatively low and the failure rate is fairly constant. Then, as degradation factors become predominant, the components enter area III, where, in spite of repairs and routine maintenance, the failure rate increases until the components become uneconomical to maintain.

For aircraft that have been in service for years, the component characteristics identified by the circle in Figure 6–4 predominate. This means that:

- Structural components have a dominant rapidly increasing wearout mechanism and a relatively low stress–strength failure rate
- Mechanical components also exhibit an increasing wearout mechanism and a relatively low stress–strength failure rate
- Electrical components have a long life (negligible wearout) and a relatively high failure rate

A stress–strength related failure, as described in Chapter 4, occurs when the instantaneous stress applied to a part exceeds the part's rated strength. The exponential reliability distribution, as defined by the part's failure rate, is used to estimate the probability of occurrence for a stress–strength failure.

The term 'operational aging' is used to refer to the combination of an undetected defect coupled with the increasing number of take-offs and landings and other aerodynamic stress cycles which, after repeated application, could damage the aircraft and result in a failure condition leading to a flight safety incident.

It was assumed that defects may be induced in components during manufacturing and maintenance operations. Defects due to the manufacturing operations arise from hidden flaws in a part or component or minute cracks in the structure which are introduced by the particular process involved (welding, stamping, assembly, etc.), and, as indicated, these defects combined with fatigue and corrosion during use may result in aircraft failure. Many of these are eliminated through the quality control process, but inspections are not 100% effective and the manufacturing facility may not be equipped to detect all hidden or latent defects. The application of environmental stress screening (ESS), described in Chapter 4, will significantly reduce the number of latent defects escaping manufacturing.

Maintenance work is carried out in accordance with the depot maintenance work requirement (DMWR) described in Chapter 5. However, as during initial manufacturing, the requirements may not be followed precisely or may not be adequately specified. The quality control (QC) inspection and testing process cannot be assumed to be perfect and defects may escape. Also, as a necessary expedient during maintenance, items not completely conforming to their procurement specifications may be used. In such cases, a material review board determines whether discrepant items can be used without loss of integrity or function of the end item. It must also be assumed that this process is not perfect: the material review boards are not always properly constituted and motivated, and unacceptably discrepant material does occasionally come into use during maintenance. Therefore, maintenance induced defects were assumed to exist, arising from a number of causes. ESS can also be applied to reduce the number of latent defects escaping depot maintenance.

During maintenance, components are often replaced by spares, or are repaired with parts or materials, all drawn from storage. If these items have been stored in accordance with strict application of completely adequate storage serviceability standards which establish depot quality control and reliability management procedures for assuring materiel readiness, virtually no failures due to storage-induced defects or deterioration would occur. These standards, described in Chapter 4, contain mandatory instructions for the inspection, testing and/or restoration of items in storage. They encompass storage criteria, preservation, packaging, packing and marking requirements, and inspection schedules to determine the materiel serviceability and the degree of degradation that has occurred. However, it must be assumed that the standards may be inadequate, or misapplied, and that spares defects, or deterioration due to inadequate storage control, may occur as a result of cumulative storage stresses.

Figure 6–5 presents a standard fault tree logic diagram for component failure reflecting the life-cycle influences and logic definitions discussed above. As shown in the figure, failure may be either stress–strength or wearout related. As previously indicated, stress–strength failures occur randomly and their occurrence probabilities can be estimated from the exponential distribution using standard part failure rates derived from field experience data.

A wearout failure is due to a defect in the component which was not detected by preflight inspection. Procedures for preflight inspection are given in a checklist format and included in the appropriate operator's technical manual. The checklist includes procedures for day, night and instrument flights, with annotative indicators immediately preceding the check to which they are pertinent. Pilots and crew members do not rely on memory for accomplishment of the inspection checks. The checklist items are confirmed by the pilot or crew members.

The upper inhibit gate, shown in Figure 6–5, indicates that the repeated application of aerodynamic stress cycles (operational aging) may lead to a worsening of an undetected defect such that it can cause cumulative damage to the aircraft and a flight safety incident. The gate condition (open or closed), for a particular aircraft, is dependent on its total operating time, which is a measure of number of aerodynamic stress cycles that the aircraft was subjected to, relative to the aircraft service life or the time between overhauls (TBO) for individual components. The gate condition is defined by:

$$P(D \rightarrow F) = \frac{\text{operating hours}}{\text{service hours}}$$

where $P(D \rightarrow F)$ is the probability of a defect becoming an actual component failure.

A defect may be induced during manufacture, or during maintenance

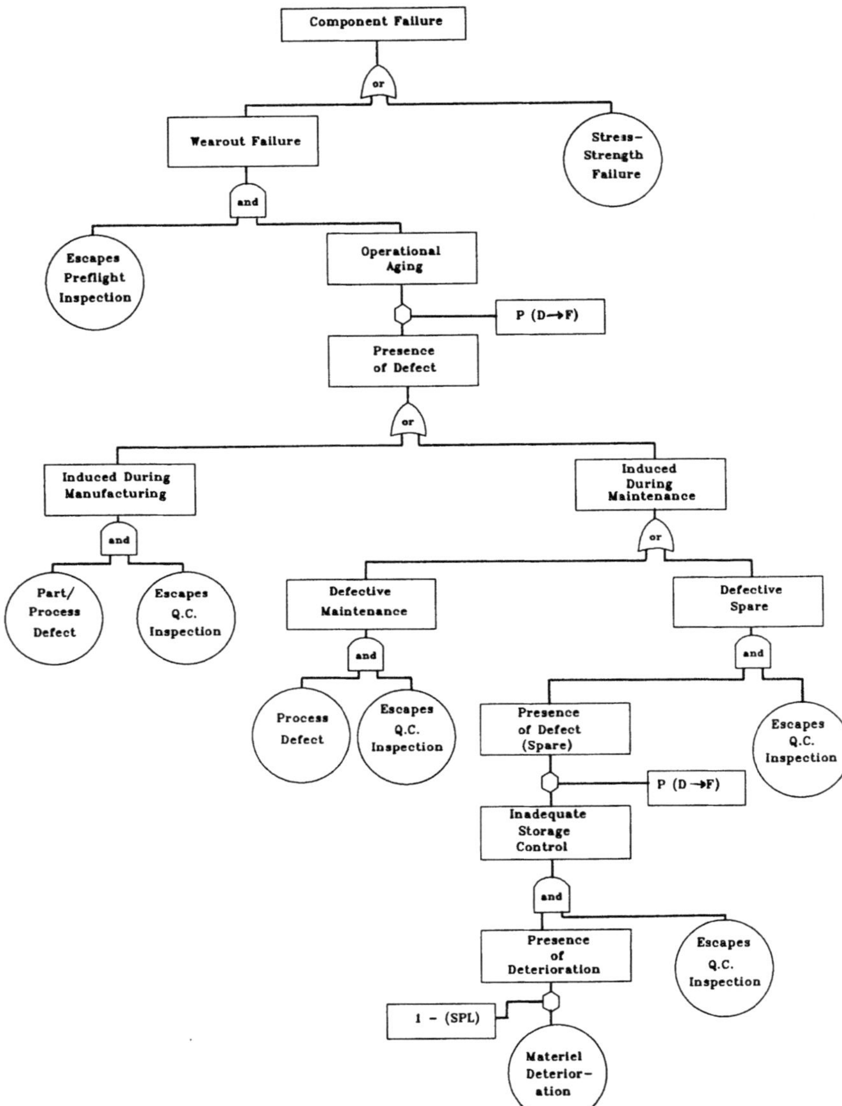

Figure 6–5 Standard Component Failure Fault Tree Diagram

(and not detected by quality control inspection). If the component is reconditioned or repaired during maintenance, the reconditioning or repair (including spare replacement) may be defective, with the defect not detected by inspection. If the component is replaced by a spare, the spare may be defective, with the defect made worse by material deterioration due to cumulative handling and long-term storage stresses. It then may have escaped inspection prior to issue. Such material deterioration results from an inadequate storage protection level of the spare and inadequate inspection of the spare while in storage.

An inhibit gate was also used to account for these cumulative non-operational stresses due to long-term storage. The gate condition is dependent on the ratio of actual storage time to total allowed storage time or shelf life, if specified. It is defined by:

$$P(D \rightarrow F) = \frac{\text{active storage time}}{\text{total storage time}}$$

where $P(D \rightarrow F)$ is the probability of material deterioration becoming a defective spare.

An inhibit gate was used to account for the storage protection level (SPL) against material deterioration of a spare. If the material is highly sensitive to storage deterioration with time, but is controllable by the SPL, the gate condition is taken to be dependent on the extent of the protection provided. SPL is defined by:

$$\text{SPL} = \text{PPC} \times \text{TSC}$$

where:

PPC is the Preservation Packing Code. Its value is dependent on the level of protection established by the storage serviceability standard (SSS); it is generally assigned in accordance with the following practice:

Maximum Military Protection:	1.0
Minimum Military Protection:	0.8
Commercial Protection:	0.6

and:

TSC is the Type Storage Code. Its value is dependent on the type of storage; it is generally assigned as follows:

Controlled Humidity Warehouse:	1.0
Heated Warehouse:	0.9

Unheated Warehouse:	0.8
Open, Improved:	0.4
Open Space:	0.00

The gate condition is defined by 1–SPL. When SPL is one, the gate is closed and the 'presence of deterioration' probability becomes zero regardless of the deterioration sensitivity of the material. On the other hand, when SPL is zero, the gate is open and the 'presence of deterioration' probability is a value directly related to the deterioration sensitivity of the material.

The component logic failure mechanism, shown in Figure 6–5, are applied at appropriate points in the helicopter FTD's, thus providing a means for determining the impact of component failure on flight safety in terms of basic engineering and management actions.

Actual reliability experience data, if available, should be used to establish the stress–strength component failure rates. However, during initial fielding when few or no data exist, the representative failure rates presented in Table 6–1 may be used as a first approximation.

Pilot error is either a judgment error in carrying out a flight operation or a failure to react, if possible, when a system or component fails during flight operations (lack of intervention). A lack of intervention pilot error is included within the fault trees for the appropriate component involved. Basic pilot judgment error is treated in a separate fault tree diagram.

Pilot error probabilities can be estimated from the following equation:

$$P_E = P_e P_t P_n P_c P_f$$

where:

- P_E is the pilot error probability
- P_e is an intrinsic error probability (under ideal or normalized conditions)
- P_t is a factor to account for level of training or experience
- P_n is a factor to account for type of mission flown
- P_c is a factor to account for the complexity of the aircraft
- P_f is a factor to account for the flight environment

The instrinic probability and the adjustment factors for a particular error are derived from specific experiences or from generic data sources on human reliability. Table 6–2 presents basic human error probabilities for typical tasks and uncertainty bounds associated with the tasks. Generic human error data should be used with some caution to assure that the assumptions on which they are based are valid and that the factors for the specific application are compatible with the data. Since no central data

TABLE 6-1
Part Failure Rates (per hour)

	10^{-6}
Accumulator	500·0
Actuator	0·3–405·4
Battery	676·8
Bearing	12·6–53·2
Brake (magnetic)	241·5
Circuit protect device	28·5
Control/instrument (gauge)	37·5–269·6
Fan	9·1
Filter	26·0–49·5
Gasket/seal	2·4–31·6
Generator (DC)	205·9
Gyroscope	300·0
Heat exchanger	38·4
Hose and fittings	3·9–32·9
Lamp, incandescent	18·6
Mechanical device	1·7–986·6
Pump	1·7–395·0
Regulator	3·0–136·2
Relap	1·0–31·0
Sensor	76·6
Solenoid	65·6
Switch	18·6–95·0
Tank	108·8–159·3
Transducer (tech gen.)	57·9–100·0
Valve	10·1–133·5
Structural elements	0·00004–0·004 (depending on complexity)
Avionics	500·0–1000·0 (depending on part technology/complexity)

bank of human performance data exists and until such a bank can be developed, human reliability prediction will depend to a large extent on judgment.

Inspection is treated explicitly throughout the fault trees. Inspection efficiencies were established from the following model:

$$E = E_1 E_2 E_3 \ldots E_n$$

where the number of inspection parameters, n, is a function of the level and complexity of each inspection. For example, inspection performed after manufacturing (or maintenance) includes the following parameters:

E_1 is the probability that all functions to be tested are incorporated in the test procedure (hardware complexity)
E_2 is the reliability of the inspection test
E_3 is the probability of no inspection error (experience and training)

Aircraft Fault Tree Diagrams (Figures 6–6 through 6–12)

The *overall helicopter FTD*, presented in Figure 6–6, identifies 12 basic functional groups [numbered (02) through (13)]. The figure indicates that a flight safety incident can occur as a result of three basic failure mechanisms: (1) a functional group failure, (2) a pilot error in flight operations, or (3) an uncontrollable mishap (i.e., lightning strike, etc.). As shown, the FTD's focus on functional group failure where basic faults (including lack of pilot intervention) leading to a flight safety incident during preflight inspection/ground operation and flight operation are identified.

The *airframe* (02) components whose failure can cause a flight safety incident are the pylon support, the engine mount, the tailboom mount and support, or the windshield. *Alighting* (*landing*) *gear* (03) failure can be caused either by failure of a landing skid or cross tube, or by failure of a fitting or fastener holding the landing gear together or attaching it to the fuselage structure. In each case, as identified in the figure, the standard component FTD given in Figure 6–5 applies.

The *power plant* (04) FTD, presented in Figure 6–7, indicates that system failure may be performance related or due to component failure (defined by the standard component FTD, Figure 6–5), foreign object damage (FOD) or a pilot induced error. As shown, a performance failure may be due to a loss of power, the engine's quitting or erratic speed (over or under speed), a compressor stall or a starting malfunction.

A 'loss of power' is defined as an engine/power turbine rpm (N_2) of less than 98%, i.e., there is less power available than that required for flight. A loss of power may lead to a flight safety incident if the pilot fails to land as soon as possible. A loss of power can be caused by failure in an engine component including:

(1) compressor (chipped blades, rubbing blade tips, case leakage, air bleed valve leakage, anti-ice valve leakage or stator vane deformity)
(2) turbine (burned blades, deformed blades or warped vanes)
(3) combustion chamber (cracked or warped liner, cracked fuel nozzle boss, cracked outer case or air tube leakage)
(4) fuel system (low pressure in the fuel system regulator valve,

TABLE 6-2
Basic Human Error Probabilities

Task	*Human error probabilities*
1. Walk-around inspections: recognize incorrect status, using checklist correctly	0·01 (0·005–0·05)
2. Walk-around inspections: recognize incorrect status, using checklist incorrectly	0·1 (0·05–0·5)
3. Walk-around inspections: recognize incorrect status, no checklist, first walk-around	0·9 (0·5–0·99)
4. Use checklist correctly	0·5 (0·1–0·9)
5. Follow established policies or procedures	0·01 (0·003–0·03)
6. Passive inspection	0·1 (0·05–0·5)
7. Respond to an annunciator (one of one)	0·0001 (0·00005–0·001)
8. Read annunciated lamp	0·001 (0·0005–0·005)
9. Read digital display	0·001 (0·0005–0·005)
10. Read analog meter	0·003 (0·001–0·01)
11. Read analog chart recorder	0·006 (0·002–0·02)
12. Read graph	0·01 (0·005–0·05)
13. Read printing recorder (cluttered)	0·05 (0·01–0·2)
14. Record more than 3 digits	0·001 (0·0005–0·005)
15. Detect a deviant meter with limit marks during initial unit	0·05 (0·01–0·1)
16. Check–read specific meters with limit marks	0·001 (0·0005–0·005)
17. Check–read specific meters without limit marks	0·003 (0·001–0·01)
18. Check wrong indicator lamp in a group of similar lamps	0·003 (0·001–0·01)
19. Note incorrect status of an indicator lamp (in a group)	0·99 (0·98–0·998)
20. Note incorrect status of a legend lamp (in a group)	0·98 (0·96–0·996)
21. Remember oral instructions, one of one	0·001 (0·0005–0·005)
22. Select wrong panel control:	
a. Among a group of similar controls	0·003 (0·001–0·01)
b. If functionally grouped	0·001 (0·0005–0·005)
c. If part of a mimic-type panel	0·0005 (0·0001–0·001)
23. Set a multiposition switch	0·001 (0·001–0·1)
24. Mate a connector	0·01 (0·005–0·05)
25. Turn control in wrong direction:	
a. If no violation of population stereotype	0·0005 (0·0001–0·001)
b. If populational stereotype is violated	0·05 (0·01–0·1)
26. Check each item on a short list, (10 items or less) using checkoff	0·001 (0·0005–0·005)
27. Check each item on a long list, (more than 10 items) using checkoff	0·003 (0·001–0·01)
28. Check each item on a short list, (10 items or less) not using checkoff	0·003 (0·001–0·01)
29. Check each item on a long list, (more than 10 items) not using checkoff	0·01 (0·005–0·5)

Adapted from A.D. Swain and H.E. Guttman, 'Handbook of Human Reliability Analysis with Emphasis on Nuclear Power Plan Applications' (Ref. 20).

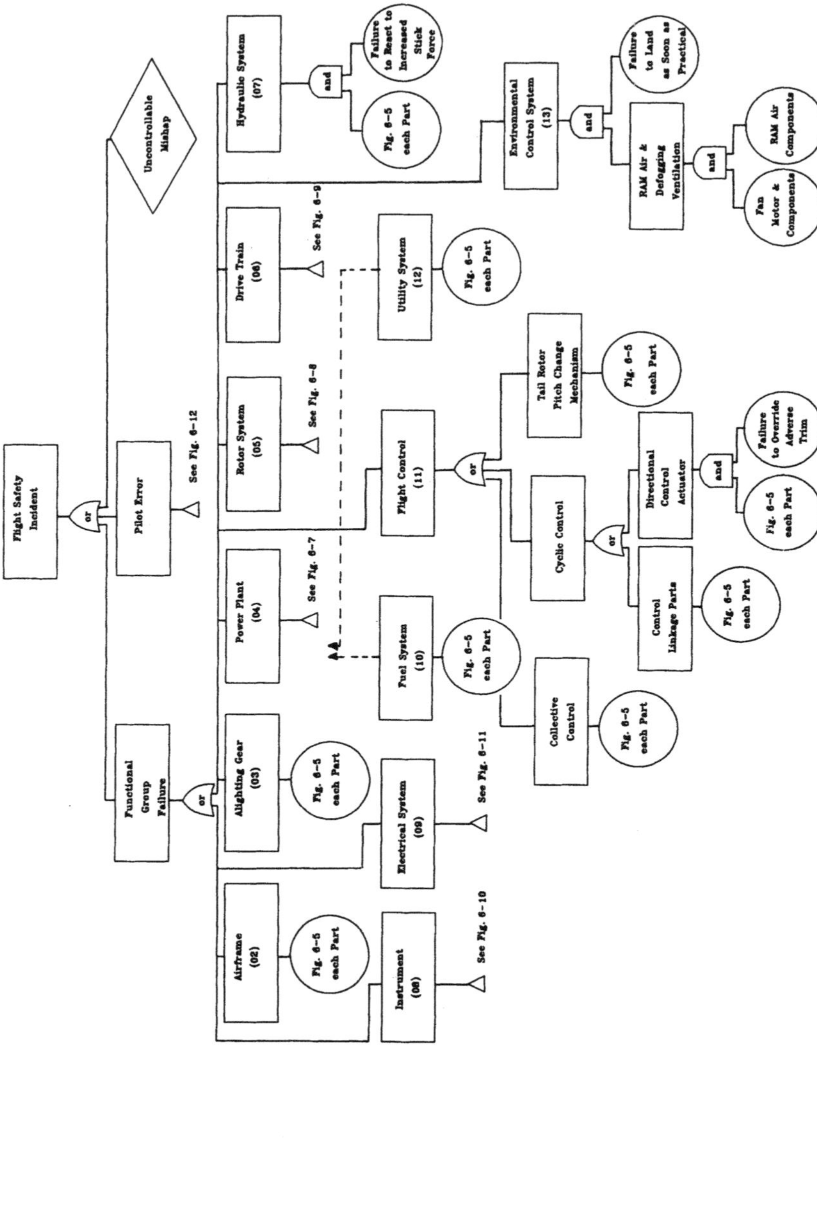

Figure 6–6 Overall FTD

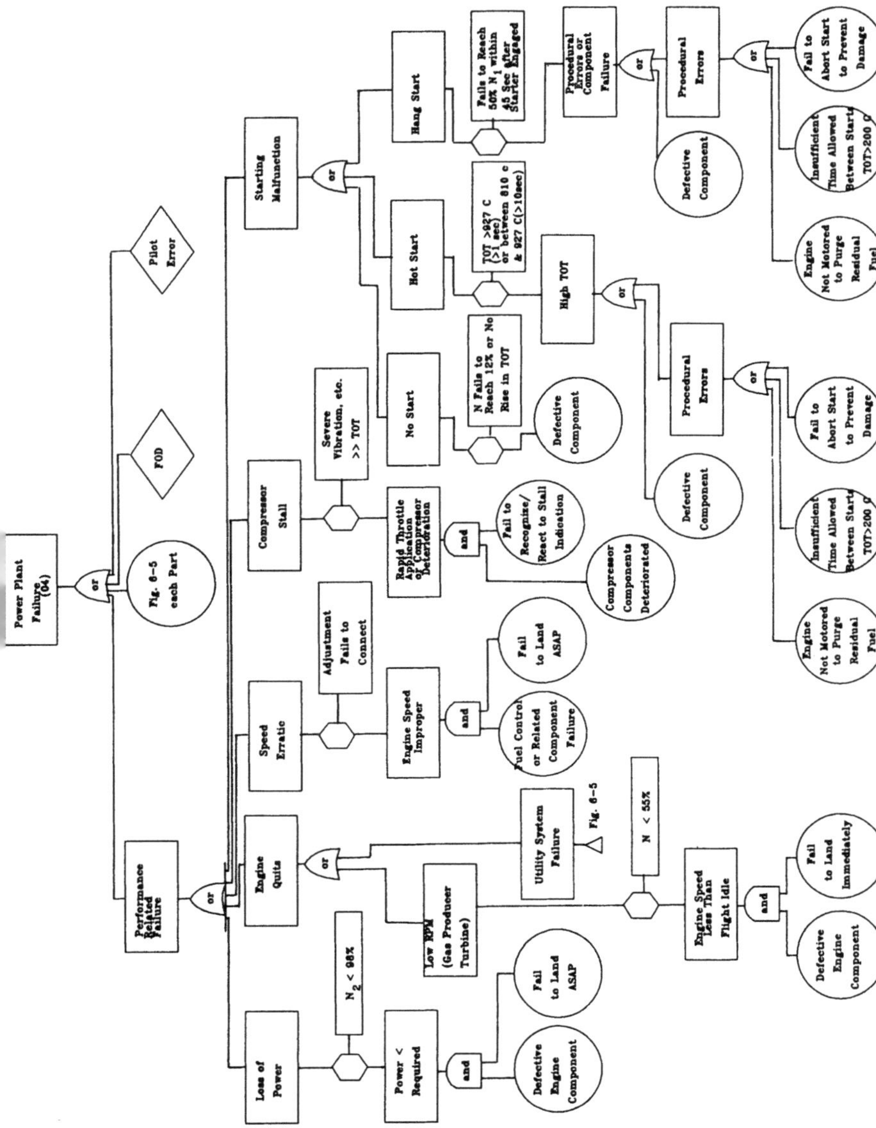

Figure 6–7 Power Plant FTD

restricted fuel lines, worn fuel pump gears combined with fuel cell boost pump failure, fuel control failure, governor failure or air system leakage)

(5) other, non-engine-mounted, fuel system components
(6) gears (binding) or the lubrication system (oil leakage, low oil supply or bearings binding)

The engine is defined to have 'quit' when the gas producer turbine rpm (N_1) is less than 55% of normal. When N_1 is less than 55%, engine speed is less than flight idle. The engine's quitting may lead to a flight safety incident if the pilot fails to 'land immediately' which is defined in the emergency procedures as 'execute a landing without delay to assure the survival of occupants'.

The engine's quitting may be due to utility system failure or to a defective engine component including:

(1) compressor (damage from a foreign object, a cracked diffuser scroll, a fractured blade or vane, a cracked air tube, a seized bearing, a cracked case or flame-out due to failure of the anti-ice valve and resultant ingestion of water or ice)
(2) turbine (a fractured blade or vane, blade damage due to overheating, a seized bearing or damage from a foreign object)
(3) the combustion chamber (a cracked case or a deformity in the chamber)
(4) the fuel system (contaminated fuel, a failing-closed of the fuel nozzle valve, worn fuel pump gears combined with fuel cell boost pump failure, fracture of the fuel pump shaft, failure of fuel control, a clogged filter combined with a failed-closed filter by-pass valve, failure of the fuel pressure regulating valve or blocked fuel lines)
(5) other non-engine-mounted fuel system components
(6) gearbox (fractured gear shafts or teeth, seized gears, a cracked case or a seized bearing)
(7) the lubrication system (a clogged filter combined with the failing-closed of the filter by-pass valve, fractured oil pump gears or gear shaft, blocked oil cooler leading to high oil temperature, blocked internal oil passages or ports, a ruptured oil cooler, low oil supply, failed chip detector, a failing-open of the check valves, a ruptured tank, a failing-open of the oil pressure sensor, contaminated oil, rupture of the external sump, or rupture or leakage of the oil lines)

Erratic speed occurs upon an improper engine speed which is not corrected by governor switch or throttle adjustment. Erratic speed leads to a flight safety incident only if the pilot fails to land as soon as possible (as defined in the paragraph on loss of power). Erratic speed can be caused by failure in fuel control or related components. High engine speed may be due to a binding of the fuel control fuel by-pass valve, a defective double check valve, wrong settings on the fuel control idle, a defective governor, incorrect rigging of the throttle linkage or dirt in the fuel control. Low engine speed may be due to accumulator leaks, leaks in the air lines to the fuel control, a failing-open of the linear actuator, an excessive generator load at idle, a dirty compressor or a wrong rigging on the droop compensator. Variable engine speed may be due to a clogged fuel control fuel filter, defective fuel control or dirty ports in the fuel control.

Compressor stall, which is indicated by severe engine vibration, rumbling, banging and a rapid rise in turbine outlet temperature, may be caused by rapid throttle application or by compressor deterioration. Compressor stall may lead to a flight safety incident if the pilot fails to recognize the stall indicator and to react properly. Compressor component deterioration may be a defective air bleed valve, dirt or foreign matter build-up on the blades, rubbing blades or blade erosion.

A starting malfunction may be a 'no start', a 'hot start' or a 'hang start'. A 'no start' occurs when the gas producer turbine rpm (N_1) fails to reach 12% of normal or when there is no rise in turbine outlet temperature. This may be caused by failure in an engine starting component such as the compressor rotor (binding), the ignition system (low/no voltage from ignition exciter, an open ignition lead or switch, or a fouled or damaged igniter plug), the battery circuit (a low or dead battery or loose or open battery cables), the starter circuit (reversed leads on the starter/generator terminal block or an open starter winding) or the fuel system (no fuel in tanks, a clogged fuel nozzle orifice, air in the fuel lines, crushed or restricted fuel lines, low fuel pressure, a stuck fuel nozzle valve, the fuel control linkage in the cut-off position or an inoperable fuel pump).

A 'hot start' occurs when turbine outlet temperature is greater than 927°C for more than 1 second or between 810 and 927°C for more than 10 seconds. This engine failure due to high temperature may be due to battery/starter failure (low battery or a defective starter), compressor failure (compressor air leakage, dirty compressor blades or a jammed air bleed control valve) or a fuel or fuel control related failure (a fuel nozzle jammed fully open, a defective throttle linkage, a too-rich fuel control start enrichment or a shift in the fuel control calibration).

A 'hang start' occurs when the engine fails to reach 50% of normal gas producer turbine rpm (N_1) within 45 seconds after the starter is engaged. This may occur due to procedural errors or to equipment component failure. Component failures include starter/battery failure (low battery or defective starter), compressor failure (cracked air discharge tube or leaking air discharge tube seals or dirty compressor blades), a cracked outer combustion case, fuel system failure (a defective fuel pump leading to low pressure, the fuel nozzle jammed partly open or a partly blocked fuel line) or a fuel control related failure (a clogged PC filter, leaking fuel control air tubing, a defective power turbine governor, a too-lean starting enrichment, a cracked accumulator, a jammed-open fuel control by-pass valve, shifts in fuel control calibration, a sticking double check valve).

The *rotor system* (05) *FTD*, presented in Figure 6–8, indicates that system failure may be due to a main rotor blade failure, a pylon failure, a main rotor hub failure, or a tail rotor hub or blade failure. A failure of either the pylon or the main rotor hub is defined by the standard component FTD.

A main rotor blade failure may be a material failure in the blade or a failure induced by the pilot or the ground crew, i.e., the rotor may be allowed to overspeed during autorotation or the ground crew may damage the blade during tie-down. A material failure may be due to a defect in the nose block, spar or one of the other components making up the blade.

A defective component may be either a failure of the component structure itself or a delamination failure if the component is attached to the rest of the blade by adhesives. Component structural failures may be due to cracking, corrosion and erosion. Erosion, of course, applies only to those components exposed to the air during operation of the blade.

Cracks may be present in components and can be induced by manufacturing or maintenance as shown in the standard component FTD (Figure 6–5). It should be emphasized that corrosion occurs when moisture is present due to entrapment near the component and this moisture acts on the component due to cumulative storage stresses. Moisture entrapment may be induced during: (1) manufacturing and not detected by QC inspection, (2) maintenance and not detected by inspection or (3) storage.

Storage induced moisture entrapment is due to a combination of factors: inadequate preservation, packaging and packing of the blade;

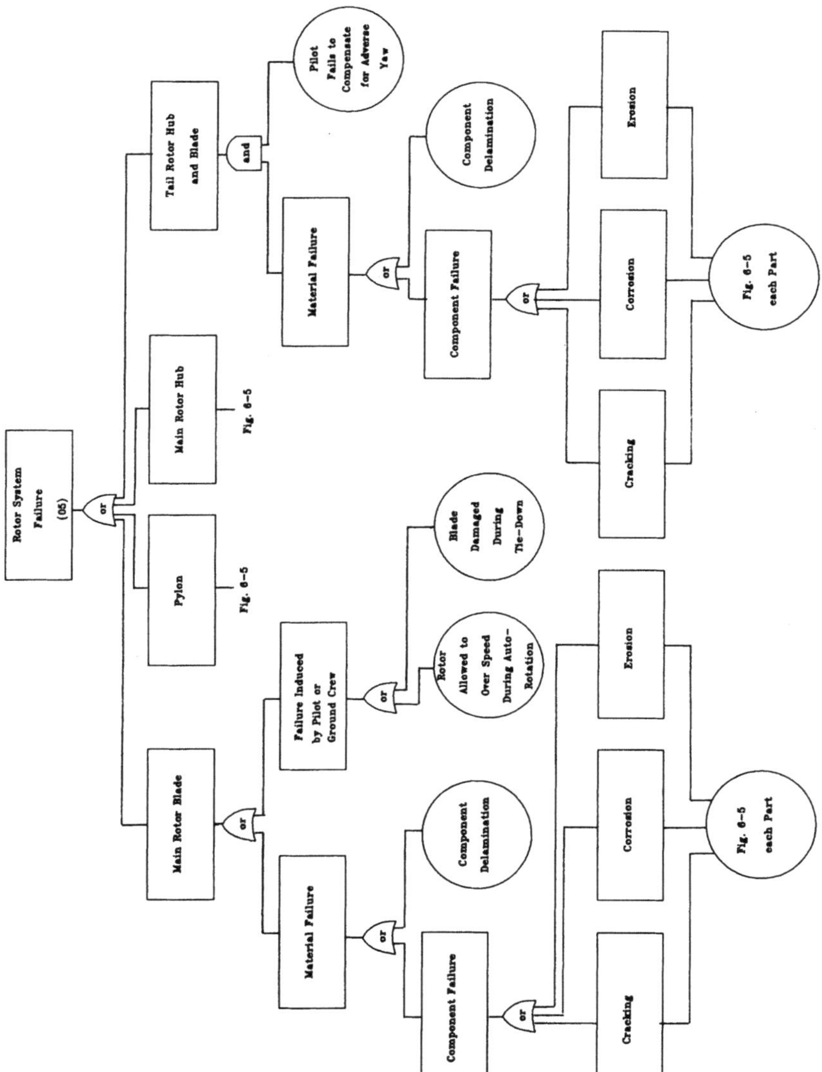

Figure 6–8 Rotor System FTD

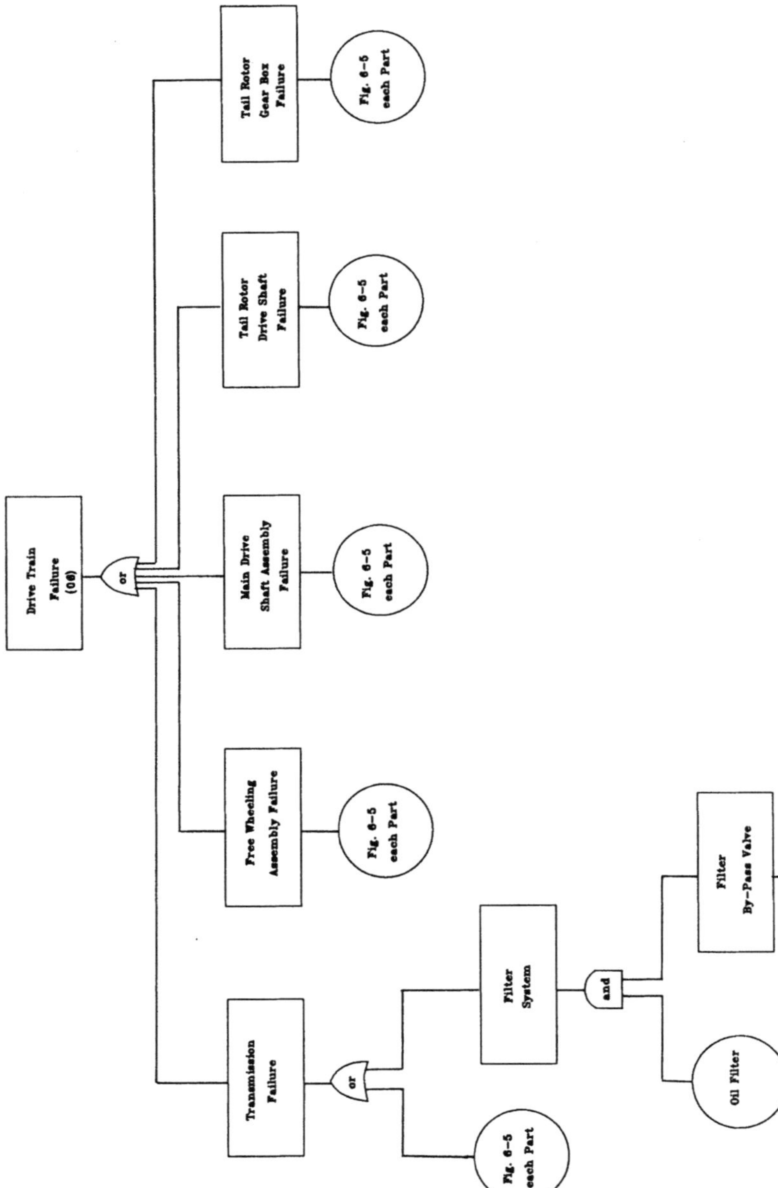

Figure 6–9 Drive Train FTD

improper storage environment and inadequate inspection during storage; and presence of a defect in the blade that allows moisture entrance to the component. Such defects may be induced during manufacture, and not detected by QC inspection; they may be induced during previous operation, for instance, by overstressing the blade or flying in dusty, sandy conditions; they may be induced due to improper maintenance and not detected by inspection; as a result of storage itself. As with other failure mechanisms, corrosion can lead to blade failure if not detected by inspection, and corrected, in time. Cracks, of course, occur in many different sizes and degrees of severity, but a crack of any size can lead to blade failure if not detected by inspection in time.

Erosion occurs in those components exposed to the air during operation of the blade and can lead to blade failure if excessive erosion of a component is not detected by inspection in time and operating stresses are then applied.

Delamination of components attached to the rest of the blade by adhesive may occur when there is a void present in the adhesive. A void may be induced by adhesive degradation resulting from moisture entrapment in the adhesive due to cumulative storage stress. The moisture entrapment failure mechanism is the same as that discussed above on corrosion.

Failure of either the pylon or the main rotor hub is defined by the standard component FTD given in Figure 6–5.

Tail rotor hub or blade failure leads to a flight safety incident or mission abort only if the pilot fails to compensate for adverse yaw resulting from the failure. Tail rotor hub failure may be a failure of the yoke or of the retaining bolt. Failure mechanisms for the tail rotor blade itself are the same as those discussed for the main rotor blade for components that are common. Also the tail rotor blade does not have a pilot/ground crew induced failure mode.

The *drive train* (06) *FTD*, presented in Figure 6–9, indicates that system failure may be due to the transmission, the free wheeling assembly, the main driveshaft assembly, the tail rotor driveshaft or the tail rotor gearbox. In each case the standard component FTD applies.

Hydraulic system (07) components whose failure can cause a flight safety incident, if the pilot fails to compensate for the resultant increased stick forces, are the fittings, hoses, the reservoir, the check valve, the relief valve, the pump, the solenoid valve, the boost switch, the filter or the boost solenoid circuit breaker.

The *instrument system failure* (08) *FTD*, presented in Figure 6–10, indicates that system failure may be due to the avionics equipment or the analog gauges (instruments) or the warning/caution indicators. For purposes of this model, it is assumed, as shown on the FTD, that an avionics equipment failure will cause a flight safety incident.

Avionics equipment failure may be failure of the communication equipment, the navigation equipment or the radar equipment. Communication equipment failure occurs either when all four types of radio equipment in the cockpit, the UHF radio, the VHF (AM) radio and the two units of VHF (FM) radio, fail or when interphone panel communication control equipment which is common to all four types of radios fails. Navigation equipment failure occurs when both the ADF receiver and the CONUS navigation receiver fail. Radar equipment failure occurs either when the transponder fails or when one or more of the three radarsets (altimeter, proximity warning or threat detector), fails.

In the case of each item of avionics equipment the failure may be due to failure of the set or unit itself, associated wiring or the circuit breaker for the item or, for the communication equipment, of interphone panel communication control equipment dedicated to the item or of any associated antenna or switches.

Additionally, the VHF (AM) radio and the two FM radios are connected to the non-essential bus, and the CONUS navigation receiver is connected to AC power, so failure of these electrical systems leads to failure of the connected avionics equipment.

For the analog gauges (instruments) and the warning/caution indicator, failure of a given instrument or indicator leads to a flight safety incident only when the pilot reacts incorrectly due to the failure. For each of the gauges and indicators a failure may be due to failure in the gauge unit or lamp, in the associated sensor in associated wiring or in the circuit breaker controlling the device. In addition to the failure of individual indicators, the entire caution system may fail due to failure of the caution panel lights circuit breaker.

The *electrical system* (09) *FTD*, presented in Figure 6–11, indicates that system failure may be due to the battery, the charging circuit, the starting circuit, the non-essential bus circuit, the AC power circuit, or instrument or landing light circuit. For battery, charging circuit, or instrument or landing light circuit component failure, the failure leads to a flight safety incident only if the pilot fails to 'land as soon as practicable', which is defined in the emergency procedures as 'execute a landing to the nearest suitable airfield/heliport'. Non-essential bus circuit component failure and AC power circuit component failure each lead to a flight safety

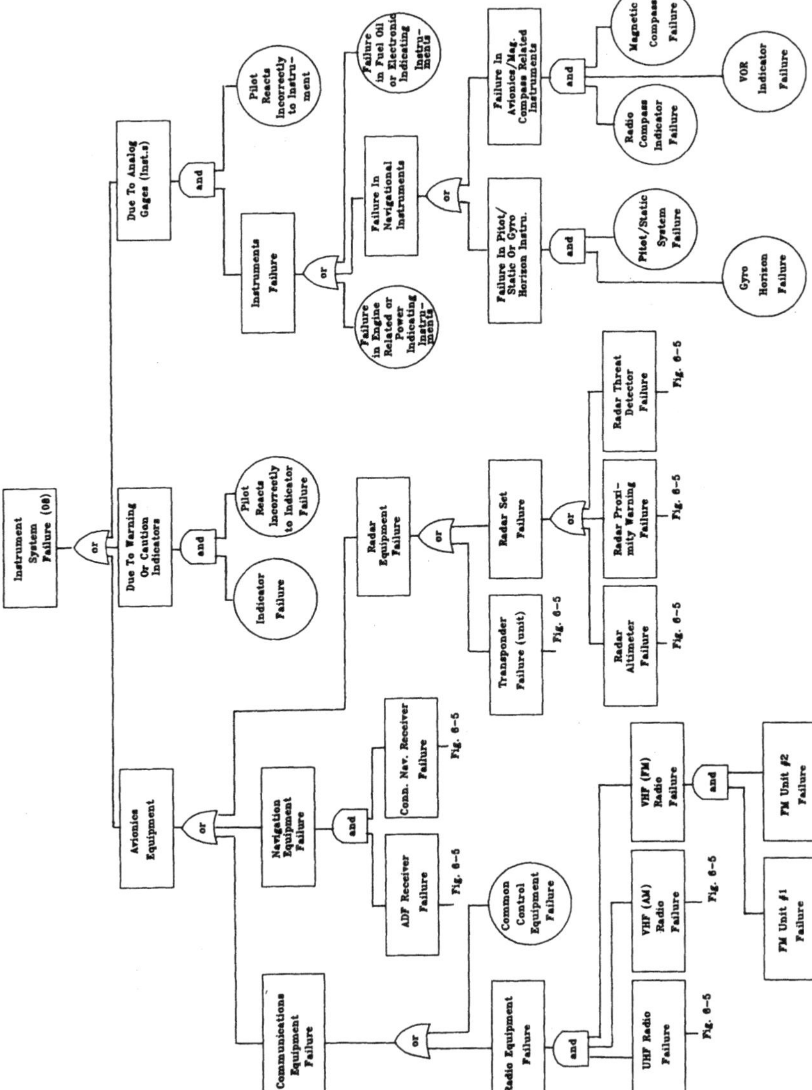

Figure 6–10 Instrument Failure FTD

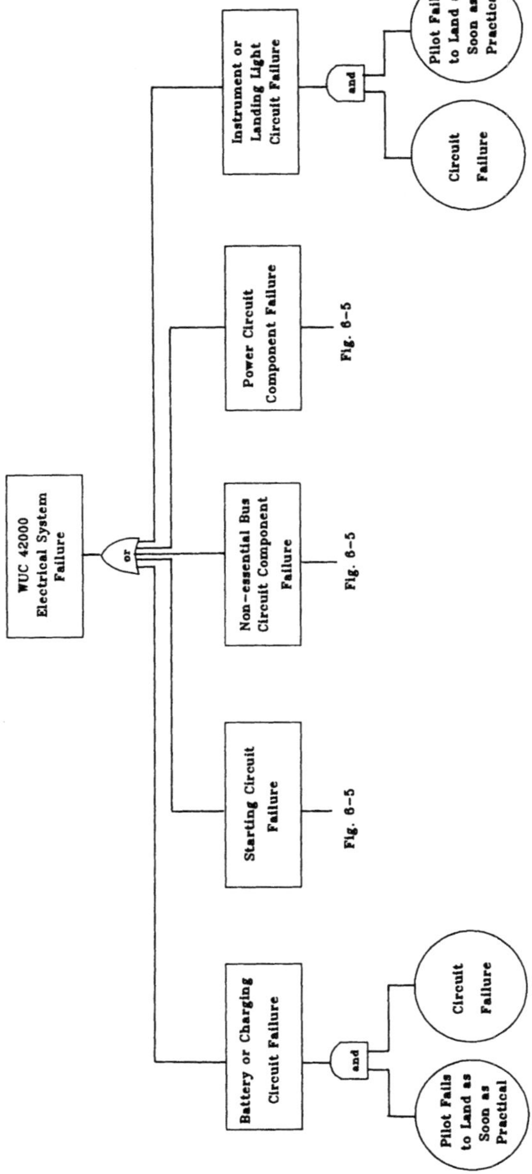

Figure 6–11 Electrical System FTD

incident in that their respective failures may cause an avionic equipment failure.

A *fuel system* (10) failure, including fuel cell boost pump failure, leads to a loss of engine power as shown in Figure 6–6. The failure may be a failure of the fuel cell itself; a failure in the lines (hoses or tubing), fittings or seals of the system; a failure in the shutoff valve or its associated linkage; or a failure of the boost pump or its associated circuit breaker or wiring. In each case the standard component FTD applies.

Flight control (11) system failures are either cyclic or collective control failures or tail rotor pitch change mechanism failures which in turn are failures in the pedal assembly, the control linkage or the pitch change element and are defined by the standard component FTD. Control linkage failure may be either serial element failures or redundant element (primary and backup control linkage) failures. Also, the cyclic control can fail, leading to adverse trim which could cause a flight safety incident if the pilot fails to override the adverse trim.

Utility engine anti-icing (12) system failure leads to a flight safety incident in that its failure causes the engine to 'quit'. System failure may result from failure of the engine anti-icing switch, wiring, circuit breaker, actuator, lever, tube or clevis.

Environmental control (13) system failure leads to a flight safety incident only when there is both a fan motor related failure (failure in the defogging fan motor, fan wiring, or the defogging and ventilation switch) and a ram air related failure (failure in the ventilation control cable, the plenum valve control rod or the duct door control rod) and the pilot fails to 'land as soon as practicable', which as previously stated is defined in the Army aircraft emergency procedures as 'execute a landing to the nearest suitable airfield/heliport'.

The *pilot error FTD*, presented in Figure 6–12, indicates that a flight safety incident can occur when the pilot misreads the instruments in flight operation. Pilot error may be due to the misreading of the instruments or to physical, psychological or training factors which impede appropriate pilot action.

6.3 APPLICATION OF THE FLIGHT SAFETY PREDICTION MODEL WITH THE RCM LOGIC

This section illustrates how the helicopter flight safety prediction model can be used with the RCM logic to select those life-time maintenance tasks

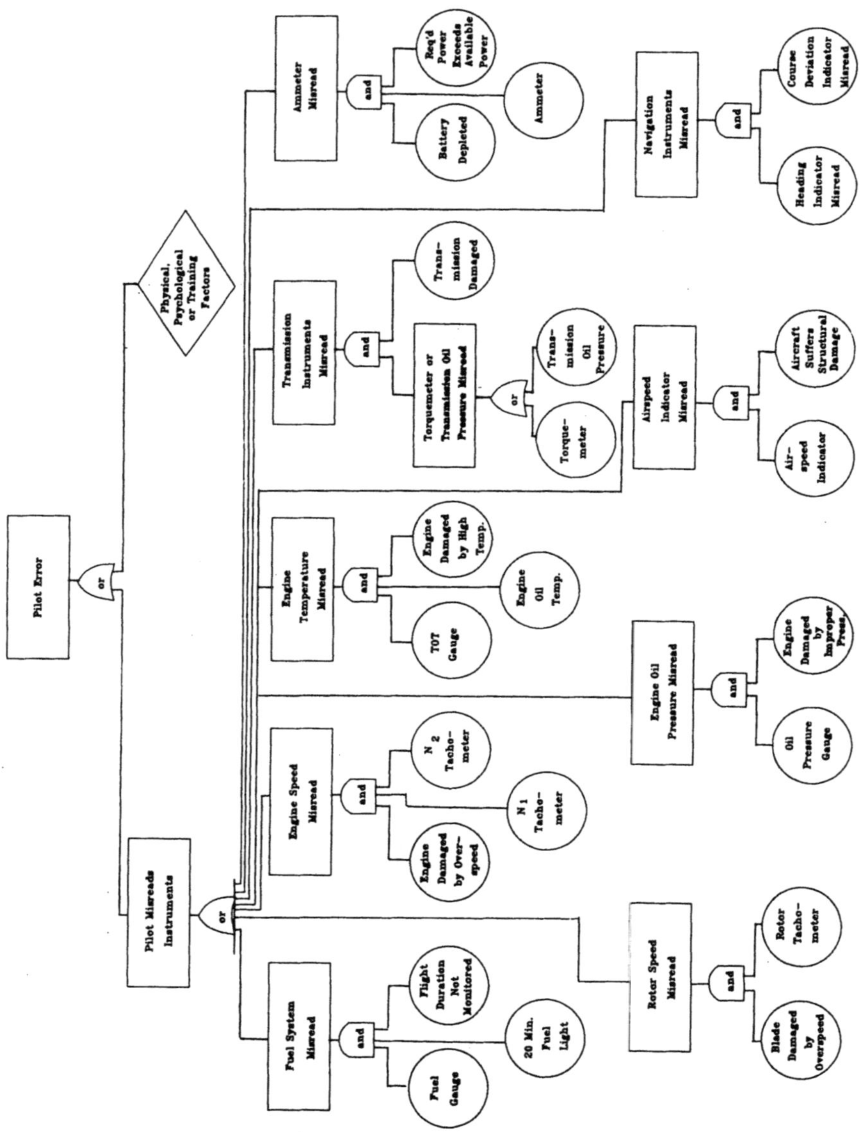

Figure 6–12 Pilot Error FTD

most effective in preventing a decrease in reliability and flight safety. Sample part criticality data were developed for a particular turbine engine using the standard FTD given in Figure 6–7. The potential part failure modes that would make up each basic fault for the engine were defined and probability of occurrence and criticality numerics were computed for each.

A fault matrix was prepared for this example and is presented in Table 6–3. Listed are individual failure modes applicable to the standardized basic component faults and associated probability of occurrence and criticality data. For purposes of this example, a worst-case situation was assumed for the basic faults associated with the pilot's reaction to failure, i.e., $P(x)$'s for these faults were set at one. Also, for this example, it was assumed that the probability of a utility system failure is zero.

The probability of occurrence numerics were computed from estimated failure rates, derived from Table 6–1, and an operating time of 2.5 hours. This operating time takes into account preflight inspection; the flight or mission, which could involve intermediate landings and restarts, through to final landing; and the tie-down of the aircraft. Defects discovered during the pilot or crew chief walk-around preflight inspection or initial engine start which result in mission cancellation, were not included.

The criticalities for all failure modes are arranged in the fault matrix in descending order (the most critical failure mode is listed first, while the least critical failure mode is listed last). Associated with each ranked criticality value is a cumulative sum of all previously ranked criticalities. These data were then used to plot a criticality curve as previously described and shown in Figure 6–2. The critical region of the curve, for this example, contains nine failure modes listed below:

1. Compressor rotor seal leakage
2. Carrier and gear cylindrical roller bearing cage fracture
3. Rear compressor bearing spalling/cage/pin fracture
4. Forward turbine rotor bearing face seal leakage
5. Centrifugal impeller fracture
6. First stage gas producer turbine rotor blade/disc fracture
7. Compressor rotor front shaft (first stage) blade fracture/ separation from disc
8. Air diffuser housing vane fracture
9. Front compressor ball bearing spalling/cage fracture

The RCM decision logic was then applied to each critical part failure mode, and cost-effective maintenance tasks and requirements were selected from a review of the resultant maintenance process analysis data. The

TABLE 6-3
Fault Matrix (Sample Data)

	Basic fault	*Failure effect*	$P(X_i)$	$P(H/X_i)$	*CR*
1	Compressor rotor seal leakage	Loss of power	176·8	1·0	176·8
2	Carrier and gear cylinder roller bearing cage fracture	Engine quits	59·3	1·0	59·3
3	Rear compressor bearing spalling/cage/pin fracture	Loss of power	32·8	1·0	32·8
4	Forward turbine rotor bearing face seal leakage	Loss of power	23·8	1·0	23·8
5	Centrifugal impeller fracture	Engine quits	22·3	1·0	22·3
6	Second stage gas producer turbine rotor blade/disc fracture	Engine quits	21·5	1·0	21·5
7	Compressor rotor front shaft (1st stage) blade fracture/separation from disc	Engine quits	21·0	1·0	21·0
8	Air diffuser housing vane fracture	Loss of power	21·0	1·0	21·0
9	Front compressor rotor ball bearing spalling cage fracture	Loss of power	17·8	1·0	17·8
10	Rear compressor rotor shaft fracture	Engine quits	5·0	1·0	5·0
11	Axial compressor rotor front shaft disc fracture/separation from compressor	Loss of power	5·0	1·0	5·0
12	Power turbine shaft fracture	Engine quits	5·0	1·0	5·0
13	Compressor rotor bolts fracture	Compressor stall	5·0	1·0	5·0
14	Accessory gearbox annular ball bearing cage fracture	Engine quits	3·8	1·0	3·8
15	Second stage power turbine rotor annular ball bearing fracture	Loss of power	3·3	1·0	3·3
16	Second stage power turbine rotor annular ball bearing cage fracture	Engine quits	3·3	1·0	3·3
17	Accessory gearbox ball bearing cage fracture	Engine quits	3·3	1·0	3·3
18	Accessory gearbox annular ball bearing cage fracture	Speed erratic	3·3	1·0	3·3
19	Accessory gearbox annular ball bearing cage fracture	Engine quits	2·8	1·0	2·8
20	Accessory gearbox annular ball bearing cage fracture	Speed erratic	2·8	1·0	2·8
21	2nd–5th stage compressor or rotor blade fracture/separation from disc	Engine quits	2·3	1·0	2·3
22	Second stage gas producer turbine rotor blade/disc fracture	Engine quits	2·3	1·0	2·3

TABLE 6-3—*contd.*
Fault Matrix (Sample Data)

	Basic fault	*Failure effect*	$P(X_i)$	$P(H/X_i)$	*CR*
23	Second stage power turbine rotor nut and locking cap shear failure	Engine quits	2·3	1·0	2·3
24	First stage power turbine rotor disc and hub fracture	Engine quits	2·3	1·0	2·3
25	Fuel control shaft fracture	Engine quits	2·3	1·0	2·3
26	Fuel pump shaft/rotating element fracture	Engine quits	2·3	1·0	2·3
27	Oil pump drive spur gear teeth/hub fracture	Engine quits	2·3	1·0	2·3
28	Accessory gearbox spur gearshaft drive teeth/hub fracture	Speed erratic	2·3	1·0	2·3
29	N11 idler spur gear teeth/hub fracture	Speed erratic	2·3	1·0	2·3
30	Start fuel manifold clogged	Starting malfunction	2·3	1·0	2·3
31	Accessory drive inner bevel spur gear teeth fracture	Starting malfunction	2·3	1·0	2·3
32	Start fuel manifold clogged	Starting malfunction	2·3	1·0	2·3
33	Engine speed low main fuel manifold clogged	Speed erratic	2·3	1·0	2·3
34	First stage power turbine rotor blade fracture	Engine quits	1·8	1·0	1·8
35	Output shaft fracture	Engine quits	1·8	1·0	1·8
36	Accessory drive carrier annular ball bearing cage fracture	Engine quits	1·8	1·0	1·8
37	Carrier and gear cylinder roller bearing cage fracture	Engine quits	1·8	1·0	1·8
38	IGV actuator assembly stuck closed	Compressor stall	1·8	1·0	1·8
39	Accessory drive carrier annular ball bearing cage fracture	Starting malfunction	1·8	1·0	1·8
40	Accessory drive inner bevel, spur gear teeth spalling/hub failure	Loss of power	1·75	1·0	1·75
41	Accessory drive carrier annular ball bearings cage fracture	Engine quits	1·75	1·0	1·75
42	N11 driver idler cluster gear teeth spalling	Loss of power	1·0	1·0	1·0
43	Compressor rotor fracture	Engine quits	1·0	1·0	1·0
44	Second stage power turbine rotor blade fracture	Engine quits	1·0	1·0	1·0
45	Accessory drive shaft gear teeth/hub/shaft/pin fracture	Engine quits	1·0	1·0	1·0
46	Accessory drive annular ball bearing cage fracture	Engine quits	1·0	1·0	1·0
47	Carrier and gear drive bearing cage fracture	Speed erratic	1·0	1·0	1·0
48	Drive idler cluster gear teeth/hub fracture	Speed erratic	1·0	1·0	1·0

TABLE 6-3—*contd.*
Fault Matrix (Sample Data)

	Basic fault	*Failure effect*	$P(X_i)$	$P(H/X_i)$	*CR*
49	Accessory drive carrier annular ball bearing cage fracture	Starting malfunction	1·0	1·0	1·0
50	Axial compressor rotor nut/lockring fracture	Loss of power	0·5	1·0	0·5
51	Second stage turbine rotor disc and hub fracture	Engine quits	0·5	1·0	0·5
52	Carrier and gear nut and locking cup shear failure	Engine quits	0·5	1·0	0·5
53	Accessory drive carrier bevel gear hub fracture	Engine quits	0·5	1·0	0·5
54	Accessory gearbox cylindrical roller bearing cage fracture	Engine quits	0·5	1·0	0·5
55	Accessory gearbox annular ball bearing cage fracture	Engine quits	0·5	1·0	0·5
56	Carrier and gear assembly bearing cage fracture	Speed erratic	0·5	1·0	0·5
57	Drive idler cluster gear teeth/hub fracture	Speed erratic	0·5	1·0	0·5
58	Accessory drive carrier bevel gear teeth fracture	Starting malfunction	0·5	1·0	0·5
59	Start fuel solenoid valve spring failure	Starting malfunction	0·5	1·0	0·5
60	Flow divider and dump valve plunger failed open/spring failure	Starting malfunction	0·5	1·0	0·5
61	Failure in fuel system; main fuel nozzle clogged	Loss of power	0·2	1·0	0·2
62	Aging failure not detected during inspection	Component failure (wearout/stress–strength)	0·35	0·006	0·0021
63	Inadequate quality control inspection	Component failure (wearout/stress–strength)	0·3	0·006	0·0018
64	Inadequate repair inspection	Component failure (wearout/stress–strength)	0·3	0·006	0·0018
65	Manufacturing induced spare defect	Component failure (wearout/stress–strength)	0·3	0·006	0·0018
66	Inadequate quality control inspection	Component failure (wearout/stress–strength)	0·3	0·006	0·0018
67	Inadequate inspection prior to issue	Component failure (wearout/stress–strength)	0·3	0·006	0·0018
68	Spare inadequate reconditioning prior to installation	Component failure (wearout/stress–strength)	0·3	0·006	0·0012
69	Inadequate storage inspection (cyclic)	Component failure (wearout/stress–strength)	0·3	0·006	0·0012
70	Material deterioration	Component failure (wearout/stress–strength)	0·2	0·006	0·0012
71	Used beyond specified operating life period (wearout)	Component failure (wearout/stress–strength)	0·001	1·0	0·001
72	Replacement period (operating life not adequate or not specified)	Component failure (wearout/stress–strength)	0·001	0·006	< 0·000

TABLE 6-3—*contd.*
Fault Matrix (Sample Data)

Basic fault	*Failure effect*	$P(X_i)$	$P(H/X_i)$	*CR*
73 Manufacturing induced engine defect	Component failure (wearout/stress–strength)	0·03	0·006	0·0001
74 Faulty repair	Component failure (wearout/stress–strength)	0·03	0·006	0·0001
75 Defect induced due to exceeding design limits during previous operations	Component failure (wearout/stress–strength)	0·001	0·006	< 0·0001
76 Spare part defective due to faulty design	Component failure (wearout/stress–strength)	0·001	0·006	< 0·0001
77 Accessory drive gear teeth/hub fracture	Engine quits	0	1·0	0
78 Accessor gearbox annular ball bearing cage fracture	Engine quits	0	1·0	0
79 Rotary oil vane fracture	Engine quits	0	1·0	0
80 Rotary oil pump shaft fracture	Engine quits	0	1·0	0
81 Engine speed high interstage air bleed actuator stuck closed	Speed erratic	0	1·0	0
82 Contamination of main turbine fuel control computer passage and elements	Speed erratic	0	1·0	0
83 Fuel filter clogged	Speed erratic	0	1·0	0
84 Interstage air bleed actuator stuck closed	Compressor stall	0	1·0	0
85 Bleed band crack/leakage	Compressor stall	0	1·0	0
86 Low/no output from ignition exciter	Starting malfunction	0	1·0	0
87 Ignition lead short circuit	Starting malfunction	0	1·0	0
88 Ignition plug short circuit	Starting malfunction	0	1·0	0
89 Ignition exciter wiring shorted/open	Starting malfunction	0	1·0	0
90 Ignition lead open circuit	Starting malfunction	0	1·0	0
91 Ignition plug open circuit	Starting malfunction	0	1·0	0
92 Engine not motored to purge residual fuel	Starting malfunction	0	1·0	0
93 Insufficient time allowed between starts	Starting malfunction	0	1·0	0
94 Pilot fails to abort start to prevent damage	Starting malfunction	0	1·0	0
95 Engine component failure leading to 'Hang start' flow divide and dump valve plug failed closed	Starting malfunction	0	1·0	0

maintenance process analysis worksheet, described in Chapter 2, was used to record the logic data for the critical parts. The worksheet for the compressor rotor centrifugal impeller fracture failure mode (item 5 above) is given in Figure 6–13. The answers to the first four questions (step 1) indicate that this part failure mode falls in the 'safety hidden' consequent

MAINTENANCE PROCESS ANALYSIS WORKSHEET				1
MAJOR ITEM *COMPRESSOR ROTOR*	PREPARED BY *DH*		PREPARING ORGANIZATION *RTA*	
NOMENCLATURE *CENTRIFUGAL IMPELLER*	PART NO. *XXXX*	DATE *8/29/84*	REVISION NO. *0*	
FAILURE MODES	A *FRACTURE*	B	C	D

NO.	LOGIC QUESTION	FM	Y	N	INFORMATION SUMMARY
1	Can Pilot/Crew Detect Failure?	A		X	
		B			
		C			
		D			
2	Does Failure Cause a Mission Abort or Flight Safety Incident?	A			
		B			
		C			
		D			
3	Does Failure Alone Cause a Mission Abort or Flight Safety Incident?	A	X		*LOSS OF POWER*
		B			
		C			
		D			
4	Does Failure Adversely Effect Operational Performance?	A			
		B			
		C			
		D			

FAILURE MODES CONSEQUENCE CATEGORY	A *FRACTURE*	B	C	D

NO.		FM	Y	N	INFORMATION SUMMARY
* 5, 11, (16), 22	A Servicing Task?	A		X	
		B			
		C			
		D			
6, 12*	A Crew Monitoring Task?	A			
		B			
		C			
		D			
(17) 23	Verify Operation?	A		X	
		B			
		C			
		D			
7, 13, (18), 24	An On-Condition Task?	A	X		*PSA – VISUAL, FLUORESCENT PENETRANT,*
		B			*DIMENSIONAL INSPECTION*
		C			
		D			
8, 14, (19), 25	A Rework Task?	A	X		*REPAIR RUBBING, NICKS, DENTS, BURRS ON VANES*
		B			
		C			
		D			
9, 15, (20), 26	Replacement?	A	X		*REPLACE IF LIMITS ARE NOT MET FOR VANE*
		B			*TIP RUBBING, VANE EROSION, WEAR AND FIT.*
		C			*ALWAYS REPLACE IF CRACKED.*
		D			
10, (21)	A Combination of Tasks?	A	X		*SEE QUESTION 18,19, 20*
		B			
		C			
		D			

* Identify Applicable PM Task Question Numbers

Figure 6–13 Maintenance Process Analysis Worksheet

TABLE 6-4
Maintenance Task Profile (Critical Parts)

Part description	*Failure mode*	*Service*	*Crew monitor*	*On-condition*	*Rework*	*Replace*
Compressor rotor seal	Leakage		×	×		×
Carrier and gear cylindrical roller bearing cage	Fracture			×		×
Rear compressor bearing spalling/cage/pin	Fracture		×	×		×
Forward turbine rotor bearing face seal	Leakage	×	×	×		×
Centrifugal impeller	Fracture			×		×
First stage gas producer turbine rotor disc	Fracture			×	×	×
Compressor rotor front shaft (first stage)	Blade fracture/ separation from disc			×	×	×
Air diffuser housing vane	Fracture			×	×	×
Front compressor ball bearing	Spalling/cage fracture		×	×	×	×

category. Therefore all subsequent logic questions were asked. Similar worksheets were prepared for each of the other failure modes. Table 6–4 provides a profile of the maintenance task requirements for this sample analysis.

As shown in Tables 6–3 and 6–4, the RCM process provides complete correlation of the maintenance tasks to the specific parts and their failure modes (and criticalities) which the maintenance tasks are to prevent. It thus increases the probability that all safety critical parts and their failure modes are considered in the development of the maintenance support program. It also increases the probability that the level and content of the maintenance program are optimally specified.

Appropriate intervals for each of the hard-time replacement and on-condition maintenance tasks can then be defined based on an evaluation of component reliability–age data or data from other hardware systems, particularly if it has been shown that the accomplished maintenance tasks are cost-effective. The maintenance task requirements and intervals are then entered into the LSAR system, as was described in Chapter 2.

References

1. MSG-3 AIRLINE/MANUFACTURERS MAINTENANCE PLANNING DOCUMENT, Air Transport Association, Washington, DC, October 1980.
2. US AMC PAMPHLET, 750-2 *Guide to Reliability-Centered Maintenance*, June 1985.
3. US MIL-STD-1388-1A, *Logistic Support Analysis—Program Requirements.*
4. US MIL-STD-1388-2A, *DoD Requirements for a Logistic Support Analysis Record.*
5. US MIL-STD-785, *Reliability Program for Systems and Equipment—Development and Production.*
6. US MIL-STD-470, *Maintainability Program Requirements.*
7. US MIL-HDBK-217, *Reliability Prediction of Electronic Equipment.*
8. US MIL-HDBK-472, *Maintainability Prediction.*
9. NONELECTRONIC PARTS RELIABILITY DATA NPRD-3, *Reliability Analysis Center, Rome Air Development Center*, Griffiss AFB, Rome, New York, 1985.
10. IEEE STD-500-1984, *IEEE Guide to the Collection and Presentation of Electrical, Electronic, Sensing Component, and Mechanical Equipment Reliability Data for Nuclear-Power Generating Stations*, IEEE, New York, 1983.
11. HAUGEN, E. B., *Probabilistic Approaches to Design*, John Wiley, New York, 1968.
12. US MIL-STD-1629A, *Procedures for Performing a Failure Mode, Effects and Criticality Analysis.*
13. US MIL-STD-810, *Environmental Test Methods.*
14. MIL-STD-781, *Reliability Qualification and Production Acceptance Test—Exponential Distribution.*
15. E. D., CODIER, Reliability Growth in Real Life, *Proceedings, 1968 Annual Symposium on Reliability*, IEEE, New York, January, 1968.

16. NAVMAT, *Navy Manufacturing Screening Program*, Department of the Navy, May 1979.
17. US MIL-STD-883, *Test Methods and Procedures for Microelectronics.*
18. US DOD-HDBK-344, *Environmental Stress Screening of Electronic Equipment.*
19. US MIL-M-63041C (TM), *Preparation of Depot Maintenance Work Requirements*, October 1984.
20. SWAIN, A. D. and GUTTMAN, H. E., *Handbook of Human Reliability with Emphasis on Nuclear Power Plant Applications*, NUREC/CR-1278, National Technical Information Service, US Department of Commerce, Springfield, VA, August 1983.

APPENDIX A

Glossary

Age — The measure of a unit's total exposure to stress, expressed as the number of operating hours, flight cycles, or other stress units since new or since the last shop visit.
Age Exploration — A systematic evaluation of an item based on analysis of collected information from in-service experience. It assesses the item's resistance to a deterioration process with respect to increasing age.
Availability — The probability that a material, component, equipment, system or process is in its intended functional condition at a given time and therefore is either in use or capable of being used in a stated environment.
Bathtub Curve — A conditional probability curve which represents the age-reliability relationship of certain items, characterized by an infant mortality region, a region of relatively constant hazard rate, and an identifiable wearout region.
Burn-in — A common form of reliability screen where items (parts, assemblies or products) are operated prior to their ultimate application to stabilize their characteristics and to identify early failures.
Conditional Probability of Failure — The probability that an item will fail during a particular age interval, given that it survives to enter that interval.
Corrective Maintenance — The actions performed, as a result of failure, to restore an item to a specified condition.
Crack Initiation — The first appearance of a fatigue crack in an item

subject to repeated loads; usually based on visual inspection, but sometimes based on the use of nondestructive testing techniques.

Crack Propagation Characteristics — The rate of crack growth, and the resulting reduction in residual strength, from the time of crack initiation to a crack of critical length.

Critical Crack Length — The length of a fatigue crack at which the residual strength of the item is no longer sufficient to withstand the specified damage tolerant load.

Critical Failure — A failure involving a loss of function or secondary damage that could have a direct adverse effect on operating safety.

Criticality Analysis — A procedure by which each potential failure mode is ranked according to the combined influence of severity and probability of occurrence.

Damage Tolerant — A qualification standard for aircraft structure. An item is judged to be damage tolerant if it can sustain damage and the remaining structure can withstand reasonable loads without structural failure or excessive structural deformation until the damage is detected.

Default Answer — In a binary decision process, the answer to be chosen in case of uncertainty; employed in the development of an initial preventive maintenance program to arrive at a course of action in the absence of complete information.

Defect — A characteristic which does not conform to applicable specification requirements and which adversely affects or potentially affects the quality of a device.

Degradation — A gradual deterioration in performance as a function of time.

Derating — The intentional reduction of the stress–strength ratio in the application of an item, usually for the purpose of reducing the occurrence of stress related failures.

Design Life — The expected time or cycles, based on the design of the item, during which the item remains operationally effective and economically useful before wearout.

Deterioration — Degradation in quality, mission accomplishment and/or reliability due to age, usage or environment.

Durability — An element of reliability, defined as the probability that an item will successfully survive to its projected service life or rebuild point (whichever is the more appropriate durability measure for the item) without experiencing a durability failure. A durability failure is considered to be a malfunction that precludes further operation of the item under consideration and is of such consequence (in terms of cost and/or time to restore) that the item must be replaced or completely rebuilt.

Economic Life Limit — A life limit imposed on an item on the basis of cost-effectiveness to reduce the frequency of age related failures.

End Item — A final combination of end products, component parts, and/or materials which is ready for its intended use; e.g., ship, tank, mobile machine shop, aircraft.

Environmental Stress Screening — The process or method whereby a group of like items are subjected to the application of physical climatic stresses or forces (or combinations thereof) to identify and eliminate defective, abnormal or marginal parts and manufacturing defects.

Environments — The conditions, circumstances, influences, stresses and combinations thereof, surrounding and affecting systems or equipment during storage, handling, transportation, testing, installation, and use in standby status and mission operation.

Error — Any discrepancy between a computed, observed or measured quantity and the true, specified or theoretically correct value or condition. A conceptual, syntactic or clerical discrepancy which causes one or more faults in the software.

Fail Safe System — A system whose function is replicated, so that the function will still be available to the equipment after failure of one of its sources.

Failure — Any deviation from the design-specified, measurable tolerance limits that causes either a loss of function or reduced capability.

Failure Cause — The physical or chemical processes, design defects, quality defects, part misapplication or other processes which are the basic reason for failure or which initiate the physical process by which deterioration proceeds to failure.

Failure Effect — The consequence(s) a failure mode has on the operations, function or status of an item. Failure effects are classified as local effect, next higher level and end effect.

a. Local effect — The consequence(s) a failure mode has on the operation, function or status of the specific item being analyzed.
b. Next higher level effect — The consequence(s) a failure mode has on the operation, functions or status of the items in the next higher indenture level above the indenture level under consideration.
c. End effect — The consequence(s) a failure mode has on the operation, function or status of the highest indenture level.

Failure Free Criteria — An acceptance requirement that is imposed during the later part of a screen test that requires no failure for a specified period or number of cycles. Successful completion of the failure free period (or number of cycles) provides an indication that the screening has been effectively completed depending on the expected failure rate and length of the failure-free period.

Failure Mechanism — The physical process or occurrence that caused a failure (e.g., stress corrosion cracking, operator error, equipment malfunction, relay contacts welded by overload and bearings frozen by contamination with foreign material).

Failure Mode — A particular way in which failures occur, independent of the reason for failure; the condition or state which is the end result of a particular failure mechanism.

Failure, Random — Any failure whose cause and/or mechanism make its time of occurrence unpredictable, but which is predictable only by probabilistic or statistical methods.

Failure Rate — The number of failures of an item per unit measure of life. The failure rate is considered constant during the useful life period.

Failure Symptom — An identifiable physical condition by which a potential functional failure can be recognized.

Fatigue — Reduction in resistance to failure of a material over time, as a result of repeated or cyclic applied loads.

Fatigue Life — For an item subject to fatigue, the total time to functional failure of the item.

Functional Failure — Failure of an item to perform its normal or characteristic actions within specified limits.

Hard Time — Scheduled removal of all units of an item before some specified maximum permissible age limit, in order to preclude functional failure.

Hazard Rate — The probability that a failure will occur at any point in the item's life-cycle (the instantaneous failure rate).

Hidden Failure — A failure which is undetectable during operation by the operator/crew.

Incipient Failure — A deteriorated condition that indicates that a failure is about to occur.

Infant Mortality — The relatively high conditional probability of failure during the period immediately after an item enters service. Such failures are due to defects in manufacturing not detected by quality control.

Inherent Reliability Level — The level of reliability of an item or of equipment that is derived from its design, is characterized by a near-constant conditional probability of failure, and cannot be improved by maintenance.

In-Service Reliability — That characteristic of design and installation that will ensure a system's (equipment's) capability to operate satisfactorily under given conditions for a specified period of time.

Inspection Task — A scheduled task requiring testing, measurement or visual inspection for explicit failure evidence by maintenance personnel.

Integrated Logistic Support (ILS) — A unified and iterative approach to the management and technical activities necessary to: (a) cause support considerations to influence requirements and design; (b) define support requirements that are optimally related to the design and to each other; (c) acquire the required support; and (d) provide the required support during the operational phase at minimum cost.

Item — Any level of the equipment or its sets of parts (including the equipment itself) isolated as an entity for study.

Life — The elapsed time (in flight hours, calendar time, cycles, etc.) since an item was newly manufactured (or zero timed) at which it suffers wearout failure or is removed from service to prevent in-service wearout failure.

Logistic Support Analysis (LSA) — The selective application of scientific and engineering efforts to (a) influence the system performance parameters and system configuration from a supportability standpoint, and (b) determine the logistic support resource requirements for the system through the use of an iterative process of definition, synthesis, trade-off, test and evaluation.

Logistic Support Analysis Record (LSAR) — That portion of LSA documentation consisting of detailed data pertaining to the identification of logistic support resource requirements of a system. These data are normally referred to as the A-J sheets.

Maintainability — A measure of the ease and rapidity with which a system or equipment can be restored to operational status following a failure, expressed as the probability that an item will be retained in or restored to a specified condition within a given period of time when the maintenance is performed in accordance with prescribed procedures and resources.

Maintainability Engineering — The engineering discipline which formulates an acceptable combination of design features, repair policies, and maintenance resources, to achieve a specified level of maintainability, as an operational requirement, at optimum life-cycle costs.

Maintenance — All actions necessary for retaining an item in a specified condition before failure or breakdown (preventive maintenance) or the process of restoring an item to return it to a workable condition (corrective maintenance).

Maintenance Analysis — The process of identifying required maintenance functions by analysis of the design, to determine the most effective means to accomplish these functions.

Maintenance Capabilities — The facilities, tools, test equipment, drawings, technical publications, trained maintenance personnel, engineering support and spare parts required to restore a system to serviceable condition.

Maintenance Concept — A description of the planned general scheme for maintenance and support of an item in the operational environment. The maintenance concept provides the practical basis for design, layout and packaging of the system and its test equipment, and establishes the scope of maintenance responsibility for each level (echelon) of maintenance and the personnel resources (maintenance manning and skill levels) required to maintain the system.

Maintenance Levels — The basic levels of maintenance into which all maintenance activity is divided. The scope of maintenance performed within each level must be commensurate with the personnel, equipment, technical data and facilities provided.

Maintenance Planning — A principal element of ILS; includes development of the maintenance concept, reliability and maintainability parameters, repair level determinations, maintenance requirements, and supply support essential to adequate and economical support of the system/equipment. Planning becomes more detailed as the system/equipment progresses through the acquisition cycle.

Maintenance Tasks — An action or set of actions required to achieve a desired outcome which restores an item to or maintains an item in serviceable condition, including inspection and determination of condition.

Manufacturing Defect — A flaw caused by in-process errors or uncontrolled conditions during assembly, rest, inspection or handling.

Mean Time Between Failures (MTBF) — Total operating time (frequently stated in hours) divided by the total number of failures.

Mean Time To Repair (MTTR) — The mean time required to complete a maintenance action, i.e., total active maintenance downtime (i.e., fault isolation, fault correction, calibration and checkout) divided by the total number of maintenance actions, over a given period of time, excluding those time elements which are related to preparation and delay, and administrative and supply delay downtime.

Mean Time To Restore (MTR) — That time associated with reinitiation of the system's functional capabilities. For non-redundant systems, this time is usually equivalent to MTTR. In the case of standby redundant systems, or systems where a different hardware type can provide backup service, system restoration time is equal to the time required to switch operation to the backup unit. It is computed by dividing the total system outage time by the number of system outages, over a given period of time.

Minor Repair — The level of repair required to restore serviceability to an item by correcting specific damage, fault, malfunctions or failure in part, subassembly, module (component or assembly), end item or system.

Multiple Failure — A failure event consisting of the sequential occurrence of two or more independent failures, which may have consequences that would not be produced by any of the failures occurring separately.

Nondestructive Testing — A test that is neither functionally nor potentially destructive. It is performed to establish acceptability; e.g., X–ray analysis, leak tests, ultrasonic tests, etc.

On-Condition Maintenance — Scheduled inspections, tests or measurements to determine whether an item is in, and will remain in, a satisfactory condition until the next scheduled inspection, test or measurement.

Overhaul — The level of repair required to restore an item to a completely serviceable/operational condition as prescribed by the maintenance standard (for example, DMWR). Overhaul is normally the highest degree of maintenance performed.

Phase Inspections — A series of related inspections that are performed sequentially at specific intervals. These inspections are the results of dividing the maintenance requirements into small packages containing approximately the same workload.

Potential Failure — A quantifiable failure symptom which indicates that a functional failure is imminent.

Preventive Maintenance — The care and servicing by personnel for the purpose of maintaining system/equipment safety and reliability levels through systematic inspection, detection, lubrication, cleaning, etc. Preventive maintenance includes scheduled maintenance.

Provisioning — The process of determining and acquiring the range and quantity (depth) of spares and repair parts, and support and test equipment required to operate and maintain an end item of material for an initial period of service.

RCM Analysis — Use of the RCM decision concepts to devise a preventive maintenance program by evaluating maintenance required for an item according to the consequences of each significant failure possibility, the inherent reliability characteristics of each item, and the applicability and

effectiveness of possible preventive maintenance tasks.
RCM Program — A preventive maintenance program consisting of a set of tasks each generated by RCM analysis.
RCM Task — A preventive maintenance task which satisfies the specific applicability criteria for that type of task.
Rebuild — The level of repair required to restore unserviceable equipment to a like-new condition in accordance with original manufacturing standards. Rebuild is the highest degree of material maintenance applied to equipment. The rebuild operation includes the act of returning to zero those age measurements (i.e., hours) considered in classifying hardware systems and components.
Redundancy — The design practice of replicating the sources of a function so that the function remains available after the failure of one or more items.
Reliability — The characteristic of an item expressed by the probability that it will perform a required function under a stated condition for a stated period of time.
Reliability–Age Characteristics — The characteristics exhibited by the relationship between the operating age of an item and its conditional probability of failure.
Reliability – Centered Maintenance (RCM) — A disciplined logic or methodology used to identify preventive maintenance tasks to realize the inherent reliability of equipment at a minimum expenditure of resources.
Reliability Engineering — The engineering discipline which formulates an acceptable combination of design features, repair philosophy and maintenance resources, to achieve a specified level of reliability as an operational requirement, at optimum life-cycle costs.
Reliability Growth — The improvement in the reliability of a new item as a result of product improvement after the equipment enters service.
Reliability Growth Testing — The improvement process during which hardware reliability increases to an acceptable level.
Reliability, Operational — The assessed reliability of an item based on operational data.
Repair — The level of repair required to restore serviceability to an item by correcting specific damage, fault, malfunctions or failure in a part, subassembly, module (component or assembly), end item or system.
Safe-Life Limit — A life limit imposed on an item that is subject to a critical failure, established as some fraction of the average age at which test data show that failures will occur.

Safety Consequences — A loss of a function or secondary damage resulting from a given failure mode which produces a direct adverse effect on safety. One of the four consequence branches of the RCM decision diagram.
Scheduled Maintenance — Periodic prescribed inspection and servicing of equipment accomplished on a calendar, mileage or hours of operation basis.
Scheduled Removal — Removal of a serviceable unit at some specified age limit to prevent an in-service functional failure due to an explicit wearout failure mode, or to inspect for an incipient functional failure.
Secondary Damage — The immediate physical damage to other parts of items that results from a specific failure mode.
Service Life — The period of time during which an item is expected to perform in a satisfactory manner under specified operational conditions prior to wearout or obsolescence, and subsequent removal from service.
Servicing Tasks — Scheduled tasks to replenish fluid levels, pressures and consumable supplies.
Significant Item — An item whose failure or hidden functions whose part in a multiple failure has safety, operational or major economic consequences.
Simple Item — An item whose functional failure is caused by only one or a very few failure modes.
Storage Life (Shelf Life) — The length of time an item can be stored under specified conditions and still meet specified requirements.
Teardown Inspection — The complete disassembly of a serviceable item that has survived to a specified age limit to examine the condition of each of its parts as a basis for judging whether it would have survived to a proposed higher age limit.
Unscheduled Maintenance — Those unpredictable maintenance requirements that had not been previously planned or programmed but which require prompt attention and must be added to, integrated with, or substituted for previously scheduled workloads.
Wearout — The process of attrition which results in an increase in hazard rate with increasing age (cycles, time, miles, events, etc., as applicable for the item).

APPENDIX B

Bibliography

AMC PAMPHLET 750–2, 'GUIDE TO RELIABILITY-CENTERED MAINTENANCE', June 1985.

This pamphlet is a guide for Army representatives and contractors who write and develop a detailed maintenance plan for system/equipment using the reliability-centered maintenance (RCM) philosophy. It explains in detail how to use the RCM logic and the failure mode, effects and criticality analysis (FMECA) to develop a scheduled maintenance plan which includes the maintenance task and the maintenance interval for preventive maintenance checks and services (PMCS), and provides information for overhaul, age exploration, economic analysis and redesign.

ANDERSON, R. T. AND BASS, S., 'HOW TO CONTROL RELIABILITY FROM DESIGN THROUGH BURN-IN', Evaluation Engineering, March 1981, pp. 86–93.

The increasing number of products which incorporate microprocessors and other complex and sensitive electronic components present a particularly difficult reliability problem. Special handling techniques are required during assembly, and manufacturing induced defects are often difficult to detect with ordinary inspection techniques. This paper describes an overall cost-effective program to control reliability based on screening and burn-in procedures, coupled with a well designed system of failure reporting and analysis as well as periodic process modification.

ARSENAULT, J. E. AND ROBERTS, J. A. (EDS) RELIABILITY AND MAINTAINABILITY OF ELECTRONIC SYSTEM, Computer Sciences Press, Inc., Rockville, Maryland, 1980.

The book is intended for engineers, managers and academics engaged in system engineering and concerned with reliability and maintainability. Accordingly, it takes a broad approach to the subject. The editors have tried to select sufficient theoretical and practical information to solve those reliability and maintainability problems frequently encountered. In addition, they give a comprehensive set of reference techniques required for those special problems which inevitably appear.

BAZOVSKY, I., RELIABILITY THEORY AND PRACTICE, Prentice–Hall, Inc., Englewood Cliffs, New Jersey, 1961.

The objective of this book is to develop reliability concepts and methods in a logical way, from simple components to complex systems, to give the reader a thorough understanding of the subject and show him how to solve reliability problems by analysis, design and testing. There is an abundance of useful reliability formulae in the book, which will help the reader predict system reliability, establish reliability goals and determine the procedures necessary to achieve them. Also included is a quantitative treatment of system maintainability, availability and safety, and outlined methods which have to be followed.

BEST PRACTICES, DEPARTMENT OF THE NAVY, NAVSO P–6071, March 1986, Superintendent of Documents, US Government Printing Office, Washington, DC.

As follow-on to the efforts of the US Defense Science Board Task Force on the Transition from Development to Production, this manual attempts to enhance the enlightenment of both Government and industry by identifying specific practices in current use and their potentially adverse consequences in terms of cost, schedule, performance and readiness. It then describes proven best practices which avoid or alleviate these consequences, and provides enough background information to understand their rationale.

BLANCHARD, B. S., 'LOGISTICS ENGINEERING AND MANAGEMENT', Prentice–Hall, Inc., Englewood Cliffs, New Jersey, 1974.

This book provides an introduction to logistics engineering and management. It focuses on the language, principles and some of the quantitative measures used for prediction and assessment. It covers cost-effectiveness, system-effectiveness, reliability and maintainability, and the application of statistical techniques for logistics, as well as the application of logistics to the system/equipment life-cycle, commencing with identification of a need and extending through operational use and ultimate equipment phase-out.

CUNNINGHAM, C. E. AND COX, W. 'APPLIED MAINTAINABILITY ENGINEERING', Wiley–Interscience Publication, New York, 1972.

This book provides implementation guidance for setting up and pursuing a maintainability program in accordance with US Department of Defense specifications and in particular MIL-STD – 470 'Maintainability Program for Systems and Equipment'.

DA-P – 750–40 GUIDE TO RELIABILITY CENTERED MAINTENANCE FOR FIELDED EQUIPMENT (DARCOM FINAL REPORT DAG39–77–C – 0169), Darcom, Alexandria, VA, February 1980.

This guide illustrates how the elements of reliability-centered maintenance (RCM) are planned, developed and incorporated into maintenance plans/programs for material systems. Individual material developers are expected to tailor the techniques to fit their particular item/system needs.

FRANKLIN RESEARCH CENTER, A REVIEW OF EQUIPMENT AGING THEORY AND TECHNOLOGY (EPRI NP – 1558), Philadelphia, Pennsylvania, 1980.

The theory and technology of equipment aging is reviewed, particularly as it relates to qualification of safety-system equipment for nuclear power generating stations. A fundamental degradation model is developed, and its relation to more restricted models (e.g., Arrhenius and inverse stress models) is shown. The most common theoretical and empirical models of aging are introduced, and limitations on their practical application are analyzed. Reliability theory and its application to the acceleration of aging are also discussed. The difficulty of accelerating the aging of an assembly of materials and components in a scientifically rigorous manner, through the application of aging models, is demonstrated. A compendium of aging data for materials and components, including degradation mechanisms, failure modes and activation energies, is included.

GREEN, A. E. AND BOURNE, A. J., 'RELIABILITY TECHNOLOGY', John Wiley and Sons, New York, 1972.

The purpose of this book is to examine the problems of reliability against a background of cost, efficiency and safety. It describes the techniques for solving them, and deals with applications over a wide range of technological products. The opening chapter formulates a definition of reliability to make it a measurable quantity and the following chapters show how reliability criteria based on this definition can be applied to items in many branches of technology. Contents include: reliability concepts; an approach to reliability assessment; the performance requirements; the performance achievement; variations in the performance achievement, the transfer characteristic; properties of distribution; sampling, estimation and confidence; reliability considerations for systems, synthesis of system reliability; synthesis of complex systems; and the application of reliability assessment.

Institute of Environmental Sciences (IES), ESSEH, ENVIRONMENTAL STRESS SCREENING GUIDELINES.

Environmental Stress Screening of Electronic Hardware (ESSEH) is a process performed on all 'items' at various levels of assembly. It is intended to identify, force and/or segregate those items (part, module, unit or system) defined as defective. This document provides guidelines to plan and implement an ESSEH program. It is a compendium of information relative to the state-of-the-art of environmental stress screening of electronic hardware. This information has been derived from hard data solicited from companies which have developed and implemented successful stress screening programs in support of their product lines. All levels of assembly have been addressed. In addition, cost–benefit models have been included, which are easily understood and which may be used to assess the cost-effectiveness of various stress screens being considered for implementation. The document supports the decision making process, as it relates to the technical and economic aspects of environmental stress screening of electronic hardware.

LAKNER, A. A. AND ANDERSON, R. T., 'AN ANALYTICAL APPROACH TO DETERMINING OPTIMUM RELIABILITY AND MAINTAINABILITY REQUIREMENTS', The Radio and Electronic Engineer, 48, July/August 1978.

This paper describes the methodology, its approach and the specific life cycle cost (LCC) models used in the computation of optimal levels of R&M. It provides criteria, guidelines, rationale and formulae which the Federal Aviation Authority (FAA) is applying to its procurements in order to determine optimum mean time between failures (MTBF) and mean time to repair (MTTR) design values for use in its hardware specification. The methodology is based upon LCC principles which are fully in accord with FAA needs.

LAKNER, A. A. AND ANDERSON, R. T., 'COST-EFFECTIVE RELIABILITY TESTING', Proceedings Annual Reliability and Maintainability Symposium, IEEE, New York, 1978.

This paper addresses the design and implementation of cost-effective reliability tests and is based on work performed by the airways facilities of the Federal Aviation Administration in their development of a total methodology and database for reliability and maintainability and cost.

LAKNER, A. A. AND ANDERSON, R. T., 'RELIABILITY ENGINEERING FOR NUCLEAR AND OTHER HIGH TECHNOLOGY SYSTEMS', Elsevier Applied Science Publishers, Barking, Essex, 1985.

This book describes a systems approach to reliability/safety engineering and provides guidelines on how to integrate reliability into the hardware specification, design and development process; safety aspects peculiar to nuclear power plant design and operational procedures are given special consideration. The book is designed for the program manager, project engineer, design engineer, reliability/safety engineer of industrial concerns, as well as others who are involved in hardware/system acquisition and are concerned with procuring and developing reliable and cost-effective equipment.

LLOYD, D. AND LIPOW, M., 'RELIABILITY: MANAGEMENT, METHODS AND MATHEMATICS', 2nd edn., Redondo Beach, California, 1977.

This book describes the management, methods and mathematics of reliability, with a comprehensive treatment suitable for both graduate engineering students and practicing engineers. All activities of a reliability department are described, including failure and operating time reporting systems, and a thorough treatment of the mathematics of reliability is given. This includes basic probability theory and statistics, reliability point and confidence limit estimation, reliability demonstration methods based on various discrete and continuous statistical distributions, reliability growth models and various forms of system models. Binomial and exponential sampling plans and the latest methods of industrial experimentation are described. Added in this second edition is a chapter on computer software describing techniques for design and production of reliable computer programs and methods for measuring their reliability. A feature of the book is that in the mathematical chapters an introductory discussion is presented in non-mathematical language. Many examples are given from the authors' technical and management experience. Also included are numerous exercises in the statistical methods contained within the text, and useful charts and tables of sample sizes and confidence levels.

LOCKS, M. O., 'RELIABILITY, MAINTAINABILITY & AVAILABILITY ASSESSMENT', Hayden Book Company Inc., Rochelle Park, New Jersey, 1973.

The material is organized around the subject of confidence assessment, the measure of the quality of estimated reliability. The assumption is made that the component success or failure data used for reliability estimation are governed by some parametric probability distribution. The distribution groups treated include: binomial and other Bernoulli-type distributions; exponential, Poisson and gamma; normal and lognormal; and Weibull. Simplified graphical goodness-of-fit analysis is included as well as the use of optimum (maximum likelihood or linear) or Bayesian models for point and interval estimation and Monte Carlo simulation. The book covers reliability, maintainability and availability analysis of both repairable and non-repairable systems and components. This book is written for quality, reliability and safety analysts, engineers and operations researchers. This volume features a textbook style, with problems, to facilitate classroom use as well as self-study. Models include both point and interval values (confidence levels) for assessing the reliability, maintainability or availability using attributes data, and time-to-failure or time-to-repair data.

MOSS, M. A., 'DESIGNING FOR MINIMAL MAINTENANCE EXPENSE: THE PRACTICAL APPLICATION OF RELIABILITY AND MAINTAINABILITY', Marcel Dekker, Inc., New York, 1985.

This book describes, in a practical fashion, how maintenance expenses on consumer, commercial, and industrial hard-goods products can be sharply reduced by applying the principles of reliability/maintainability/availability (RMA) to their designs.

MSG−3 AIRLINE/MANUFACTURERS MAINTENANCE PLANNING DOCUMENT, Air Transport Association, Washington, DC, October 1980.

Provides an extension of logical decision processes which focuses on the consequences of failure and which identifies a maintenance program which provides the specified levels of safety and reliability at the lowest possible overall cost. Unlike its predecessor (MSG−2), this revision catalogs maintenance decisions via a task orientation in lieu of the maintenance process approach.

NAVMAT P−9492, 'NAVY MANUFACTURING SCREENING PROGRAM', Department of the Navy, Washington, DC, 1979.

This report outlines, primarily for US Navy contractors, an adapted and effective manufacturing screening program consisting of temperature cycling and random vibration. With the recognition that test facility cost has been a major obstacle to the use of random vibration, a technical report, which describes in detail a proven means to generate random vibration at low cost, is included as an appendix. Together, temperature cycling and random vibration provide a most effective means of decreasing corporate costs and increasing fleet readiness.

NONELECTRONIC PARTS RELIABILITY DATA NPRD−3, Reliability Analysis Center, Rome Air Development Center, Griffiss AFB, Rome, New York, 1985.

This document, organized in four major sections, presents reliability information based on field operation, dormant state and test data for more than 380 major nonelectronic part types. The four sections are Generic Data, Background Information, Detailed Data, and Failure Modes and Mechanisms. Each device type contains reliability information in relation to the specific operation environments.

NOWLAN, F.S. AND HEAP, H.F., 'RELIABILITY CENTERED MAINTENANCE,' DDC NO. AD-A066579, Defence Documentation Center, Defense Logistics Agency, Alexandria, VA, December, 1978.

Explains basic concepts, principles, definitions and applications of a logical discipline for development of efficient scheduled (preventive) maintenance programs for complex equipment, and the on-going management of such programs. These programs are called Reliability−Centered Maintenance (RCM) programs because they are centered on achieving the inherent safety and reliability capabilities of the equipment at minimum cost. A DoD objective in sponsoring preparation of this document was that it serve as a guide for application to a wide range of military equipment.

O'CONNOR, P. D. T., 'PRACTICAL RELIABILITY ENGINEERING', Heyden, London, 1981.

The mathematical concepts described are limited to those necessary for the solution of the problems covered. Practical approaches to problem solving such as probability plotting techniques and computer programs are stressed throughout. Full coverage is given to major national and international standards and specifications on reliability engineering. This is a vital aspect of the practical approach since so much engineering development is now governed by such documents. The effects of current engineering, commercial and legislative developments, such as microelectronics, software-based systems, consumerism and product liability, are covered in detail.

'RADC RELIABILITY ENGINEER'S TOOLKIT', Systems Reliability and Engineering Division, Rome Air Development Center, Griffiss AFB, Rome, New York, July 1988.

This publication was designed for easy use with quick reference indices tied to questions that need to be answered by reliability engineers, such as 'How do I evaluate contractor proposals?' It is not intended to be a tutorial, or complete technical treatment, but rather a compendium of useful R&M information to be used in everyday practice.

RDH – 376, ANDERSON, R. T., 'RELIABILITY DESIGN HANDBOOK', Reliability Analysis Center, USAF/RADC, Rome, New York, 1976.

The purpose of the handbook is to provide information and direction to the designer which will help him engineer reliability into an equipment during its basic design stage. To this end, it provides design data and guidelines for those safety, mission, maintenance and cost factors which together form the working elements of reliability engineering, system engineering and cost-effectiveness. This handbook is primarily intended for use in the design of new equipment or systems which are largely composed of electronic parts and components. However, it can also be used for the design of systems which encompass both nonelectronic and electronic parts, as well as for the modification of existing systems. The handbook embodies a preventive approach to reliability.

SELBY, J. AND MILLER, S., 'RELIABILITY PLANNING AND MANAGEMENT—RPM,' Symposium for Reliability and Maintainability Technology for Mechanical Systems, AOA, Washington, 1982.

This paper presents a new approach to the reliability planning and management of complex weapon systems. RPM is essentially a management tool for bridging the gap between stated reliability requirements and implementation planning. The RPM methodology, equally usable by buyer and contractor, is applicable to establishing plans, projecting effort, evaluating proposals and monitoring contract performance.

SWAIN, A.D. AND GUTTMAN, H.E., 'HANDBOOK OF HUMAN RELIABILITY WITH EMPHASIS ON NUCLEAR POWER PLANT APPLICATIONS', US Nuclear Regulatory Commission. Reproduced by National Technical Information Service, Department of Commerce, Springfield, VA, Aug. 1983.

This handbook aids qualified persons in evaluating the effects of human error on the availability of engineered safety features and systems in nuclear power plants. The handbook expands the human error analysis presented in WASH – 1400 and includes principles of human behavior and ergonomics, analytical procedures mathematical models, and human error probabilities derived from related performance measure and experience. The derived probabilities should be adequate to determine the relative merits of different configurations of equipment, procedures, and operating practices within a plant, and for gross comparisons among plants. Limitations of the handbook and cautions to be observed in its use are explicitly stated.

US MIL-HDBK – 189 RELIABILITY GROWTH MANAGEMENT.

Provides an understanding of the concepts and principles of reliability growth, advantages of managing reliability growth, and guidelines and procedures to be used in managing reliability growth.

US MIL-HDBK – 217 RELIABILITY PREDICTION OF ELECTRONIC EQUIPMENT.

Establishes uniform methods for predicting the reliability of military electronic equipment and systems. Provides a common base for reliability prediction during acquisition programs of military equipment and serves as a means of comparing the reliability of related or competitive designs. This document provides two methods of reliability prediction: (a) parts stress analysis and (b) parts count. Mathematical expressions for part failure rates are provided for use in computer programming. Tables, rather than curves, are used for base failure rates to improve ease of manual application of the prediction methods. These prediction methods will be continually updated as new information becomes available. This handbook includes information relating to part stress analysis prediction in the areas of: (a) microelectronic devices; (b) discrete semiconductors; (c) tubes, electronic vacuums; (d) lasers; (e) resistors; (f) inductive devices; (g) rotating devices; (h) relays; (i) switches; (j) connectors; (k) wire and printed wiring boards; and (l) miscellaneous parts. Also covered is parts count reliability prediction. Appendices dealing with system reliability modeling and approximation for reliability calculation, and a comprehensive bibliography, are also included.

US MIL-HDBK – 266 APPLICATION OF RELIABILITY CENTERED MAINTENANCE TO NAVAL AIRCRAFT, WEAPONS SYSTEMS AND SUPPORT EQUIPMENT.

Provides basic and fundamental information to apply the principles of Reliability-Centered Maintenance (RCM) covered by DoD report AD-A066579 (29 December 1978). Also provides the procedures for developing preventive maintenance requirements as part of maintenance planning and analysis process as specified in MIL-STD – 2080.

US MIL-HDBK – 472 MAINTAINABILITY PREDICTION.

Establishes uniform methods for predicting the maintainability of military equipment and systems. Provides a common basis for maintainability prediction during acquisition programs of military equipment and serves as a means of comparing the maintainability of related or competitive designs. Includes procedures dependent on the use of recorded R&M data and experience which have been obtained from comparable systems and components under similar conditions of use and operation. Prescribes four maintainability prediction procedures. Procedure I—system downtime of airborne electronic and electromechanical systems involving modular replacement at the flight-line. Procedure II—methods and techniques used to predict corrective, preventive and active maintenance parameters. Procedure III—method of performing a maintainability prediction of ground electronic systems and equipment by utilizing the basic principles of random sampling. Procedure IV—historical experience, subjective evaluation, expert judgment and selective measurements for predicting the downtime of a system/equipment; uses existing data to the extent available; provides an orderly process by which the prediction can be made and integrated preventive and corrective maintenance; task time to perform various maintenance actions are estimated and then combined to predict overall system/equipment maintainability. Procedures I and III are solely applicable to electronic systems and equipment. Procedures II and IV can be used for all systems and equipments.

US MIL-STD-470 MAINTAINABILITY PROGRAM REQUIREMENTS.

This standard provides requirements for establishing a maintainability program and guidelines for the preparation of a maintainability program plan that encompasses: (a) analysis; (b) design criteria; (c) design trade-offs; (d) parameter values; (e) subcontractor and vendor contract specifications; (f) design reviews; (g) data collection, analysis and corrective action systems; and (h) inputs and status reports.

US MIL-STD-756 RELIABILITY PREDICTION.

Establishes uniform procedures for predicting the quantitative reliability of aircraft, missiles, satellites and electronic equipment. Graphically portrays the effects of system complexity on reliability, to permit the ready prediction of tolerance and interaction problems. Provides appropriate *k* factors by which to adjust MIL-HDBK-217 predictions for airborne, missile and space environments.

US MIL-STD-785 RELIABILITY PROGRAM FOR SYSTEMS AND EQUIPMENT—DEVELOPMENT AND PRODUCTION.

Establishes uniform criteria for reliability programs and provides guidelines for the preparation of reliability program plans. Lists detailed requirements as program elements including: (a) reliability management (reliability organization, management and control, subcontractor and supplier reliability program, program review); (b) reliability design and evaluation (design techniques, reliability analysis, effects of storage, design reviews); (c) reliability testing and demonstration (reliability test plans, development testing, reliability demonstration); (d) failure data (failure data collection analysis and corrective action, failure summaries); (e) production reliability (transition from development reprocurement); and (f) status reports.

US MIL-STD-810 ENVIRONMENTAL TEST METHODS.

Describes environmental test procedures and criteria for military equipment/systems.

US MIL-STD-882 SYSTEM SAFETY PROGRAM REQUIREMENTS.

Defines those elements of a system safety program which are required during the development, production and initial deployment of systems and equipment.

US MIL-STD-1388-1A LOGISTIC SUPPORT ANALYSIS.

Establishes criteria governing performance of a Logistic Support Analysis (LSA), integral to the engineering process, to define support system requirements and inject support criteria into system/equipment design and acquisition.

US MIL-STD-1388-2A DoD REQUIREMENTS FOR A LOGISTIC SUPPORT ANALYSIS RECORD.

Identifies and describes the logistic support analysis (LSA) record system.

US MIL-STD-1629A PROCEDURES FOR PERFORMING A FAILURE MODE, EFFECTS AND CRITICALITY ANALYSIS.

Establishes requirements and procedures for performing a failure mode, effects, and criticality analysis (FMECA) to systematically evaluate and document, by item failure mode analysis, the potential impact of each functional or hardware failure on mission success, personnel and system safety, system performance, maintainability and maintenance requirements. Each potential failure is ranked by the severity of its effect in order that appropriate corrective actions may be taken to eliminate or control the high risk items.

US MIL-STD-1635 RELIABILITY GROWTH TESTING.

Establishes the requirements and procedures for reliability development (growth) tests. These tests are conducted during the hardware development phase on commitment. These tests provide engineering information on the failure modes and mechanisms of a test item under natural and induced environmental conditions of military operations. Reliability improvement (growth) results when failure modes and mechanisms are identified and their recurrence prevented through implementation of corrective action.

Index

MIX
Papier aus verantwortungsvollen Quellen
Paper from responsible sources
FSC® C105338

If you have any concerns about our products,
you can contact us on
ProductSafety@springernature.com

In case Publisher is established outside the EU,
the EU authorized representative is:
Springer Nature Customer Service Center GmbH
Europaplatz 3, 69115 Heidelberg, Germany

Printed by Libri Plureos GmbH
in Hamburg, Germany